Fisheries Compatibility Disputes

Agreeing to Disagree, Committed to Conserve

Alexandros X.M. Ntovas

University of Southampton Law School, UK

Edward Elgar
PUBLISHING

Cheltenham, UK • Northampton, MA, USA

Published by

Edward Elgar Publishing Limited
The Lypiatts
15 Lansdown Road
Cheltenham
Glos GL50 2JA
UK

Edward Elgar Publishing, Inc.
William Pratt House
9 Dewey Court
Northampton
Massachusetts 01060
USA

Authorised representative in the EU for GPSR queries only: Easy Access System Europe – Mustamäe tee 50, 10621 Tallinn, Estonia, gpsr.requests@easproject.com

A catalogue record for this book is available from the British Library

Library of Congress Control Number: 2025950492

This book is available electronically in the **Elgar**online
Law subject collection
https://doi.org/10.4337/9781800378612

MIX
Paper | Supporting responsible forestry
FSC® C014540

ISBN 978 1 80037 860 5 (cased)
ISBN 978 1 80037 861 2 (eBook)
ISBN 978 1 0353 9104 2 (ePub)

Printed and bound in the UK by Short Run Press Ltd, Exeter EX2 7LW

Contents

Acknowledgements

To my life partner and wife *Αἰκατερίνη*, for all the quiet days, early mornings, and late evenings you graciously allowed me to lose myself in writing this; *σε εὐχαριστῶ* for your love, patience, and unfailing encouragement.

My special thanks go to Ben Booth, Elizabeth Ruck, Amber Watts, and Emily White at Edward Elgar Publishing for making this book a reality. I am deeply grateful to the many others, within and beyond my family, academic and professional circles, whose support, in various forms, proved invaluable in bringing this endeavour to completion. I hesitate to name you individually, for fear of inadvertently omitting someone.

I lift my humble prayer in thanksgiving to the Virgin Mary, Most Holy *Μῆτερ τοῦ Θεοῦ*, for her intercession to the One God Almighty—Father, Son, and Holy Spirit—Who reigns eternal and sovereign over all. May His everlasting blessing be upon all.

Southampton, 27 June 2025

List of abbreviations and acronyms

Term	Definition
A/AC.138/...	Numbered document of the United Nations Committee on the Peaceful Uses of the Seabed and the Ocean Floor Beyond the Limits of National Jurisdiction
A/CN.4/...	Numbered document of the United Nations International Law Commission
A/Conf.13/...	Numbered document of the United Nations Conference on the Law of the Sea
A/Conf.19/...	Numbered document of the Second United Nations Conference on the Law of the Sea
A/Conf.62/...	Numbered document of the Third United Nations Conference on the Law of the Sea
A/Conf.164/...	Numbered document of the United Nations Conference on Straddling Fish Stocks and Highly Migratory Fish Stocks
A/Res.	United Nations General Assembly Resolution
AJIL	*American Journal of International Law*
AJIL *Suppl.*	*Special Supplement of the American Journal of International Law*
Annuaire Français	*Annuaire Français de Droit International*
Asia Pacific LPJ	*Asia Pacific Law and Policy Journal*
Asian JIL	*Asian Journal of International Law*
Aust. YbIL	*Australian Yearbook of International Law*
BBNJ Agreement	*United Nations Agreement under the United Nations Convention on the Law of the Sea on the Conservation and Sustainable Use of Marine Biological Diversity of Areas Beyond National Jurisdiction*, 2023

Term	**Definition**
BYBIL	*British Yearbook of International Law*
Canad. YbIL	*Canadian Yearbook of International Law/Annuaire Canadien de Droit International*
CCAMLR	Commission on the Conservation of Antarctic Marine Living Resources
CCSBT	Commission for the Conservation of the Southern Bluefin Tuna
CJIL	*Chinese Journal of International Law*
CMM	Conservation and Management Measure(s)
CN	Communication of Depositary Notification by the United Nations Secretary–General
CPPS	Comisión Permanente del Pacífico Sur
CS	Continental Shelf
CSC	*United Nations Convention on the Continental Shelf*, 1958
CUP	Cambridge University Press
DOALOS	United Nations Division for Ocean Affairs and the Law of the Sea
DWFS	Distant Water Fishing State(s)
EEP	Edward Elgar Publishing
EEZ	Exclusive Economic Zone
EFP	Experimental Fishing Programme(s)
EFZ	Exclusive Fishing Zone
EJIL	*European Journal of International Law*
ELQ	*Ecology Law Quarterly*
ENB	*Earth Negotiations Bulletin*
FAO	Food and Agriculture Organization of the United Nations
FAO COFI	Committee on Fisheries of the Food and Agriculture Organization of the United Nations
FAO/TD/TRE	Treaties Database of the Food and Agriculture Organization of the United Nations

Term	Definition
FCC	*United Nations Convention on Fishing and Conservation of the Living Resources of the High Seas*, 1958
FFA	Forum Fisheries Agency of the Pacific Islands
Fish Stocks Conference	*United Nations Conference on Straddling Fish Stocks and Highly Migratory Fish Stocks*
FSA	*United Nations Agreement for the Implementation of the Provisions of the United Nations Convention on the Law of the Sea of 10 December 1982 relating to the Conservation and Management of Straddling Fish Stocks and Highly Migratory Fish Stocks*, 1995
FSA Review Conference	*Review Conference on the 1995 United Nations Agreement for the Implementation of the Provisions of the United Nations Convention on the Law of the Sea of 10 December 1982 relating to the Conservation and Management of Straddling Fish Stocks and Highly Migratory Fish Stocks*
Geneva Off. Records	Official Records of the United Nations Conference on the Law of the Sea, including Preparatory Documents, Summary Records of Plenary Meetings and of Meetings of the 5 Committees, with Annexes and Final Act; published by the United Nations Publications
GFCM	General Fisheries Commission for the Mediterranean
GIELR	*Georgetown International Environmental Law Review*
GöJIL	*Göttingen Journal of International Law*
GYIL	*German Yearbook of International Law*
HSC	*United Nations Convention on the High Seas*, 1958
IATTC	Inter-American Tropical Tuna Commission
ICCAT	International Commission for the Conservation of Atlantic Tunas
ICJ	International Court of Justice
ICJ Reports	Annual Reports of Judgments, Advisory Opinions and Orders of the International Court of Justice
ICLQ	*International and Comparative Law Quarterly*

Term	Definition
ICSP	Informal Consultations of States Parties to the *1995 United Nations Agreement for the Implementation of the Provisions of the United Nations Convention on the Law of the Sea of 10 December 1982 relating to the Conservation and Management of Straddling Fish Stocks and Highly Migratory Fish Stocks*
IJMCL	*International Journal of Marine and Coastal Law*
ILA	International Law Association
ILC	International Law Commission of the United Nations
ILM	*International Legal Materials* reproduced by the American Society of International Law, and published by Cambridge University Press
IOTC	Indian Ocean Tuna Commission
ITLOS	International Tribunal for the Law of the Sea
ITLOS Reports	Annual Reports of Judgments, Advisory Opinions and Orders of the International Tribunal for the Law of the Sea
IUU	Illegal, Unreported and Unregulated
LCP	*Journal of Law and Contemporary Problems*
LJIL	*Leiden Journal of International Law*
LoN	League of Nations
LOSC	*United Nations Convention on the Law of the Sea*, 1982
LOSI UH	The Law of the Sea Institute, University of Hawaii
MCS	Monitoring, Control and Surveillance
MLR	*Malaya Law Review*
MPA	Marine Protected Area(s)
MSY	Maximum Sustainable Yield
MUP	Manchester University Press
NAFO	Northwest Atlantic Fisheries Organization
NEAFC	North-East Atlantic Fisheries Commission
NJIL	*Nordic Journal of International Law*
nm	Nautical miles
NPFC	North Pacific Fisheries Commission

Term	**Definition**
NYUJ *Int'l LP*	*New York University Journal of International Law & Politics*
OCM	*Journal of Ocean & Coastal Management*
ODIL	*Ocean Development & International Law*
OJEU	*Official Journal of the European Union*
OJL	Series 'L' in the *Official Journal of the European Union*
OLDEPESCA	Organización Latinoamericana de Desarrollo Pesquero
OLR	*Oregon Law Review*
OUP	Oxford University Press
PCA	Permanent Court of Arbitration
PCIJ	Permanent Court of International Justice
Platzöder	Documents of the Third United Nations Conference on the Law of the Sea; compiled and edited by Renate Platzöder, and published by Oceana Publications
PRL&PJ	*Pacific Rim Law & Policy Journal*
Recueil des Cours	*Recueil des Cours de l'Académie de Droit International de la Haye*
RFMO/A	Regional and Subregional Fisheries Management Organization(s) and Arrangement(s)
RIAA	Reports of International Arbitral Awards
SDG	Sustainable Development Goal(s)
SDLR	*San Diego Law Review*
SEAFO	South East Atlantic Fisheries Organisation
SIDS	Small Island Developing States and Territories
SIOFA	Southern Indian Ocean Fisheries Arrangement
SPFF	South Pacific Forum Fisheries Agency of the Pacific Islands
SPRFMO	South Pacific Regional Fisheries Management Organisation
SRFC	Sub-Regional Fisheries Commission
t	Tonnes (measured in megagram units)
TAC	Total Allowable Catch
TAFE	Total Allowable Fishing Effort
UN	United Nations

Term	Definition
UNC	*Charter of the United Nations and Statute of the International Court of Justice*, 1945
UNCED	United Nations Conference on Environment and Development
UNCLOS/I	United Nations Conference on the Law of the Sea
UNCLOS/II	Second United Nations Conference on the Law of the Sea
UNCLOS/II Off.Records	Official Records of the Second United Nations Conference on the Law of the Sea, including Summary Records of Plenary Meetings and of Meetings of the Committee of the Whole, Annexes and Final Act; published by the United Nations Publications
UNCLOS/III	Third United Nations Conference on the Law of the Sea
UNCLOS/III Off.Records	Official Records of the Third United Nations Conference on the Law of the Sea, including Documents and Summary Records of the Plenary, General and of the First, Second and Third Committees Meetings, published by the United Nations Publications
UNGA	United Nations General Assembly
UNTS	United Nations Treaty Series
USC	United States Code (Official version of the Federal Statutory Code)
VCLT	*Vienna Convention on the Law of Treaties*, 1969
Virginia Commentary	Commentaries on the *United Nations Convention on the Law of the Sea*, 1982 based on formal and informal documentation of the Third United Nations Conference on the Law of the Sea; produced by the University of Virginia Center for Oceans Law and Policy, and published by Brill
VJIL	*Virginia Journal of International Law*
VMS	Vessel Monitoring System(s)
VUWLR	*Victoria University Wellington Law Review*
Washington LRSBJ	*Washington Law Review and State Bar Journal*
WCPFC	Western and Central Pacific Fisheries Commission

Term	**Definition**
WCPFC MHLC	Multilateral High-Level Conference for the Conservation and Management of Highly Migratory Fish Stocks in the Western and Central Pacific
WTO	World Trade Organization
WTO DSB	Dispute Settlement Panels and Appellate Bodies of the World Trade Organization
YbIEL	*Yearbook of International Environmental Law*
YbILC	*Yearbook of the United Nations International Law Commission*
Yearbook UN Law	*Max Planck Yearbook of United Nations Law*
YUP	Yale University Press
ZaöRV	*Zeitschrift für ausländisches öffentliches Recht und Völkerrecht*

International cases

International instruments

1. General introduction: fisheries disputes and the 1995 United Nations Fish Stocks Agreement

1.1 THE INEXORABLE RISE OF FISHERIES DISPUTES IN AN ERA OF ELUSIVE SUSTAINABILITY

It is no longer surprising to witness clashes over the exploration or exploitation of oil and gas deposits, and soon, similar contention may follow in the race for rare-earth elements. Nor should it be unexpected for hostilities to flare up by the diversion of watercourses or the management of transboundary groundwater reservoirs. The finiteness of natural sources has become an undeniable factor in shaping new perceptions of geoeconomics, which, in turn, have forged a radical conception of State interests through the lens of environmental sovereignty.[1]

This predicament finds its most striking expression in the context of fisheries.[2] For example, in the period immediately following the post-Cold War era, both the Russian Federation ('Russia') and the United States of America ('United States') identified the deteriorating condition of marine resources in the Pacific Ocean, rather than any other matter, as a grave threat to their shared

[1] *see* Haas, 'Constructing Environmental Conflicts from Resource Scarcity', *Global Environmental Politics* (2002) 1; Hulme, 'Environmental Security: Implications for international law', YbIEL (2009) 3; and Dalby, *Rethinking Environmental Security* (2022) 34*ff.*

[2] *see* Peterson–Teal, 'Ocean Fisheries as a Factor in Strategic Policy and Action', in Westing (Ed.) *Global Resources and International Conflict* (1986) 127; Barnes–Rosello, 'Fisheries and Maritime Security: Understanding and enhancing the connection', in Evans–Galani (Eds) *Maritime Security and the Law of the Sea* (2020) 48; and Menzel–Otto, 'Understanding Security and Sustainable Development in the Maritime Domain', in Otto–Menzel (Eds) *Global Challenges in Maritime Security* (2024) 7 *et seq.*

vital national interest.[3] Fishing, as a mutually exclusive activity—where one State's harvesting of stock units precludes others from doing so—exemplifies an environmental conflict arising from non-cooperative access to common-pool resources.[4] In such antagonistic circumstances, where unresolved disagreements have repeatedly led to disputes—and even to direct naval confrontations—the competing interests are expressed, *grosso modo*, between the coastal States of the region and the Distant Water Fishing States ('DWFS') operating therein. That said, the latter often lack clear identification as a stable or cohesive group beyond the confines of international political negotiations.[5] Fisheries, however, differ markedly from other marine natural sources in that they actively respond to conservation and management measures ('CMMs'), making them a true renewable resource. Given appropriate conditions, fish populations can remain available to humanity *ad infinitum* through sustainable regulation.[6] Countering the once-dominant narrative of a fated 'tragedy of the commons', recent decades have brought forth new intellectual insights highlighting the merits of cooperative regimes; those founded on scientifically sound measures and rigorously upheld.[7]

As a living resource, fish naturally cross human-made maritime boundaries. Achieving effective CMMs has proven challenging, even when stock migration is confined to adjoining sea areas under national jurisdiction and involves only a limited number of bordering coastal States. In such cases, disagreements over harvesting rights and equitable participation in decision-making processes have often hindered quota-sharing arrangements.[8] Far more troubling is

---

[3] Gjerde *et al.*, 'Living Resource Problems', in Broadus–Vartanov (Eds) *The Oceans and Environmental Security* (1994) 53–58.

[4] *see* Soroos, 'Conflict in the Use and Management of International Commons', in Käkönen (Ed.) *Perspectives on Environmental Conflict and International Politics* (1992) 32; Van Dyke, 'Sharing Ocean Resources – In a Time of Scarcity and Selfishness', in Scheiber (Ed.) *Law of the Sea* (2000) 35; and Palacios-Abrantes *et al.*, 'The Transboundary Nature of the World's Exploited Marine Species', *Nature Scientific Reports* (2020) 17668.

[5] FAO, *Mapping Distant-Water Fisheries Access Arrangements* (2022) 1*ff.*

[6] *see* Sumalia, 'A Review of Game-Theoretic Models of Fishing', *Marine Policy* (1999) 1 and Longhurst, *Mismanagement of Marine Fisheries* (2010) 291*ff.*

[7] *compare* Hardin, 'The Tragedy of the Commons', *Science* (1968) 12 and indicatively for the Law of the Sea; *see* Anand, 'Tyranny of the Freedom of the Seas Doctrine', *International Studies* (1973) 416, *with* Ostrom, *Governing the Commons* (1990) and Taylor–Lynch–Schechter (Eds) *Sustainable Fisheries* (2011); and *see* Finus–Schneider, 'Scope and Compatibility of Measures in International Fisheries Agreements', *Oxford Economic Papers* (2015) 865.

[8] *see* Ørebech, 'The "Lost Mackerel" of the North East Atlantic – The Flawed System of Trilateral and Bilateral Decision-Making', IJMCL (2013) 343

the situation in which fish stocks migrate into high seas areas beyond national jurisdiction. For instance, a substantial portion of *Boreogadus saida* (Arctic cod) stocks—originally shared by Norway and the Union of Soviet Socialist Republics ('USSR') since the 1970s, and subsequently by Russia following the USSR's dissolution in 1992—shifted into a high seas enclave in the early 1990s. Heightened interest in the straddling stock by DWFS, especially Iceland, prompted an acrimonious dispute that was abated in 1999 through the establishment of a trilateral cooperative arrangement.[9] In other ocean regions, jurisdictional complexities are further perplexed by long-standing maritime claims, which have resulted in either 'glaring regulatory gaps'[10] or fragmented regimes based on 'self-restraint', 'intermediate', or 'provisional' areas,[11] as well as by emerging institutional challenges linked to new fishing opportunities in an evolving geopolitical landscape.[12]

Notwithstanding internationally concerted efforts steered by the United Nations ('UN') since the 1940s, especially through the UN Food and Agriculture Organization ('FAO'),[13] intensive industrial fishing on a global scale has led to the overexploitation of fish populations. In most fishing grounds worldwide, stocks were targeted at biologically unsustainable levels, triggering collapses that had tremendous environmental, economic, and social repercussions. In

---

and Østhagen–Spijkers–Totland, 'Collapse of Cooperation? The North-Atlantic Mackerel Dispute and Lessons for International Cooperation on Transboundary Fish Stocks', *Maritime Studies* (2020) 155.

9 Stokke, 'Managing Fisheries in the Barents Sea Loophole: Interplay with the UN Fish Stocks Agreement', ODIL (2001) 241.

10 One pertinent case is the territorial dispute between Argentina and the United Kingdom in the southwest Atlantic, which has stalled efforts to strengthen cooperation and formalize the South Atlantic Fisheries Commission into a regulatory body; *qv.*, Overfield–Yllemo, 'Options for Sustainable High Seas Fisheries Management in the Southwest Atlantic', *International Law Studies* (2025) 87–88.

11 *see* Kim, 'Governing Fishing Stocks in Northeast Asia's Disputed Waters: Preventing a "tragedy of the commons"?', IJMCL (2018) 495 and Van Logchem, *The Rights and Obligations of States in Disputed Maritime Areas* (2021) 33–38.

12 *e.g.* the United Kingdom, following its withdrawal from the European Union, reassumed full sovereignty in its own right, as declared on 31/12/2020 over all matters covered by the Convention [CN.577.2020:XXI-6] and by the Agreement [CN.578.2020:XXI-7]; for the uncertainty that ensued in the context of the 'new and special relationship' for fisheries *see* Bjørndal–Munro, *Brexit and the Consequences for Fisheries Management in the North Sea* (2020) and Churchill, 'Fisheries Management in European Union and United Kingdom Waters after Brexit: A change for the better?', *Ocean Yearbook* (2022) 287.

13 Henriksen, 'The FAO and Ocean Governance', in Attard–Fitzmaurice–Ntovas (Eds) *UN Specialized Agencies and Global Ocean Governance* (2018) 3.

addition to aggressive—and frequently destructive—harvesting pressure, the persistence of illegal, unreported and unregulated ('IUU') fishing, and the lack of enforceable obligations for transparent fleet capacity control by FAO,[14] as well as at the macroeconomic level, through the World Trade Organization ('WTO'), by means of a comprehensive treaty addressing subsidies that contribute to excessive capacity,[15] remain root causes of ineffective regulation. Even though aquaculture exhibits an alternative potential,[16] ocean capture fisheries will remain at least by 2050 the main and most important supply to assure global food security, alleviate poverty in developing States and meet the nutritional demands for consumption of fish by the continuously growing human population and future generations.[17]

Looking ahead, the United Nations General Assembly ('UNGA') in the form of political aspirations, establishing as well some normative legal value, has therefore attached high importance to the protection and preservation of marine resources as part of the Sustainable Development Goals ('SDG') to be attained by 2030.[18] In particular, target (4) of the SDG 14 on 'Life Below Water' has called upon States to sustainably regulate fishing and restore stocks in the shortest time feasible, at least to levels that can produce Maximum Sustainable Yields ('MSY') as determined by their biological characteristics. To support the timely delivery of the goals, UN has proclaimed the period leading up to 2030 as both the *Decade of Ocean Science for Sustainable Development* to promote the role of knowledge and technology in achieving especially SDG 14 and the *Decade on Ecosystem Restoration* in stressing the need for international cooperation at large.[19]

---

[14] Currently implemented on the basis of action plans within the framework of the voluntary *Code of Conduct for Responsible Fisheries*, which was unanimously adopted by FAO on 31/10/1995.

[15] As evidenced by the long-awaited adoption of the *Agreement on Fisheries Subsidies* (2022), although limited in scope; *qv.* Hoekman–Mavroidis–Sasmal, 'Managing Externalities in the WTO: The Agreement on Fisheries Subsidies', *Journal of International Economic Law* (2023) 266.

[16] Hilborn–Hilborn, *Ocean Recovery* (2019) 163*ff.*

[17] FAO, *The State of Food Security and Nutrition in the World* (2023).

[18] A/Res.70/1 'Transforming our World, The 2030 Agenda for Sustainable Development' (25/9/2015) and A/Res.71/312 'Our Ocean, Our Future' (6/7/2017); *qv.* Telesetsky, 'The Sustainable Development Goals and Informal Lawmaking Processes', in Klein (Ed.) *Unconventional Lawmaking in the Law of the Sea* (2022) 271.

[19] A/Res.72/73 'Oceans and the Law of the Sea' (5/12/2017), ¶292 and A/Res.73/284 'United Nations Decade on Ecosystem Restoration (2021–2030)' (1/3/2019).

Yet, progress still falls significantly short of what is necessary to achieve SDGs. At the 2023 halfway mark, most targets were still out of reach, underscoring the urgent need for renewed high-level political commitment and fresh institutional guidance on transformative and accelerated actions—including those essential for the sustainability of fishery resources—in the years ahead.[20] This is not the first instance in which political ambitions, pursued under the banner of globalization, have failed to come to fruition. Commitments undertaken at the World Summit on Sustainable Development to urgently restore depleted fisheries—ideally by 2015—to MSY levels were ultimately unfulfilled.[21] Data compiled and verified by FAO indicate that, despite noticeable improvements in several regions, the proportion of global marine stocks fished within biologically sustainable levels continues to decline. Since 1980, the share of ocean capture fisheries subjected to unsustainable harvesting has more than doubled. As reported at the UN Ocean Conference 'Our Ocean, Our Future, Our Responsibility', held in Nice from 9 to 13 June 2025, 35.5% of stocks are classified as overfished; this is an alarming figure that continues to rise at a rate of approximately 1% per year.[22] At the same time, per capita consumption has been steadily rising.[23] States must now urgently address the need to rigorously implement fisheries laws and reform related policies in accordance with the best available scientific evidence, in order to rebuild depleted stocks and prevent further degradation of those already exploited at, or near, their MSY levels.[24] A more dire reality is conveyed by the expert scholarship positing that fisheries may have been declining at an even faster rate than FAO data suggest.[25]

Even with appropriate international measures to address overexploitation, climate change—one of the most formidable challenges of our time—along with ongoing ecosystem degradation and biodiversity loss caused by multiple interacting human-induced stressors (e.g. marine pollution and contamination,

---

[20] *see* UN, 'Global Sustainable Development Report' (2023), 1 *et seq* and 'Sustainable Development Goals Report' (2024) 34–37.

[21] UN 'Plan of Implementation of the World Summit on Sustainable Development' (4/9/2002), ¶31(a).

[22] FAO, *Review of the State of World Marine Fishery Resources* (2025) 21 *et seq*; and *see* A/Res.78/68 'Sustainable Fisheries' (5/12/2023), 5*ff* and A/Res.79/68 'Oceans and the Law of the Sea' (19/3/2024), ¶8.

[23] FAO, *The State of World Fisheries and Aquaculture* (2024) 42–53.

[24] A/Res.79/145 'Sustainable Fisheries' (16/12/2024), 5.

[25] Pauly–Zeller, 'Catch Reconstructions Reveal that Global Marine Fisheries Catches are Higher than Reported and Declining', *Nature Communications* (2016) 10244 and Cheung–Pauly–Sumaila, 'Hope or Despair Revisited: Assessing progress and new challenges in global fisheries', *Fish and Fisheries* (2025) 257.

ocean acidification and eutrophication, invasive and predatory species) already hamper progress towards the delivery of SDGs and pose major hurdles to cooperative regulatory regimes.[26] Fisheries disputes are bound to be driven by dynamic circumstances that already exert tremendous environmental pressure, causing spatial shifts in stock migration patterns. These shifts may represent long-term directional changes rather than temporary or random variations. Under such conditions, prevailing assumptions, legal concepts, and the interpretation and application of conservation and management rules will undoubtedly be tested, given the far-reaching economic, political, and social consequences. The International Tribunal for the Law of the Sea ('ITLOS') has already emphasized that the impacts of climate change and ocean acidification on fisheries negatively affect the 'income, livelihoods, and food security of marine resource-dependent communities' and put 'key cultural dimensions of lives and livelihoods at risk'.[27] As we are becoming increasingly aware of how the 'whole-planet' and 'global ocean' effects impact human living conditions, our foremost concern regarding fisheries will not be only about sustaining quantity (i.e. the abundance and productivity of stocks), but importantly also their comestible quality (i.e. the health of the fish populations) as aquatic habitats are already experiencing pervasive changes in their physical, chemical and biological conditions.[28]

Fisheries can be collectively utilized, but only within an enduring framework of sustainable regulation. This demands a comprehensive suite of CMMs to ensure that fishing activities remain biologically productive, ecologically responsible, economically viable, and socially inclusive. Such outcomes are achievable only through robust international cooperation. Without it, increasingly complex fisheries disputes will assuredly take centre stage in

---

26 UN Environment Programme, 'Strengthening Ocean Efforts to Tackle Climate Change, Marine Biodiversity Loss and Pollution' (1/3/2024).

27 *Climate Change and International Law* (Advisory Opinion) 2024, §66.

28 *see* Zalasiewicz–Williams, 'The Anthropocene Ocean in Its Deep Time Context', in Vidas–Schei (Eds) *The World Ocean in Globalisation* (2011) 19; Molenaar, 'Integrating Climate Change in International Fisheries Law', in Johansen–Busch–Jakobsen (Eds) *The Law of the Sea and Climate Change* (2020) 263; Stokke *et al.*, 'Climate Change and Resilient Fisheries Management', in Stokke–Østhagen–Raspotnik (Eds) *Marine Resources, Climate Change and International Management Regimes* (2022) 3*ff*; and Lennan, 'Fisheries Redistribution under Climate Change: Rethinking the law to address the "governance gap"?', in Platjouw–Pozdnakova (Eds) *The Environmental Rule of Law for Oceans* (2023) 163.

contemporary global affairs.[29] Where cooperation falters, compulsory and binding procedures for the peaceful resolution of disputes will become an indispensable means of realizing fair and ethical ocean governance, guided by the rule of law.

1.2 PROGRESSIVE DEVELOPMENT OF INTERNATIONAL FISHERIES LAW BY THE UNITED NATIONS AND THE ACQUIRED NORM OF PEACEFUL SETTLEMENT

The normative development of the law of the sea is deeply rooted in the jurisprudence of courts and tribunals. In particular, since the establishment of its modern foundations in the 1893 *Rights of Jurisdiction in the Bering's Sea and the Preservation of Fur Seals* arbitration,[30] fisheries law has been characterized by the institutionalization of dispute settlement procedures.[31] Being an open resource of natural wealth and ineluctably a source of potential international conflict, the importance of fisheries is reflected in the abundance of case law, even if one were to begin simply by recounting the extensive record of cases starting with the 1910 *North Atlantic Coast Fisheries* arbitration, which marked an emblematic turn in the course of progressively developing conservation rights and obligations upon the doctrine of maritime jurisdictions.[32] Not long after, the imperative necessity for international cooperation to regulate the biological wealth of seas was foregrounded by the League of Nations ('LoN') Committee of Experts for the Progressive Codification of International Law.

[29] UN 'World Ocean Assessment II' (2021), Chapters 6(C) & 15; A/Res.78/339 'Oceans and the Law of the Sea' (1/9/2023), 2¶1; UN, 'Intergovernmental Panel on Climate Change Report' (2023); UN, 'United Nations Convention on the Law of the Sea at 40: Successes and future prospects' (2023); and ILA, *Paris White Paper 17 Ocean* (2023), 61*ff.*

[30] This was a seminal case concerning the preservation of marine living resources, in which the United States unsuccessfully attempted to assert coastal State jurisdiction over conservation matters in high seas areas adjacent to its territorial sea; *qv.* Williams, 'Reminiscences of the Bering Sea Arbitration', AJIL (1943) 562.

[31] *see* Orrego-Vicuña, 'From the 1893 *Bering Sea Fur-Seals* Case to the 1999 *Southern Bluefin Tuna* Cases: A century of efforts at conservation of the living resources of the high seas', YbIEL (2000) 40 and Rothwell, 'The Law of the Sea, International Courts, and Judicialization', AJIL *Unbound* (2021) 373.

[32] In the *North Atlantic Coast Fisheries Case* (Award) 1910, the peculiar prominence of the fisheries question in the treaty arrangements was the cause of further unexpected differences between the United States and the United Kingdom; *qv.* Anderson, 'The Final Outcome of the Fisheries Arbitration', AJIL (1913) 1.

On the question of whether it was possible to establish by way of international agreement rules regarding the exploitation of the products of the sea, Rapporteur José León Suárez (Argentina) noted that the 'biologico-geographical solidarity of species … should find its counterpart in legal solidarity in the sphere of international law'. Nonetheless, acknowledging the disparity of views expressed among States, he concluded that any prospective regulatory scheme should consider not only the applicable principles and measures but also the most effective method for ensuring their implementation.[33]

As fisheries law primarily derives from a mix of rules and principles, and a combination of practices, standards and guidelines aimed at the rational and optimal utilization—and equitable allocation—of marine resources, it has remained fragmented and non-hierarchical in nature.[34] This legal patchwork renders it susceptible to a wide array of disputes, from genuine disagreements over the exercise of rights and fulfilment of corresponding obligations by States, often set against the backdrop of overlapping jurisdictional claims, to more strategic maritime assertions intended to contest sovereignty over sea areas. Given that many of these disputes remain unresolved, and others are bound to be precipitated by deteriorating environmental conditions, there is a growing expectation of an intensifying trend towards further judicialization of law of the sea disputes,[35] especially those concerning fisheries.[36]

1.2.1 United Nations Conference on the Law of the Sea, 1958

The field of law of the sea, particularly the legal regime pertaining to territorial waters and high seas, was among the first topics selected for codification by the International Law Commission ('ILC') in 1949.[37] Later that year, a dispute

---

[33] Committee of Experts for the Progressive Codification of International Law, 'Exploitation of the Products of the Sea' (8/12/1925) [(1926) AJIL *Suppl.* 20(5):232–240]; and *see* Hudson, 'The Progressive Codification of International Law', AJIL (1926) 667.

[34] *see* Christy, 'Fisheries and the New Conventions on the Law of the Sea', SDLR (1970) 457 and Molenaar–Caddell, 'International Fisheries Law: Achievements, limitations and challenges', in Caddell–Molenaar (Eds) *Strengthening International Fisheries Law in an Era of Changing Oceans* (2019) 5.

[35] McLaughlin-Mitchell–Owsiak, 'Judicialization of the Sea: Bargaining in the shadow of UNCLOS', AJIL (2021) 584.

[36] *see* Harrison, 'Emerging Dispute Settlement Principles and Practices in RFMOs', in Kunoy–Heidar–Yiallourides (Eds) *International Fisheries Law* (2024) 88.

[37] A/CN.4/13*Corr.*1–3 'Report of the International Law Commission on the work of its First Session' (12/4/1949) [1949*YbILC*(I) 279].

that had been brewing for nearly four decades between the United Kingdom of Great Britain and Northern Ireland ('United Kingdom') and Norway, over the latter's unilateral claims to exclusive fishery jurisdiction—allegedly exceeding limits generally accepted in international practice—reached its peak. The United Kingdom referred the matter to the International Court of Justice ('ICJ'), asking whether the notional base system used by Norway to delimit its territorial sea, which both parties agreed to be 4 nautical miles (nm) in breadth, was the correct method. The Court ruled in favour of Norway, noting that, in addition to the geographic realities specific to its coastline, the exercise of historic fishery rights, as well as considerations of economic interest and dependence on marine resources, should be taken into account when defining baselines, provided they remained moderate and reasonable.[38]

The Commission concluded its study in 1956 by adopting 73 draft *Articles Concerning the Law of the Sea* ('*ILC Draft Articles*'), which were presented to UNGA with a recommendation to call an international conference to examine the law of the sea in all its legal, technical, biological, economic, and political aspects.[39] The Commission admitted that the distinction between progressive development and codification had been peculiarly difficult to maintain in the domain of the law of the sea.[40] A year later, UNGA convened the United Nations Conference on the Law of the Sea,[41] which met at Geneva from 24 February to 27 April 1958 ('UNCLOS/I'). The conference culminated its negotiations with the adoption of the Final Act containing four separate conventions (commonly known as the '*Geneva Conventions*'), namely: the *Convention on the Territorial Sea and the Contiguous Zone*; the *Convention on the High Seas* ('HSC'); the *Convention on Fishing and Conservation of the Living Resources of the High Seas* ('FCC' or the '*Fishing & Conservation Convention*' (1958)); and the *Convention on the Continental Shelf* ('CSC'). In addition, the conference adopted the *Optional Protocol of Signature concerning the Compulsory Settlement of Disputes* ('*Optional Protocol*' (1958)) and nine resolutions.[42]

---

38 *Fisheries Case* (Judgment) 1951, pp.128–133.

39 A/CN.4/104 'ILC Report Covering the Work of its Eighth Session' (23/4 to 4/7/1956) [1956*YbILC*(II) 253].

40 *ibid.*, 255–256; a statement that attracted warranted criticism, as noted in Johnson, 'The Preparation of the 1958 Geneva Conference on the Law of the Sea', ICLQ (1959) 128–129.

41 A/Res.1105/XI 'International Conference of Plenipotentiaries to Examine the Law of the Sea' (21/2/1957).

42 A/Conf.13/L.58 'Final Act of the United Nations Conference on the Law of the Sea' (27/4/1958) [*Geneva Off.Records* II:146]. Of relevance to this book are the resolutions on 'International Fishery Conservation Conventions'; the

The *Fishing & Conservation Convention* (1958) was the first instrument with universal reach to progressively develop fisheries law. Nevertheless, it failed to garner widespread ratification and was the last to enter into force, receiving considerably less attention than the other three conventions. To better understand the convention's development, it is worthwhile to revisit the history of the negotiations to highlight aspects of the working methods through which the treaty text was adopted. The question of conservation of fisheries, which had been recognized as a matter of common concern and overriding interest to all States, was allocated to the Third Committee of UNCLOS/I. The Committee was tasked with examining and negotiating *ILC Draft Articles* 49–60, which covered fishing and conservation of marine living resources on the high seas. After considering the recommendations of the General Committee on the organization of the conference's work, the Committee decided to begin by holding a series of 10 meetings—from the 3rd to the 12th meeting—dedicated to a general debate on the first reading of the articles. In total, 42 substantive meetings were held between 26 February and 22 April 1958. At the 13th meeting, Chairman Carlos Sucre (Panama) put the articles to a provisional vote. Based on the deliberations and the outcomes of the votes, the articles were referred to a subcommittee for drafting improvements. The subcommittee held a single meeting on 16 April, making only a few changes before returning the articles to the Third Committee for a second reading and a definitive vote on 23 April. The articles were accepted and subsequently referred to the plenary of the General Committee. These corresponded to the numbering of the *ILC Draft Articles*, plus a new draft article 60A.[43] The latter, so-called 'Icelandic clause', had been added by the Third Committee to grant preferential rights to a State with overwhelming dependence on its coastal fisheries. Following deliberations on the articles in the plenary's 15th to 18th meetings, a decision was made—based on a defeated motion to the contrary—that draft article 60A should be put to a separate vote. On 25 April, draft articles 49–60 were provisionally adopted by 44 votes to 16, with 8 abstentions. Draft article 60A, receiving 30 votes in favour and 21 against, with 18 abstentions, failed to obtain the required two-thirds majority and was not adopted.[44] On 26 April, the adopted articles—proposed to form the text of the *Fishing & Conservation Convention* (1958)—were put *en bloc* to a final vote and approved by 45 votes in favour, 1 against, and 18 abstentions.[45]

'Co-operation in Conservation Measures'; and the 'Special Situations relating to Coastal Fisheries'.

[43] A/Conf.13/C.3/L.21/Add.1 'Text of the articles adopted by the Third Committee' (23/4/1958) [*Geneva Off.Records* V:160].

[44] UNCLOS/I Plenary 15th meeting (25/4/1958) [*Geneva Off.Records* II:42*ff*].

[45] UNCLOS/I Plenary 18th meeting (26/4/1958) [*Geneva Off.Records* II:57*ff*].

Being drafted in an exceptionally legalistic manner by ILC and subsequently accepted with only minor language refinement and modifications at UNCLOS/I, the treaty text aimed to strike a strict balance of interests. Yet, it ultimately left both DWFS and coastal States dissatisfied. The provisions recognizing a special interest to coastal States in high seas areas were perceived by DWFS as encouraging an unprecedented seaward extension of national jurisdiction. Conversely, coastal States were underwhelmed by the substantive principles, which remained merely declaratory in the absence of enforcement jurisdiction authorized by the convention. Furthermore, both sides viewed the inclusion of stringent procedures for the compulsory settlement of fisheries disputes as an unwelcome necessity; albeit for different reasons. As recounted by William Herrington—a uniquely influential member of the United States delegation at Geneva on all fisheries issues—while the convention was initially met with the highest hopes for success, its principal impact did not lie in the fulfilment of its specific terms. Rather, its significance lay in its moral contribution to advancing international discourse on management policies against overexploitation, as well as in its technical influence on the normative development of standards by subsequent treaties.[46] Since then, scholarly opinion generally regards the *Fishing & Conservation Convention* (1958) as having had 'little practical significance'.[47] Even so, two seminal contributions to the progressive development of fisheries law should be acknowledged. First, as part of the evolution of extended fishing zones under exclusive coastal jurisdiction ('EFZ'), the convention stressed the need for cooperation to prevent overexploitation by coordinating CMMs for marine living resources across maritime boundaries.[48] Second, it was the only one among the *Geneva Conventions* to incorporate dispute settlement rules directly into the treaty text by securing the parties' prior acceptance of the principle of compulsory adjudication.[49] This represented a remarkable and novel feature of the convention, as it is well established in international law that no State can be compelled, without

---

46 Scheiber, 'The Law of the Sea Institute: A new forum for debate of ocean law in the 1960s "decade of uncertainty"', in Scheiber *et al.* (Eds) *Ocean Law Debates* (2018) 55–64.

47 *e.g.*, Jensen–Bankes, 'Compulsory and Binding Dispute Resolution under the United Nations Convention on the Law of the Sea: Introduction', ODIL (2017) 210 (*fn*6).

48 Sohn–Noyes, *Cases and Materials on the Law of the Sea* (2004) 737; however, the second edition by Sohn–Noyes–Franckx–Juras, *Cases and Materials on the Law of the Sea* (2014), does not explicitly address this point.

49 Chandra, *Law of the Sea* (1985) 35 and Oda, *Fifty Years of the Law of the Sea* (2003) 144*ff.*

its consent, to submit its disputes with other States to any form of settlement procedure.[50]

1.2.2 Second United Nations Conference on the Law of the Sea, 1960

In hindsight, the lack of enthusiasm from States towards the *Fishing & Conservation Convention* (1958) was chiefly due to the unresolved issue of the breadth of the territorial sea—in excess of the customary 3 nm limit, at the time—and the extent of fisheries jurisdiction to be exercised by coastal States in adjacent high seas areas. Disagreement on these crucial issues had cast its shadow over the negotiations of the fisheries provisions at UNCLOS/I. Negotiations continued at the Second United Nations Conference on the Law of the Sea, convened in Geneva from 17 March to 26 April 1960 ('UNCLOS/II'), with the specific aim of resolving the outstanding issues of the territorial sea limits and fisheries jurisdiction.[51] The conference fell short by a single negative vote of securing the two-thirds majority required to adopt a text based on a compromise formula. This formula envisioned the establishment of a zone extending up to 12 nm, comprising a fishing zone contiguous to the territorial sea (the '6 nm + 6 nm formula'), with its regime hinging on the *mutatis mutandis* application of the dispute settlement procedures provided in the *Fishing & Conservation Convention* (1958).[52] Consequently, the conference adopted two

[50] *see* in *Status of Eastern Carelia* (Advisory Opinion) 1923, p.27; *Corfu Channel* (Preliminary Objections) 1948, p.27; *East Timor* (Judgment) 1995, §26 and on the significance of the principle of consent to adjudication *see* Orakhelashvili, 'The Concept of International Judicial Jurisdiction: A reappraisal', *Law and Practice of International Courts and Tribunals* (2003) 508*ff*; Cassese, International Law (2005) 58–59; Brown, *A Common Law of International Adjudication* (2007) 37–38; Crawford, *Brownlie's Principles of Public International Law* (2019) 692*ff*; and de Brabandere, 'The Principles and Methods of International Dispute Settlement', in Evans (Ed.) *International Law* (2024) 531–532.

[51] A/Res.1307/XIII 'Convening of a Second United Nations Conference on the Law of the Sea' (10/12/1958).

[52] A/Conf.19/C.1/L.4 'Canada: Proposal' (24/3/1960) [*UNCLOS/II Off. Records* I:167] and A/Conf.19/C.1/L.10 'Canada and United States of America: Proposal' (8/4/1960) [*UNCLOS/II Off.Records* I:169]; and *see* Dean, 'The Second Geneva Conference on the Law of the Sea: The Fight for Freedom of the Seas', AJIL (1960) 774–777; Bowett, 'The Second United Nations Conference on the Law of the Sea', ICLQ (1960) 424; and Bishop, 'The 1958 Geneva Convention on Fishing and Conservation of the Living Resources of the High Seas', *Columbia Law Review* (1962) 1224.

resolutions and adjourned a final decision on these issues to further negotiations *sine die*.[53]

1.2.3 Third United Nations Conference on the Law of the Sea, 1973–1982

Mounting tensions between DWFS and coastal States brought deep divisions over access to international fisheries to the forefront. At the same time, rapid advancements in fishing technologies, which expanded the operational capacity of fleets, raised serious concerns about the sustainability of marine living resources. Unresolved legal questions surrounding conservation and management amplified the prevailing sense of urgency. In response to these growing concerns—both for the rational use of ocean resources and for the preservation of peace and stability in the global order—UNGA called the Third UN Conference on the Law of the Sea ('UNCLOS/III'). Mandated to address all matters relating to the law of the sea, the conference was an ambitious undertaking unparalleled in both scale and scope.[54] The inaugural session was held in New York from 3 to 15 December 1973, and the conference was initially expected to conclude its work by 1975.[55] All negotiations were conducted under the cardinal 'package deal' principle, which required the acceptance of a comprehensive text in its entirety, rather than as a series of separate treaties, as was the case with the *Geneva Conventions*. To this end, negotiations centred on a 'single negotiating text' for each Committee, which the chairman was permitted to revise during the early stages without any formal binding commitment from delegations. As discussions progressed, further revisions required broad support to demonstrate a substantially improved prospect of

---

53 A/Conf.19/L.15 'Final Act' (26/4/1960) [*UNCLOS/II Off.Records* I:175].

54 A/Res.2749/XXV 'Declaration of Principles Governing the Seabed and the Ocean Floor, and the Subsoil thereof, Beyond the Limits of National Jurisdiction' (17/12/1970) and A/Res.2750/XXV 'Reservation Exclusively for Peaceful Purposes of the Seabed and the Ocean Floor and the Subsoil thereof, Underlying the High Seas Beyond the Limits of Present National Jurisdiction and Use of their Resources in the Interests of Mankind, and Convening of a Conference on the Law of the Sea' (17/12/1970); *qv.* Stevenson–Oxman, 'The Preparations for the Law of the Sea Conference', AJIL (1974) 13.

55 A/Res.3067/XXVIII 'Reservation Exclusively for Peaceful Purposes of the Seabed and the Ocean Floor, and the Subsoil thereof, Underlying the High Seas Beyond the Limits of Present National Jurisdiction and Use of their Resources in the Interests of Mankind, and Convening of the 3rd United Nations Conference on the Law of the Seabed and the Ocean Floor Beyond the Limits of National Jurisdiction' (16/11/1973).

reaching consensus. In addition, the decision-making process was guided by an informal commitment to procedural etiquette, requiring 'every effort to reach agreement on substantive matters by way of consensus', with voting not permitted 'until all efforts at consensus have been exhausted'.[56] The organic nature of the drafting process fostered essential *quid pro quo* compromises and played a vital role in tempering interpersonal, inter-delegation, and inter-coalition dynamics at the longest and most expansive conference ever held.[57]

The topic of fisheries was assigned to the Second Committee, chaired by Reynaldo Galindo Pohl (El Salvador), which was tasked with addressing some of the most complex and controversial subjects, in comparison with those handled by the other main Committees and the Plenary.[58] The sheer workload was evident from the outset, as it quickly became clear that an early conclusion of UNCLOS/III was unrealistic. Even narrowing down and resolving the basic political differences, let alone formulating the numerous legal issues within the Second Committee before opening substantive debates, was expected to take years.[59] Early on, fisheries were understood to be the true 'chess game' of the negotiations. Coastal States, driven by predominant interests in offshore natural resources, had set up the board by previously enacting domestic legislation to extend national authority over adjacent high seas areas, thereby signalling their unilateral intentions, purportedly for the purpose of fisheries conservation.[60] Fisheries law was in a state of flux, as evidenced in the cases jointly brought by the Federal Republic of Germany and the United Kingdom against Iceland following the *Cod war*, which escalated at a politically opportune time amidst the UNCLOS/III negotiations.[61] The importance of fishery resources to all States had gradually led to the decoupling of questions concerning the legal

---

[56] Appendix to A/Conf.62/30/Rev.1 'Rules of Procedure' (27/6/1974) [(1974) 13(5)ILM1209] and A/Conf.62/62 'Organization of work' (13/4/1978) [*UNCLOS/III Off.Records* X:6]; and *see* Vignes, 'Will the Third Conference on the Law of the Sea Work According to the Consensus Rule?', AJIL (1975) 119 and Sebenius, *Negotiating the Law of the Sea* (1984) 12–13.

[57] *see* Miles, *Global Ocean Politics* (1998) and Koh, *Building a New Legal Order for the Oceans* (2020).

[58] A/Conf.62/29 'Organization of the second session of the Conference and allocation of items (2/7/1974) [*UNCLOS/III Off.Records* III:59].

[59] Stevenson–Oxman, 'The Third United Nations Conference on the Law of the Sea: The 1974 Caracas Session', AJIL (1975) 12.

[60] *see* Knight, 'Law of the Sea Negotiations 1971–1972: From internationalism to nationalism', SDLR (1972) 386–387; Jacobson, 'Bridging the Gap to International Fisheries Agreement: A guide for unilateral action', SDLR (1972) 455–456; and Schram, 'Self-Government and Natural Resources', NJIL (1988) 331–334.

[61] Katz, 'Issues Arising in the Icelandic Fisheries Case', ICLQ (1973) 83.

nature and extent of coastal jurisdiction from the juridical status of the territorial sea. Whereas, based on the near-miss agreement and general consensus at UNCLOS/II, the customary limits of coastal jurisdiction were already widely accepted to cover waters up to 12 nm from the coast, the concept of a fishery zone extending into adjacent high seas areas—wherein the coastal State would assume exclusive jurisdiction or claim preferential fishing rights against DWFS—remained highly contested. While UNCLOS/III was underway, ICJ balanced the preferential fishing rights claimed by a coastal State with the traditional fishing rights of DWFS, deliberately avoiding a decision *sub specie legis ferendae*. It found that neither of these rights was absolute, and the parties remained under the obligation to cooperate and negotiate an equitable solution. Nonetheless, by extending national jurisdiction over fisheries to 50 nm from its coast, Iceland had infringed upon the principle of reasonable regard for the interests of other States, as enshrined in article 2 HSC.[62]

Alongside the formal endorsement of extending the territorial sea to up to 12 nm, the consensus reached in 1977 at UNCLOS/III on the novel concept of Exclusive Economic Zone ('EEZ') was set to profoundly transform the jurisdictional paradigm in the governance of marine resources. From a historical perspective, the institution of EEZ represented the ultimate and most definitive concession by DWFS within the package deal, marking the *non plus ultra* in the legal obsession of coastal States with expanding their offshore jurisdiction and sovereign rights.[63] For the first time in over three centuries—since Huig de Groot's *Mare Liberum* (1609) gained ascendancy over John Selden's *Mare Clausum* (1652)—States revisited the traditional principle of freedom of fishing by agreeing to extend coastal jurisdiction into areas that were previously part of the high seas.[64] On account of fisheries, the regime of EEZ was driven predominantly by coastal States' conservation arguments, pointing to

---

[62] *Fisheries Jurisdiction Cases* (Merits) 1974, §§51–54 & 67–78.

[63] For a critical exposition of coastal States' incessant tendencies towards the normative territorialization of high seas areas, before and after the Convention, *see* Allen 'Control of Fisheries Beyond Three Miles' and 'Legal Limits of Coastal Fishery Protection', published respectively, in *Washington* LRSBJ (1939) 91 and (1946) 1; Scelle, 'Obsession du Territoire', in van Asbeck (Ed.) *Symbolae Verzijl* (1958) 347; Oudendijk, *Status and Extent of Adjacent Waters* (1970) *passim*; Bardonnet, 'Frontières Terrestres et Frontières Maritimes', *Annuaire Français* (1989) 46*ff*; and Oxman, 'The Territorial Temptation: A siren song at sea', AJIL (2006) 833.

[64] Christy, 'Transitions in the Management and Distribution of International Fisheries', *International Organization* (1977) 235 and Finlay, 'The Proposed New Convention on the Law of the Sea: A candid appraisal', *Syracuse Journal of International Law and Commerce* (1980) 139.

estimates that over 90% of global commercial fish catches occurred within 200 nm from shore. Although the spatial extension of coastal jurisdiction echoed a pragmatic, if not aspirational, approach to managing most types of marine living resources, the EEZ regime failed to provide a rational framework for addressing pelagic fisheries characterized by extensive oceanic migrations.[65] Of particular gravity are the remarks of Arvid Pardo (Malta), the *inspirateur* of the UNGA Resolutions 2749/1970 and 2750/1970,[66] who considered this approach seriously flawed in terms of the conservation and management of marine living resources. Early in the negotiations, he cautioned against perceiving the marine environment as an unrestricted arena for State activity, where traditionally *laissez faire* rules had instigated competitive and uneconomical fishing. In this regard, he foresaw that despite all efforts at UNCLOS/III, its aftermath would entail further depletion of the marine living resources and increasingly bitter conflict between DWFS and coastal States.[67] This has been the case with straddling fish stocks and stocks of highly migratory species (hereinafter referred to colloquially together as 'fish stocks'), for which, in the absence of an agreed legal regime coextensive with their distribution or migration range, the task of their conservation and management falls to regional cooperation efforts.[68]

Closely related to these concerns, and within the broader context of progressively developing the substantive law of the sea, the integration of dispute settlement procedures into the treaty text was regarded by UNCLOS/III President Hamilton Shirley Amerasinghe (Sri Lanka) as the pivot upon which the delicate equilibrium of the entire package deal rested.[69] This enunciated

---

[65] *see* Alexander–Hodgson, 'The Impact of the 200-Mile Economic Zone on the Law of the Sea', SDLR (1975) 569 and Brodd, 'A Common Heritage Approach to Fisheries through Regional Controls', NYUJ *Int'l LP* (1977) 191.

[66] *supra cit.* (*fn*54).

[67] Nowhere is this more clearly expressed than in his scholarly writings in 'Development of Ocean Space – An International Dilemma', *Louisiana Law Review* (1970) 49–50; 'The Emerging Law of the Sea', in Walsh (Ed.) *The Law of the Sea: Issues in ocean resource management* (1977) 63; 'The Law of the Sea – Where Now? Before and After', LCP (1983) 104; and [with Borgese] *The New International Economic Order and the Law of the Sea* (1976) *passim.*

[68] *compare* Rayfuse, 'Some Reflections on What's Wrong with the Law of the Sea', in Ryngaert–Molenaar–Nouwen (Eds) *What's Wrong with International Law?* (2015) 22 *with* Moore, 'The United Nations Convention on the Law of the Sea: One of the greatest achievements in the international rule of law', in Nordquist–Moore–Long (Eds) *Legal Order in the World's Oceans* (2017) 11 & 21–22.

[69] A/Conf.62/WP.9/Add.1 'Memorandum by the President of the Conference on document A/Conf.62/WP.9' (31/3/1976) [*UNCLOS/III Off.Records* V:122(¶6)].

the prevailing preoccupation with the concept of 'the package', which, in technical terms, guided the negotiations towards reaching consensus on a single, comprehensive text.[70] The pivotal role of dispute settlement resonates with the 'procedural architecture'[71] crafted by UNCLOS/III and the rulemaking approach of adopting, by consensus, provisions that absorbed unresolved issues—such as those relating to fisheries and the protection and preservation of the marine environment—and which, in turn, were susceptible to what has been eloquently described as the 'creeping jurisdiction syndrome'.[72] For these matters, the Convention establishes a framework of conduct to which States are referred, calling upon them to engage with each other and cooperate through formal processes. In cases of disagreement, this normative and process-oriented framework also provides for recourse to third-party dispute settlement. This is of particular importance in circumstances where the 'intentionalism paradox'[73] arises in treaty interpretation, giving rise to the doctrinal challenge of 'auto-interpretation'.[74] Briefly put, while States in disagreement may assert the right to invoke self-judging arguments—claiming to have held differing intentions during the negotiation of the treaty text—they cannot, under the guise of sovereign equality, refuse to submit their dispute to third-party adjudication while simultaneously insisting that their own interpretation must prevail over all others. This was the foundational legal norm originally laid in the *Fishing & Conservation Convention* (1958) and subsequently espoused by UNCLOS/III. The preservation of the Convention's inherently progressive nature has been entrusted to procedures entailing binding decisions, which form part and parcel of the substantive treaty text with the clear purpose of

---

[70] Allott, 'Power Sharing in the Law of the Sea', AJIL (1983) 11.

[71] Kennedy, *International Legal Structures* (1987) 201–224.

[72] Nordquist, 'The Implementation of the 200-mile Exclusive Economic Zone', *International Business Lawyer* (1978) 171.

[73] *see* Fitzmaurice, 'The Law and Procedure of the International Court of Justice, 1951–4: Treaty interpretation and certain other treaty points', BYBIL (1957) 205; Larsen, 'Between Scylla and Charybdis in Treaty Interpretation', AJIL (1969) 109; and Bianchi–Zarbiyev, *Demystifying Treaty Interpretation* (2024) 101*ff.*

[74] As classically put in Gross, 'States as Organs of International Law and the Problem of Autointerpretation', in Lipsky (Ed.) *Law and Politics in the World Community* (1953) 74–78; and *see* Sohn, 'The Importance of the Peaceful Settlement of Disputes Provisions of the United Nations Convention on the Law of the Sea', in Nordquist–Moore (Eds) *Entry Into Force of the Law of the Sea Convention* (1995) 265–266 and Kolb, 'Note sur Certaines Caractéristiques du Différend International', *The Global Community Yearbook of International Law and Jurisprudence* (2004) 227.

establishing a compulsory dispute settlement regime.[75] These procedures, themselves the product of far-reaching compromises during the negotiations, have received both high praise and sharp criticism.[76]

Although the entire treaty text was drafted throughout the negotiations wholly by consensus, it was put *in concreto* to a recorded vote at the request of the United States during the resumed part of the Eleventh Session (22–24/9/1982, New York). Following its adoption by 130 in favour to 4 against, with 17 States abstaining and 2 others not participating in the vote,[77] UNCLOS/III after nine years of arduous negotiations concluded its proceedings by opening for signature the *United Nations Convention on the Law of the Sea* ('LOSC' or the 'Convention') during the final part of the session (6–10/12/1982, Montego Bay). Since then, the Convention has frequently been described as the 'constitution for the oceans'; a polysemous metaphor first given by Canada,[78] and later echoed by the UNCLOS/III President Tommy Koh (Singapore) in his opening remarks on the final day of the proceedings. This reflects the Convention's significance as a universally acceptable order based on the rule of law for regulating the uses and natural resources of all ocean space.[79]

Yet, the vagueness enshrouding article 63(2) LOSC on straddling stocks and article 64 LOSC on highly migratory species has spurred several disputes regarding the extent of coastal jurisdiction exercised for the purpose of conservation and management, for example, in the Pacific Ocean, which to date yields the highest volume of fish captures. Soon after the conclusion of the

---

75 *Arctic Sunrise* (Merits) 2015, §366.

76 *see*, among others, Charney, 'The Implications of Expanding International Dispute Settlement Systems: The 1982 Convention on the Law of the Sea', AJIL (1996) 69; Rayfuse, 'The Future of Compulsory Dispute Settlement under the Law of the Sea Convention', VUWLR (2005) 683; Subedi, 'An Innovative Solution to an Ambitious Project: Dispute resolution in the 1982 Convention on the Law of the Sea', in Jalloh–Elias (Eds) *Shielding Humanity* (2015) 163; Anderson, 'Peaceful Settlement of Disputes under UNCLOS', in Barrett–Barnes (Eds) *Law of the Sea* (2016) 385; Pedrozo, 'Reflecting on UNCLOS Forty Years Later: What worked, what failed', *International Law Studies* (2022) 928–931; Collins, 'Navigating Choppy Waters: UNCLOS dispute settlement coming of age?', in Buchan–Franchini–Tsagourias (Eds) *The Changing Character of International Dispute Settlement* (2023) 137; and Murphy, 'Creativity in Dispute Settlement Relating to the Law of the Sea', in Brower *et al.* (Eds) *By Peaceful Means* (2024) 195.

77 A/Conf.62/121 'Final Act of the Third United Nations Conference on the Law of the Sea' (27/10/1982) [*UNCLOS/III Off.Records* XVII:139].

78 A/Conf.62/WS/23 'Statement by the delegation of Canada' (16/4/1982) [*UNCLOS/III Off.Records* XVI:262].

79 UNCLOS/III Plenary 185th meeting (6/12/1982) [*UNCLOS/III Off.Records* XVII:13].

Convention, a protracted disagreement over highly migratory species sharpened between the United States and several coastal States in the eastern subregion. The tension dated back to the 1950s, when Chile, Ecuador and Peru proclaimed 200 nm EFZs, unilaterally excluding DWFS from waters that had previously constituted high seas areas. This provoked what became known as the *Wet war* starting in 1963, during which fishing vessels of the United States were arrested for conducting unlicensed commercial fishing operations within those zones.[80] Tensions over the interpretation of article 64 LOSC were further deepened by rival claims within the regulatory area of the Inter-American Tropical Tuna Commission ('IATTC'), as DWFS pressed for collective efforts based on agreed CMMs applicable beyond the 12 nm territorial limits. In response, the coastal States of the region united in a coalition sought to develop an alternative regional institutional framework by creating the Organización Latinoamericana de Desarrollo Pesquero ('OLDEPESCA'), following the adoption of the *Convenio Constitutivo de la Organización Latinoamericana de Desarrollo Pesquero* (1982). Although the controversy was partially resolved in the early 1990s, OLDEPESCA has not yet been dissolved. Instead, it has evolved into a forum for promoting the rational exploitation of fisheries and coordinating joint actions among coastal States in the region.[81] Similarly, tensions have persisted in the western and northern Pacific Ocean, where serious fisheries claims have often been complicated by unresolved maritime boundary disputes in marginal sea areas, involving Japan, the People's Republic of China ('China'), Republic of Korea ('South Korea') and USSR/Russia, as well as other fisheries disputes in the south Pacific Ocean between DWFS and the coastal States of the South Pacific Forum Fisheries Agency ('SPFF').[82]

---

80 Wiegand, 'Seizures of United States Fishing Vessels', SDLR (1969) 428.

81 Székely, 'Yellow-Fin Tuna: A transboundary resource of the eastern Pacific', *Natural Resources Journal* (1989) 1066.

82 *see* Valencia, 'The Yellow Sea: Transnational marine resource management issues', *Marine Policy* (1988) 382; Lee, 'Korea's Claims to Maritime Jurisdiction', *Korean Journal of Comparative Law* (1990) 62; Mizukami, 'Management of Highly Migratory Species and Fisheries Relations between Japan and South Pacific States', *University of British Columbia Law Review* (1990) 127; Akaha, 'From Conflict to Cooperation: Fishery relations in the Sea of Japan', PRL&PJ (1992) 225; Yu, 'China's Coastal Ocean Uses: Conflicts and impacts', OCM (1994) 161; Akaha, 'Japan–South Korea Fishery Agreement of 1998: Pursuing pragmatic interests without compromising sovereignty', in Scheiber (Ed.) *Law of the Sea* (2000) 249; Xue, *China and International Fisheries Law and Policy* (2005) 152–203; Dupont–Baker, 'East Asia's Maritime Disputes: Fishing in troubled waters', *Washington Quarterly* (2014) 79; and Kim, *Maritime Disputes in Northeast Asia* (2017) 119–144.

Taken as a whole, the Convention comprehensively restated the law of the sea as a product of both codificatory and progressive development, the latter notably supported by a robust dispute settlement mechanism. The progressive development achieved through consensus—elicited from the package deal negotiations—together with the compulsory application of dispute settlement procedures entailing binding decisions, stands out as one of the Convention's key technical features. These features have enabled, on the one hand, the Convention's legal order to aspire to a timeless application, realized through continuous adaptation and evolution. On the other hand, they have anchored this order to a practically unamendable treaty text, whose inevitable shortcomings are addressed chiefly through UN implementation agreements, thereby preserving its universal, unified, and integral character.[83]

1.2.4 United Nations Fish Stocks Conference, 1993–1995

Following the adoption of the Convention, important international initiatives were undertaken to remedy the ecological deficit in its fisheries provisions and to promote more robust cooperation in the regulation of marine living resources.[84] Notwithstanding the impetus given to the normative development of fisheries law, the Convention remained lacking in the effective implementation of its provisions relating to straddling and highly migratory fish stocks. The UN Conference on Environment and Development ('UNCED'), held in Rio de Janeiro from 3 to 14 June 1992, acknowledged in its 'Programme of Action for Sustainable Development (Agenda 21)'—specifically in programme area of Chapter 17(C), concerning the sustainable use and conservation of marine living resources of the high seas—the urgent need for an international conference focused on the effective implementation of the Convention's provisions on fish

---

83 A/Res.79/144 'Oceans and the Law of the Sea' (16/12/2024).

84 The most impactful have been the 'Strategy for Fisheries Management and Development' adopted by the FAO World Conference on Fisheries Management and Development, held on 27/6 to 6/7/1984 in Rome; the *Convention for the Prohibition of Fishing with Long Driftnets in the South Pacific* (1989), the principles of which were widely endorsed by A/Res.44/225 'Large-scale Pelagic Driftnet Fishing and its Impact on the Living Marine Resources of the World's Oceans and Seas' (22/12/1989); and the 'Declaration of the International Conference on Responsible Fishing', adopted by the Conference on Responsible Fishing, held on 6–8/5/1992 in Cancun; and *see* further Scanlon, 'The Significance of Informal Lawmaking in International Fisheries Law', in Klein (Ed.) *Unconventional Lawmaking in the Law of the Sea* (2022) 210.

stocks.[85] Responding to this call, UNGA convened under the aegis of UN, with the scientific and technical support of FAO, the 'Conference on Straddling Fish Stocks and Highly Migratory Stocks' ('*Fish Stocks Conference*').[86] In parallel with the negotiations at the conference, ongoing global efforts at FAO led to the adoption of the *Agreement to Promote Compliance with International Conservation and Management Measures by Fishing Vessels on the High Seas* (1993), as well as the achievement of the 'Consensus on World Fisheries'.[87]

It is worth noting that the *Fish Stocks Conference* commenced its proceedings at a time when the Convention's entry into force was still uncertain. This prospect was only averted by the *Agreement relating to the Implementation of Part XI of the United Nations Convention on the Law of the Sea of 10 December 1982* (1994), following the positive results of the informal consultations that were held among States from 1990 to 1994 on outstanding issues relating to Part XI LOSC and related provisions of the Convention.[88] The governance challenge facing fisheries—should the Convention take longer to enter into force—heightened expectations for the *Fish Stocks Conference* and demanded a chairman capable of rising to the challenge by scrupulously overseeing its arrangements and expertly managing the proceedings. The role of Conference Chairman was assigned to Satya Nand Nandan (Fiji), arguably the most eminent diplomatic persona in contemporary law of the sea affairs. He brought to the position a wealth of experience, skill as well as political dexterity, gained inter alia from the heyday of the UNCLOS/III negotiations and his long service as the UN Under-Secretary-General and Special Representative of the Secretary-General for the Law of the Sea. The resolution convening the *Fish Stocks Conference* set a 2 year window for the conclusion of the multilateral negotiations. The intergovernmental forum, assembled in New York, held its First Session from 19 to 23 April 1993, during which it addressed organizational matters. From that time until 26 August 1994, it held three additional

---

85 UNCED, 'Report – Volume I: Resolutions adopted by the Conference' (12/8/1992), Annex II ¶17(50).

86 Pursuant to the eponymous A/Res.47/192 (22/12/1992), ¶¶1–3.

87 *see* Edeson, 'Sustainable Use of Marine Living Resources', ZaöRV (2003) 355 and Hayashi, 'Three Decades' Progress in High Seas Fisheries Governance: Towards a common heritage regime?', in Moore–Nordquist–Mahmoudi (Eds) *The Stockholm Declaration and Law of the Marine Environment* (2003) 378 *et seq.*

88 A reflective commentary on the evolution of legal frameworks governing the global commons under the Convention, written in the immediate aftermath of the Agreement, is offered by Joyner–Martell, 'Looking Back to See Ahead: UNCLOS/III and lessons for global commons law', ODIL (1996) 73.

sessions.[89] By the end of the Fourth Session (15–26/8/1994), the conference had not yet concluded its work and requested an extension of its mandate for further negotiations in 1995.[90] UNGA approved the convening of two additional sessions—from 27 March to 12 April 1995 and from 24 July to 4 August 1995—requiring the conference to complete its work before the opening of the Fiftieth Session of the Assembly in September that year.[91] On 4/8/1995, after 2 years of negotiations, the *Fish Stocks Conference* adopted the *Agreement for the Implementation of the Provisions of the UN Convention on the Law of the Sea of 10 December 1982 relating to the Conservation and Management of Straddling Fish Stocks and Highly Migratory Fish Stocks* ('FSA' or the 'Agreement').

The Agreement represents the most significant advancement in the edification of the UN conventional legal regime, committing States to '[t]he objective of ensuring long-term conservation and sustainable use of straddling and highly migratory fish stocks through effective implementation of the relevant provisions of the Convention'.[92] This is an extraordinary achievement, as its text—negotiated and adopted by consensus—has progressively developed conservation and management law and applies, *mutatis mutandis*, the Convention's compulsory dispute settlement procedures.[93] It is worth noting that the very process of its negotiation, and eventual adoption, served as a

---

89 *see* Doulman, *Structure and Process of the 1993–1995 United Nations Conference on Straddling Fish Stocks and Highly Migratory Fish Stocks* (1995).

90 A/Conf.164/25 'Report on the Fourth Session of the United Nations Conference on Straddling Fish Stocks and Highly Migratory Fish Stocks (Prepared by the Secretariat)' (11/10/1994), ¶¶30–31.

91 A/Res.49/121 'United Nations Conference on Straddling Fish Stocks and Highly Migratory Fish Stocks' (19/12/1994), ¶¶2–4.

92 FSA second preambular recital & 2.

93 A/Conf.164/30 'Statement made by the Chairman of the Conference at the opening of the Sixth Session, held on 24 July 1995' (1/8/1995) and A/Conf.164/38 'Final Act of the United Nations Conference on Straddling Fish Stocks and Highly Migratory Fish Stocks' (7/9/1995); and *see* further in Munro, 'The United Nations Fish Stocks Agreement of 1995: History and problems of implementation', *Marine Resource Economics* (1995) 265; Juda, *International Law and Ocean Use Management* (1996) 276*ff*; Freestone–Makuch, 'The New International Environmental Law of Fisheries: The 1995 United Nations Straddling Stocks Agreement', YbIEL (1998) 3; Burke, 'Evolution in the Fisheries Provisions of UNCLOS', in Nisuke–McWhinney–Wolfrum (Eds) *Liber Amicorum Judge Shigeru Oda* (2002) 1355; Harrison, *Making the Law of the Sea* (2011) 99–115; and Hoel, 'The Importance of Marine Science in Sustainable Fisheries: The role of the 1995 UN Fish Stocks Agreement', in Nordquist–Moore–Long (Eds) *Legal Order in the World's Oceans* (2017) 379*ff*.

catalyst for galvanizing international fisheries politics. In the less than 6 years between the *Fish Stocks Conference* and the Agreement's entry into force, disputes between DWFS and coastal States intensified across various ocean regions. This resulted in three high-profile legal cases being submitted to international adjudication, although none ultimately proceeded to the merits stage due to differing procedural or jurisdictional grounds.

The *Turbot war* arose from nearly five decades of ineffective cooperation in regulating the once-lucrative fishery resources of the Grand Banks in the north Atlantic Ocean.[94] The region had been originally regulated by the International Commission for the Northwest Atlantic Fisheries, established under the *International Convention for the Northwest Atlantic Fisheries* (1949), and which was succeeded in 1979 by the Northwest Atlantic Fisheries Organization ('NAFO'), established under the *Convention on Future Multilateral Cooperation in the Northwest Atlantic Fisheries* (1978) ('NAFO *Convention*'). Despite being the focus of regional arrangements since 1950, several straddling stocks in the regulatory area were seriously depleted, reaching historically low levels by the late 1980s.[95] NAFO proved unable to reverse their declining status, mainly owing to institutional shortcomings. The frequent resort to opt-out procedures by members weakened the implementation of CMMs, and the organization's lack of effective surveillance and enforcement mechanisms failed to deter IUU fishing in the regulatory area, leaving the burden of these responsibilities largely to Canadian authorities.[96] In July 1992, Canada closed several fisheries within its EFZ and called on NAFO to drastically reduce the fishing efforts of all members, if not to implement full moratoria, in the regulatory area.[97] Under these circumstances, the material facts of the dispute between Canada and Spain unfolded while the *Fish Stocks Conference* was still in progress. In February 1995, NAFO set the annual quotas for the straddling stock of *Reinhardtius hippoglossoides* (Greenland turbot) at 27,000 t wherefrom Canada was to receive 16,300 t and the European

[94] On the significance of northwest Atlantic Ocean fisheries for the development of the modern fishing industry and international economy *see* Lear, 'History of Fisheries in the Northwest Atlantic: The 500-year perspective', *Journal of Northwest Atlantic Fishery Science* (1998) 41.

[95] O'Reilly-Hinds, 'Crisis in Canada's Atlantic Sea Fisheries', *Marine Policy* (1995) 271.

[96] *see* the documented analysis by Douglas Day in 'Tending the Achilles' Heel of NAFO: Canada acts to protect the nose and tail of the Grand Banks', *Marine Policy* (1995) 261–263 and 'Addressing the Weakness of High Seas Fisheries Management in the Northwest Atlantic', OCM (1997) 76–83.

[97] Mitchell, 'Fisheries Management in the Grand Banks, 1980–1992 and the Straddling Stock Issue', *Marine Policy* (1997) 97.

Union 3400 t. Discontented with the allocation, the latter objected to the decision, raising its own quota to 18,600 t, as the result of what it euphemistically referred to as 'a redistributive exchange' between them.[98] On 9/3/1995, less than 3 weeks before the commencement of the penultimate session of the Conference, Canadian authorities—having monitored and determined that European Union member States had already harvested its 3400 t quota—boarded and detained the Spanish trawler *Estai* in the NAFO regulatory area, approximately 45 nm outside Canada's EFZ.[99] On 28/3/1995, Spain instituted proceedings before ICJ, asserting that Canada's conduct breached the NAFO *Convention* and general international law. Spain argued that Canada's actions exceeded the limits of fisheries enforcement, gravely undermined the principle of high seas freedom, and flagrantly violated Spain's sovereign rights. A formal settlement was reached prior to the Court's judgment through the *Agreement constituted in the form of an Agreed Minute, an Exchange of Letters, an Exchange of Notes and the Annexes thereto between the European Community and Canada on fisheries in the context of the NAFO Convention* (1995), in which both parties affirmed their commitment to closely cooperate for agreeing CMMs in the context of NAFO, including a voluntarily revised ratio of quota allocation and the implementation of an improved control and enforcement system.

Thunnus maccoyii (Southern bluefin tuna) is a prime example of a species with an extensive migratory range that exposes it to the cumulative risk of being fished at numerous locations throughout the southern hemisphere. As one of the most commercially valuable tuna species, it was intensively exploited until the 1990s, when the need for urgent regulatory measures to prevent its collapse became undeniable.[100] Negotiations among Australia, Japan and New Zealand resulted in the *Convention for the Conservation of Southern Bluefin Tuna* (1993) ('CCSBT *Convention*'), which established the Commission for the Conservation of the Southern Bluefin Tuna ('CCSBT'). This was the first treaty to give effect to article 64 LOSC by applying its provisions across the entire migration range of a species, rather than being limited to a specific geographical region. Its main objective is to ensure, through appropriate management, the conservation and optimum utilization of the species. CCSBT

---

[98] Joyner, 'On the Borderline? Canadian Activism in the Grand Banks', in Stokke (Ed.) *Governing High Seas Fisheries* (2001) 213.

[99] For an account of the incident and its immediate political aftereffect *see* Schaefer, '1995 Canada–Spain Fishing Dispute', GIELR (1996) 437; Joyner–Von Gustedt, 'The Turbot War of 1995: Lessons for the law of the sea', IJMCL (1996) 425; and Colburn, 'Turbot Wars: Straddling stocks, regime theory, and a new UN Agreement', *Journal of Transnational Law and Policy* (1997) 323.

[100] Safina, 'Bluefin Tuna: Facing extinction', *Ecologist* (2003) 47*ff.*

prioritized rebuilding the stock biomass to at least the 1980 levels and set a Total Allowable Catch ('TAC') of 11,750 t for allocation among its members. In the following years, Japan sought to increase the TAC limit and revise its national allocation. However, disagreements over the exploitation status and prospects for stock recovery prevented members from agreeing on a new limit or revised allocations. In 1998, Japan declined to confirm its annual allocation at the most recently agreed level and announced the implementation of a three-year Experimental Fishing Programme ('EFP'). After extensive consultations, they agreed to establish a working group for a joint evaluation of the stocks, which also failed to produce commonly accepted findings. While the apparent reason for disagreement on regulatory measures stemmed from irreconcilable positions regarding scientific uncertainty over stock status, the underlying economic, political, and social factors tied to domestic fisheries policies also played a crucial—although often underrated—role in polarizing the negotiations on CMMs within CCSBT.[101] Against this background, Japan resumed its EFP in June 1999. A month later, Australia and New Zealand notified Japan of the existence of a dispute and instituted proceedings under Part XV LOSC. This was the first arbitral tribunal to be formed pursuant to Annex VII LOSC. Pending its constitution, the applicants requested ITLOS to prescribe provisional measures. In their application, it was alleged that Japan had failed to comply with its obligation to cooperate in the conservation of the species, and they sought, inter alia, the immediate cessation of EFP activities and a reduction in fishing efforts to the most recently agreed TAC, as measures aligned with the principle of precautionary approach. Japan contested the applicability of compulsory procedures under Part XV LOSC but filed a counter-request, calling on the applicant parties to return to negotiations in good faith for agreeing on TAC limits and to continue the EFP on a joint basis. In August 1999, the Tribunal, having established the *prima facie* jurisdiction of the arbitral tribunal to adjudicate the dispute, ordered provisional measures appropriate to the circumstances.[102] A year later, the arbitral tribunal upheld Japan's argument that the CCSBT *Convention* precluded the application of any compulsory settlement procedure without the consent of all parties to the dispute. As a result, the tribunal found itself lacking jurisdiction to hear the merits of the case and revoked the provisional measures.[103]

The *Swordfish stocks* dispute originated in 1990 when Chile began implementing a series of national CMMs for highly migratory species, especially

---

101 *compare* Scott, 'Australia's First Tuna Negotiations with Japan', *Marine Policy* (2000) 309 *with* Sato, 'Fishy Business: A political-economic analysis of the southern bluefin tuna dispute', *Asian Affairs* (2002) 221.

102 *Southern Bluefin Tuna Cases* (Provisional Measures) 1999.

103 *Southern Bluefin Tuna Cases* (Jurisdiction) 2000.

for *Xiphias gladius* (Swordfish) stocks. These measures restricted port access for landing, transhipment, and other logistical and port services to longline fishing and fish processing vessels from European Union member States that were operating in contravention of Chilean legislation. Since the fishing activities had taken place in high seas areas beyond the Chilean EEZ, the European Union on 19/4/2000 invoked the rules and procedures of the *WTO Dispute Settlement Understanding* (1994). It sought to challenge the validity of Chilean legislation, arguing that it violated provisions of the *General Agreement on Tariffs and Trade* (1994) regarding the freedom of transit for goods and the prohibition of quantitative restrictions on imports or exports among contracting parties.[104] In response, Chile unilaterally initiated proceedings under Part XV LOSC in September 2000 by filing a case concerning the obligations and duties of the parties in the regulation of highly migratory stocks under the Convention.[105] The controversy highlighted the legal complexities of sustainably regulating common-pool marine living resources under the Convention. Yet these aspects were overshadowed by the *lis pendens* and *lis alibis* complications that could potentially ensue from the parallel forum proceedings.[106] On 25/1/2001, both parties agreed to resume bilateral negotiations and suspended the parallel legal proceedings. These diplomatic efforts concluded in October 2009 with the adoption of the *Understanding Concerning the Conservation of Swordfish Stocks in the South-Eastern Pacific Ocean* (2009), effectively settling the dispute. At the parties' joint request, ITLOS[107] and WTO DSB[108] formally terminated their respective proceedings.

While the entry into force of the Agreement on 11/12/2001 reaffirmed the commitment of States to cooperate for the long-term conservation and sustainable use of fish stocks, disagreements over CMMs, such as those discussed above, are likely to intensify. This is particularly true as environmental considerations become increasingly connected to complex legal issues, making overexploitation the 'biggest threat in our history'.[109] It is essential that States

---

104 *Measures Affecting the Transit and Importation of Swordfish–Chile* (2000).

105 *Conservation and Sustainable Exploitation of Swordfish Stocks* (Constitution of Chamber) 2000.

106 *see* Orellana, 'The Swordfish Dispute between the EU and Chile at the ITLOS and the WTO', NJIL (2002) 55 and Stoll–Vöneky, 'The Swordfish Case: Law of the sea v. trade', ZaöRV (2002) 21.

107 *Conservation and Sustainable Exploitation of Swordfish Stocks* (Order) 2009.

108 WTO DSB 'Joint Communication from the European Union and Chile' [193/4G/L/367/Add.1] (3/6/2010).

109 World Economic Forum, 'Sir David Attenborough is worried about this ocean threat—and it's not plastic' (7/9/2019).

uphold the normative framework established by the Agreement—in terms of both substantive law and the accompanying dispute settlement procedures—to ensure meaningful cooperation and resolve disagreements in a timely and objective manner. In this regard, the effectiveness of the Agreement itself is subject to a formal process conducted by the Review Conference ('*FSA Review Conference*'),[110] the designated intergovernmental forum convened by the UN Secretary-General at the request of UNGA. The timing and agenda of each review round are determined through the Informal Consultations of States Parties to the Agreement ('ICSP'). Since 2002, ICSP meetings have been convened regularly to support regional implementation of the Agreement, provide recommendations to UNGA on the scope and content of relevant annual reports, and assist in preparing the *FSA Review Conference*.[111]

1.3 THE CLAUSAL CONSTRUCTION IN ARTICLE 7 OF THE 1995 UNITED NATIONS FISH STOCKS AGREEMENT: THE ARGUMENT AND STRUCTURE OF THE BOOK

Article 7 FSA, titled 'Compatibility of conservation and management measures', sets forth a rule that represents a significant advancement in the regulation of fish stocks, having progressively developed the rights and obligations of States under the Convention.[112] Still, in terms of spatial application, it unavoidably relies on the zoning governance model of the Convention, which, for the purpose of regulating common-pool marine living resources often gives rise to situations apt to jurisdictional fragmentation. This has predominantly been the *fons et origo* of the ecological concerns under the Convention, as fish stocks do not adhere to legal boundaries but instead develop naturally across jurisdictional sea areas. The substance of the rule envisages that CMMs must be compatible throughout the full extent of stocks' distribution or migratory range. By doing so, it integrates ecosystem-based considerations into fisheries

110 FSA 36.

111 The *FSA Review Conference* was convened in New York on 22–26/5/2006 pursuant to the foundational UNGA Resolutions on 'Sustainable Fisheries' in A/Res.59/25 (17/11/2004) and A/Res.60/31 (29/11/2005); and it was subsequently resumed on 24–28/5/2010 by A/Res.63/112 (5/12/2008) and A/Res.64/72 (4/12/2009); on 23–27/5/2016 by A/Res.70/75 (8/12/2015); and on 22–26/5/2023 by A/Res.77/118 (9/12/2022).

112 Tahindro, 'Sustainable Fisheries: The legal regime of the 1995 United Nations Fish Stocks Agreement and its contribution to subsequent developments promoting sustainable fisheries', in Nordquist–Moore–Long (Eds) *Legal Order in the World's Oceans* (2017) 343.

regulation.[113] This approach has been affirmed by subsequent policy directions as essential to the long-term conservation and sustainable use of a stock, both as a discrete resource and as a biological component interacting within the broader environmental systems to which it belongs.[114] The compatibility rule, working in synergy with the general principle of the ecosystem approach, addresses the jurisdictional fragmentation left unresolved by the Convention, promoting sustainable regulation based on cooperation.[115] To that end, States have been urged to apply this principle comprehensively in the conservation and management of fish stocks.[116] However, the input to the *FSA Review Conference* has not enabled a thorough assessment of the extent to which compatible CMMs are implemented in conformity with article 7 FSA, nor of how prevalent this practice has become between DWFS and coastal States, whether through direct cooperation or via regional and subregional fisheries management organizations and arrangements.[117]

Despite its ostensible simplicity and ecologically intuitive tenet, the rule of compatible CMMs remains one of the most ambivalent regulatory concepts in fisheries law. Although the ecosystem rationale underlying the rule is catholically accepted, disagreements over the determination, adoption and implementation of CMMs throughout a stock's distribution or migration range, and therefore across jurisdictional sea areas, will draw in contentious governance issues. While the balance of interests between DWFS and coastal States under the Convention is carefully preserved, the formulation of the rule—being the outcome of progressive development—has introduced a pronounced ambiguity into article 7 FSA, particularly with respect to its spatial scope, which has, in

---

113 A/Conf.164/12 'Statement made by the Chairman of the Conference at the conclusion of the General Debate, 15 July 1993' (21/7/1993), and articles 2 & 3(1) in conjunction with the general principles of article 5 FSA.

114 *see* Commission on Sustainable Development of the UN Economic and Social Council, 'Reykjavík Declaration on Responsible Fisheries in the Marine Ecosystem' (9/11/2001) and FAO COFI, 'Declaration for Sustainable Fisheries and Aquaculture' (1/2/2021).

115 Davis–Hanich, 'Developing an Equitable and Ecosystem-Based Approach to Fisheries Management', in Scheiber–Kraska–Kwon (Eds) *Science, Technology, and New Challenges to Ocean Law* (2015) 124.

116 *see*, for instance, the long stream of annual UNGA Resolutions on 'Oceans and the Law of the Sea' and 'Sustainable Fisheries', including, A/Res.60/31 (29/11/2005), ¶4; A/Res.73/125 (11/12/2018), ¶15; A/Res.78/68 (11/12/2023), ¶22; and A/RES/79/145 (16/12/2024), ¶185.

117 A/Conf.210/2016/1 'Report of the UN Secretary-General to the Resumed Review Conference' (1/3/2016), ¶64 and *FSA Review Conference* 2023, ¶¶54, 73, 76 and Outcome A5.

turn, engendered divergent approaches to its interpretation and application.[118] This ambiguity is considered by some to be so pervasive that they claim the text of the article is confusing and poorly drafted.[119] This is a critical issue that can be further explored through a series of relevant questions. Should the coastal State be expected to relinquish its prescriptive authority by adjusting national measures within its EEZ to align with international measures adopted for the high seas, or should DWFS be required to yield to creeping jurisdiction tendencies, whereby coastal States seek to practically arrogate high seas areas to themselves? Whereas States on both sides are explicitly bound by the duty to cooperate, what threshold must be met to deem this duty fulfilled in relation to the obligation to achieve compatible CMMs when disagreement persists? Which of the two jurisdictional regimes should serve as the dominant frame of reference for resolving disagreements over the compatibility of measures? These are just a few of the normative questions likely to arise regarding the spatial scope of the rule in the broader context of a dispute over compatible CMMs (hereinafter referred to as 'compatibility dispute') between DWFS and coastal States. Although such divisive questions were identified as soon as the Agreement was concluded,[120] the subsequent literature has yet to address them with coherence. Academic views on the matter have diverged considerably, shaped by principally substantive interpretations of the compatibility rule.[121]

The issue of the rule's spatial scope is doctrinally compounded by the question of whether compatibility disputes are amenable to compulsory settlement

[118] This has been formally noted and discussed in the ICSP8/UNFSA/Rep/Inf.6 'Eighth Round of ICSP' (16/4/2009), 14¶20.

[119] *e.g.*, Rengifo-Lozano, *International Regime Theory and the Law of the Sea* (2011) 164.

[120] *see*, among others, Balton, 'Strengthening the Law of the Sea: The new agreement on straddling fish stocks and highly migratory fish stocks', ODIL (1996) 137 and Juda, 'The United Nations Agreement on Straddling Fish Stocks and Highly Migratory Fish Stocks: A critique', ODIL (1997) 154; and *see* the opposing views held in the debate on 'Coastal States' Competences over High Seas Fisheries and the Changing Role of International Law' published in ZaöRV (1995) between Francisco Orrego-Vicuña at pages 520–534 and Bernard Oxman at pages 536–543.

[121] On the persistence of such questions, *see* Tanaka, *A Dual Approach to Ocean Governance* (2008) 79–82; Cardoso, 'Considerations on Some Global Institutional Challenges within the Context of the Conservation and Management of Marine Living Resources', in Ribeiro–Bastos–Henriksen (Eds) *Global Challenges and the Law of the Sea* (2020) 112; and Murphy, 'Straddling and Highly Migratory Fish Stocks Compatibility Requirement and Scope of Rights of High Seas Fishing Nations', in Kunoy–Heidar–Yiallourides (Eds) *International Fisheries Law* (2024) 223*ff.*

procedures entailing binding decisions. This matter has likewise been the subject of inconclusive debate. The legal uncertainty on that account stems from article 32 FSA, which incorporates by reference article 297(3) LOSC. The latter provision contains a limitation on the applicability of compulsory procedures for the settlement of fisheries disputes (hereinafter referred to as the 'fisheries limitation'), which may be invoked by the coastal State to exclude disputes concerning the exercise of its sovereign rights in its own EEZ. This presents a crucial procedural question, which would feature prominently in any contested jurisdiction case concerning compatibility disputes under the Agreement and hence remains central to the present book. On the matter, various opinions have been voiced in rather cautious terms, which can essentially be condensed into the proposition that it is 'difficult to read into article 7 FSA any intention to clarify the applicability of compulsory dispute settlement'.[122] This view has been asserted despite the fact that article 7 FSA, in articulating the compatibility rule, contains a clause in paragraph (4) that expressly refers to dispute settlement procedures in the event of a disagreement over CMMs. The occurrence of the procedural stipulation in article 7(4) FSA has been cautiously assessed by some with regard to its practical effect in the light of article 32 FSA. In this connection, it has been viewed that 'the solution given in the Agreement of adopting a procedural instead of a substantive rule is in fact more cosmetic than real'.[123] These views, in turn, have influenced arguments that concede to a dichotomization effect by notionally dividing the spatial scope of the resulting compatibility dispute into EEZ and high seas components.[124]

The uncertainty surrounding the spatial scope of the rule has been further exacerbated by the interpretation of article 297 LOSC in the 2015 *Chagos MPA* arbitration. In the instant case, the tribunal read paragraph (3) of article 297 LOSC in contrast to paragraph (1), leading to the exclusion of CMMs adopted by the coastal State for sea areas under its national jurisdiction from the Convention's compulsory and binding procedures, even where such measures also fulfilled environmental purposes under the latter provision. In this context, one view holds that an '[e]nvironmental interpretation of article 297 LOSC designed to bring conservation of marine living resources within Part XV LOSC jurisdiction would overturn a clear policy choice of

---

122 Boyle, 'Problems of Compulsory Jurisdiction and the Settlement of Disputes Relating to Straddling Fish Stocks', IJMCL (1999) 22.

123 Treves, 'The Settlement of Disputes According to the Straddling Stocks Agreement of 1995', in Boyle–Freestone (Eds) *International Law and Sustainable Development* (1999) 260.

124 *e.g.* Pinto, *Fisheries Management in Areas Beyond National Jurisdiction* (2012) 93–98.

the negotiating States at UNCLOS/III', alleging that this was the 'preference reiterated and endorsed' by the *Fish Stocks Conference* in adopting 'exactly the same dispute settlement provisions, *mutatis mutandis*'.[125] A distinctly different view holds that jurisprudence has yet to resolve these issues, thereby necessitating a closer examination of the applicable legal regimes in future disputes to ensure effective ocean governance, particularly with regard to fish stocks in a critical state.[126] Academic debate of what components in a dispute concerning CMMs for straddling or highly migratory fish stocks under the Convention fall within or outside the scope of article 297 LOSC, and in particular of paragraph (3) addressing fisheries, has subsequently sparked fairly afresh interest,[127] with the Agreement nevertheless remaining enigmatic in the limelight of this renewed attention. A compatibility dispute will raise fundamental issues concerning the conservation and management of fish stocks throughout their distribution or migration range. When States reach a point of no return in negotiations—where they can only agree to disagree on CMMs—and further progress through direct, timely cooperation becomes unattainable, they may see no viable alternative but to resort to unilateral, and therefore counterproductive, measures. Such a course of action not only will frustrate the ecosystem approach but is also likely to aggravate the strained relationship on this matter between the two sides.

---

125 Boyle, 'UNCLOS Dispute Settlement and the Uses and Abuses of Part XV', *Revue Belge De Droit International* (2014) 193 followed by Nguyen, *The Development of the Law of the Sea by UNCLOS Dispute Settlement Bodies* (2023) 228.

126 *compare* Oxman, 'On the Challenges to Stability and to the Rule of Law in Implementation of UNCLOS', in Scheiber–Oral–Kwon (Eds) *Ocean Law Debates* (2018) 282–283 *with* Klein–Parlett, *Judging the Law of the Sea* (2022) 121.

127 *see* Klein, 'The Vicissitudes of Dispute Settlement under the Law of the Sea Convention', IJMCL (2017) 349–352; Scalieri, 'Discretionary Power of Coastal States and the Control of its Compliance with International Law by International Tribunals', in Del Vecchio–Virzo (Eds) *Interpretations of the United Nations Convention on the Law of the Sea by International Courts and Tribunals* (2019) 363–365; Klein, 'The Many Facets of EEZ Fisheries Disputes and their Resolution under UNCLOS', GöJIL (2023) 14; Goodman, 'Compulsory Settlement of EEZ Fisheries Enforcement Disputes under UNCLOS: *Swallowing the rule* or *balancing the equation*?', GöJIL (2023) 27; Schatz, 'The Settlement of EEZ Fisheries Access Disputes under UNCLOS: Limitations to jurisdiction and compulsory conciliation', GöJIL (2023) 82; and Juratowitch, 'Limitations on Compulsory Settlement of Disputes with Regard to Fisheries in the EEZ Arising from Article 297(3)(a) of UNCLOS', in Kunoy–Heidar–Yiallourides (Eds) *International Fisheries Law* (2024) 171.

The analysis presented in this book argues that the Agreement remedies this imbroglio by making the application of Part VIII FSA 'Peaceful Settlement of Disputes' compulsory for compatibility disputes, thereby ensuring effective cooperation for the sustainable regulation of fish stocks. To this end, article 7 FSA was formulated using a distinctive clausal construction, whereby procedural provisions concerning dispute settlement were embedded within the substantive rule. More specifically, paragraphs (4) and (5) of article 7 FSA mirror the stipulations that Part VIII FSA makes respectively under articles 30 and 31 FSA. Simply juxtaposing the two sets of provisions is sufficient to reveal this textual duplication. Paragraph (4) of article 7 FSA, in providing that, 'If no agreement can be reached within a reasonable period of time, any of the States concerned may invoke the procedures for the settlement of disputes provided for in Part VIII', corresponds to article 30 FSA, being entitled 'Procedures for the settlement of disputes', which applies to the entire Agreement. In support of the procedural stipulation in paragraph (4), paragraph (5) of article 7 FSA follows immediately with a clause providing for provisional measures. The latter recapitulates the obligation under article 31 FSA, being entitled 'Provisional measures', which requires States to enter into cooperative provisional arrangements, or failing agreement, to obtain these in the form of provisional measures following the procedures for the settlement of disputes provided for in Part VIII FSA. The structure of clausal embeddedness appears also in article 16 FSA, respecting compatible CMMs in the special case of high seas enclaves, with paragraph (2) to provide that:

> States shall act in good faith and make every effort to agree without delay on conservation and management measures to be applied in the carrying out of fishing operations in the area referred to in paragraph 1. If, within a reasonable period of time, the fishing States concerned and the coastal State are unable to agree on such measures, they shall, having regard to paragraph 1, apply article 7, paragraphs 4, 5 and 6, relating to provisional arrangements or measures.

As article 16(2) FSA directly references article 7 FSA, it will not be examined independently.

To summarize the above drafting observations, an 'embedded clause'—a term coined here for convenient reference—denotes the incorporation of procedural stipulations within substantive rules, presenting a condensed version of the treaty's general dispute settlement provisions. The replication of these provisions raises several reasonable questions. Do these procedural stipulations merely restate, in abridged form, the main dispute settlement provisions already set out in full-fledged articles that, in any case, apply across the entire treaty text? Does the construction of embedded clauses amount to textual redundancy, serving no distinct legal function? What was the drafter's intent in structuring the treaty text to include such clauses? These questions invite a deeper analysis of the drafting rationale and interpretative considerations underlying this clausal structure. As academic commentary on compatibility disputes remains scant, the structural formation of the rule against the ambit of the fisheries limitation and the remit of compulsory procedures entailing binding decisions has not been made before the subject of a meticulous study. This book closely examines the legal effect emanating from the morphology of article 7 FSA, which peculiarly amalgamates the substantive rule and the applicable dispute settlement procedure within a single article (see Box 1.1; *note: italicised lettering in the book's boxes and tables is added to highlight the textual disposition of clausal embeddedness*). Building on this general introduction, the analysis unfolds across five main chapters, each examining the evolution of this textual pattern of clausal embeddedness in both its substantive and procedural dimensions.

BOX 1.1 CLAUSAL EMBEDDEDNESS IN ARTICLE 7 OF THE 1995 UN FISH STOCKS AGREEMENT

Article 7

Compatibility of Conservation and Management Measures

1. Without prejudice to the sovereign rights of coastal States for the purpose of exploring and exploiting, conserving and managing the living marine resources within areas under national jurisdiction as provided for in the Convention, and the right of all States for their nationals to engage in fishing on the high seas in accordance with the Convention:
   (a) with respect to straddling fish stocks, the relevant coastal States and the States whose nationals fish for such stocks in the adjacent high seas area shall seek, either directly or through the appropriate mechanisms for cooperation provided for in Part III, to agree upon the measures necessary for the conservation of these stocks in the adjacent high seas area;
   (b) with respect to highly migratory fish stocks, the relevant coastal States and other States whose nationals fish for such stocks in the region shall cooperate, either directly or through the appropriate mechanisms for cooperation provided for in Part III, with a view to ensuring conservation and promoting the objective of optimum utilization of such stocks throughout the region, both within and beyond the areas under national jurisdiction.

2. Conservation and management measures established for the high seas and those adopted for areas under national jurisdiction shall be compatible in order to ensure conservation and management of the straddling fish stocks and highly migratory fish stocks in their entirety. To this end, coastal States and States fishing on the high seas have a duty to cooperate for the purpose of achieving compatible measures in respect of such stocks. In determining compatible conservation and management measures, States shall:
   (a) take into account the conservation and management measures adopted and applied in accordance with article 61 of the Convention in respect of the same stocks by coastal States within areas under national jurisdiction and ensure that measures established in respect of such stocks for the high seas do not undermine the effectiveness of such measures;
   (b) take into account previously agreed measures established and applied for the high seas in accordance with the Convention in respect of the same stocks by relevant coastal States and States fishing on the high seas;
   (c) take into account previously agreed measures established and applied in accordance with the Convention in respect of the same stocks by a subregional or regional fisheries management organization or arrangement;
   (d) take into account the biological unity and other biological characteristics of the stocks and the relationships between the distribution of the stocks, the fisheries and the geographical particularities of the region concerned, including the extent to which the stocks occur and are fished in areas under national jurisdiction;
   (e) take into account the respective dependence of the coastal States and the States fishing on the high seas on the stocks concerned; and
   (f) ensure that such measures do not result in harmful impact on the living marine resources as a whole.
3. In giving effect to their duty to cooperate, States shall make every effort to agree on compatible conservation and management measures within a reasonable period of time.
4. (P) *If no agreement can be reached within a reasonable period of time, any of the States concerned may invoke the procedures for the settlement of disputes provided for in Part VIII.* (P)
5. Pending agreement on compatible conservation and management measures, the States concerned, in a spirit of understanding

and cooperation, shall make every effort to enter into provisional arrangements of a practical nature. In the event that they are unable to agree on such arrangements, any of the States concerned may, for the purpose of obtaining provisional measures, submit the dispute to a court or tribunal in accordance with the procedures for the settlement of disputes provided for in Part VIII.

6. Provisional arrangements or measures entered into or prescribed pursuant to paragraph 5 shall take into account the provisions of this Part, shall have due regard to the rights and obligations of all States concerned, shall not jeopardize or hamper the reaching of final agreement on compatible conservation and management measures and shall be without prejudice to the final outcome of any dispute settlement procedure.
7. Coastal States shall regularly inform States fishing on the high seas in the subregion or region, either directly or through appropriate subregional or regional fisheries management organizations or arrangements, or through other appropriate means, of the measures they have adopted for straddling fish stocks and highly migratory fish stocks within areas under their national jurisdiction.
8. States fishing on the high seas shall regularly inform other interested States, either directly or through appropriate subregional or regional fisheries management organizations or arrangements, or through other appropriate means, of the measures they have adopted for regulating the activities of vessels flying their flag which fish for such stocks on the high seas.

Note: The symbol '(P)' has been used to highlight the textual disposition of clausal embeddedness.

Sustainable regulation requires agreed-upon CMMs to address each fish stock as a unified whole, regardless of the jurisdictional sea areas wherein it occurs. This renders cooperation a fundamental process for achieving compatible measures, which must be determined, adopted, and implemented across a stock's distribution or migration range at the regional, subregional, or even national level, as appropriate. While the normative implications of the compatibility rule, in principle, impose a shared responsibility between DWFS and coastal States for regulating fish stocks in line with the ecosystem approach, the primacy of the jurisdictional regime under which the measures should be agreed—and, where agreement is not possible, the basis for resolving the disagreement—remains a matter of considerable debate, giving rise to competing interpretations. Chapter 2 examines the progressive development of the law concerning 'Compatible conservation and management measures', with a focus on the substantive provisions of the compatibility rule and its applicable spatial scope across sea areas under coastal jurisdiction and the high seas. In turn, Chapter 3 explores the subject matter of disagreements that may arise over compatible CMMs and the application thereon of the 'Settlement procedures for compatibility disputes in the 1995 United Nations Fish Stocks Agreement'. Further analysis in these chapters is devoted to contrasting interpretations of the rule's substantive content, particularly as they pertain to its spatial scope and the corresponding exclusionary effect of the fisheries limitation.

The next two chapters, in pursuing a schematic argument to the textual functionality served by the construction of embedded clauses, embark on a structural analysis of treaty articles that exhibit this morphology. Chapter 4 scrutinizes this textual pattern by revisiting two 'Drafting antecedents of embedded clauses in the Law of the Sea'. First, it focuses on the articles of the *Fishing & Conservation Convention* (1958) that establish the rule recognizing the special interest of coastal State in maintaining the productivity of living resources in high seas areas adjacent to and beyond its territorial sea. To elucidate the pattern of clausal embeddedness, the chapter draws crucial perspectives from the ILC deliberations and *travaux préparatoires* during the development of the *ILC Draft Articles*, as well as from the UNCLOS/I negotiations, which led to the adoption of the convention. This is the first treaty fostering an ecosystem approach to the conservation and management of marine living resources traversing jurisdictional sea areas, and in progressively developing fisheries law devised a system of compulsory procedures entailing binding decisions for the settlement of compatibility disputes *avant la lettre* of the provisions in the Convention and their implementation by the Agreement. Second, the analysis takes a longer detour in developing the argument of the book by examining how this pattern of clausal embeddedness was employed to formulate the EEZ and Continental Shelf ('CS') delimitation rule in the Convention. Although priority is accorded to the structural formation of articles 74/83 LOSC, readers with a keen interest in the progressive development of the delimitation rule will also find valuable reflections on the history of the UNCLOS/III negotiations and the discussion of the rich jurisprudence that has emerged in this area of the law of the sea. Against this backdrop, the analysis in Chapter 5 returns to article 7 FSA to examine 'The drafting of the compatibility rule in the 1995 United Nations Fish Stocks Agreement'. This chapter explores the textual evolution of the rule's clausal construction, drawing on the official documents of the *Fish Stocks Conference* to demonstrate how the interrelation between its substantive and procedural constituent elements informed the structural formation of the article. Together, these two chapters posit that clausal embeddedness constitutes a drafting pattern employed in the context of progressive development rulemaking, which stems from doctrinal considerations aimed at intentionally linking the inherent ambiguity of substantive rules to the principle of compulsory adjudication. When disagreements arise over their interpretation or application, the designation of dispute procedures ensures that such disputes are submitted to the compulsory jurisdiction of the competent court or tribunal for a binding legal decision.

The compatibility rule, in terms of both its substantive and procedural constituent elements, is further examined by Chapter 6 in light of the 'Subsequent practice and approaches in regional regimes'. Following the adoption of the Agreement at the global level, efforts to sustainably regulate fish stocks—prioritizing their rebuilding, responsible development, and optimal utilization—have brought the negotiations undertaken in Regional Fisheries Management Organizations and Arrangements ('RFMO/As') to the forefront; the latter are equally important international cooperative regimes with the difference of not being formed as full-fledged organizations with executive or administrative organs distinct to their members. RFMO/As play a vital role in the sustainable regulation of global fish stocks, serving as regional focal points for the coordination of joint schemes that integrate law, policy, and science.[128] Those with regulatory competence, in particular, play a leading role in collective efforts to achieve compatible CMMs through cooperation between DWFS and coastal States. As a result, RFMO/As have thus become indispensable to the effective governance of sustainable fisheries. Given their crucial role and institutional accountability in addressing overexploitation, as well as growing concerns about the environmental impacts of fishing and fishing-related activities, UN has required RFMO/As to undergo regular performance reviews since 2005. The primary purpose of these reviews is to enhance their effectiveness by aligning their practices with modern regulatory principles—such as the principles of ecosystem and precautionary approach—as reflected in the Agreement and other relevant international treaties.[129] These formally instituted processes have furnished valuable findings in terms of subsequent treaty practice and approaches to the compatibility rule. The analysis in this chapter ultimately puts in a critical regional perspective the argument advanced by the

[128] *see* Molenaar, 'Regional Fisheries Management Organizations', in Ribeiro–Bastos–Henriksen (Eds) *Global Challenges and the Law of the Sea* (2020) 81 and Davis–Hanich, 'Transparency in Fisheries Conservation and Management Measures', *Marine Policy* (2022) 104088.

[129] FAO, 'Report of the Twenty-sixth Session of the Committee on Fisheries' (2005), ¶¶111–112; A/Res.60/31 (2005), *supra cit.* (*fn*111), ¶60 and *FSA Review Conference* 2006, Outcome II-B32(j); and *see* Haas–Haward–McGee–Fleming, 'The Influence of Performance Reviews on Regional Fisheries Management Organizations', *International Council for the Exploration of the Sea Journal of Marine Science* (2019) 2082.

book in favour of applying compulsory and binding procedures to compatibility disputes.

The book concludes with a set of focused 'Final remarks' on the procedural referral of compatibility disputes under the terms of article 7 FSA and on its interpretation in accordance with the general rule spelled out in article 31 *Vienna Convention on the Law of Treaties* (1969) ('VCLT'). Jurisprudence relating to the Convention has principally relied on the general rule of interpretation, and this has often been applied to avoid endorsing unilateral interpretations, as well as to lay stress on the progressive emergence of new rules that reflect a change in the interests of the international community as a whole. As remarked by a commentator, one can most certainly commend a jurisprudence that clearly seeks to restrain new forms of creeping jurisdictionalism that, in recent years, have unfortunately informed the agenda of an ever-increasing number of States.[130] Still, what constitutes the task of interpretation is itself a subject of prolific scholarly debate in general international law, reflecting a broad spectrum of epistemological standpoints and doctrinal concerns.[131] Whether interpretation as an act should distinguish between discovery and justification of legal propositions, or whether as a process it should be influenced—methodologically or otherwise—by surrounding circumstances, values and interests at stake, specific to a dispute or not, are questions that bear directly on the analysis developed throughout this book. The present work advances a procedural paradigm of treaty interpretation that reconciles the positivist objectivity rooted in textualism with the evolutive adaptability characteristic of a teleological and systemic integration approach. This paradigm facilitates interpretations that preserve the normative coherence of the compatibility rule by remaining consistent with a strict conception of the legal text while at the same time allowing dynamic reasoning to embody a rule-utilitarian flexibility. Limiting the scope of the book in this way inevitably narrows the generality of its conclusions, yet it affords the advantage of enabling

---

130 Virzo, 'The "General Rule of Interpretation" in the International Jurisprudence Relating to the United Nations Convention on the Law of the Sea', in Del Vecchio–Virzo (Eds) *Interpretations of the United Nations Convention on the Law of the Sea by International Courts and Tribunals* (2019) 15.

131 *see*, among others, Dupuy, 'Evolutionary Interpretation of Treaties: Between memory and prophecy', in Cannizzaro (Ed.) *The Law of Treaties Beyond the Vienna Convention* (2011) 123; d'Aspremont, 'Towards an International Law of Brigandage: Interpretative engineering for the regulation of natural resources exploitation', *Asian* JIL (2013) 1; and *compare* Linderfalk, 'Is Treaty Interpretation an Art or a Science? International Law and Rational Decision Making', EJIL (2015) 169 *with* Allott, 'Interpretation – An Exact Art', in Bianchi–Peat–Windsor (Eds) *Interpretation in International Law* (2015) 373.

an in-depth examination of the structure of article 7 FSA and offers a critical analysis that has been largely absent from existing scholarship. Given the sparse academic attention devoted to compatibility disputes, the structural formation of the rule—particularly as it relates to the scope of the fisheries limitation and the remit of compulsory procedures entailing binding decisions—has not previously been the subject of detailed scholarly examination.

The law is stated as at 31 March 2025. All electronic sources were last accessed on that date. Subsequent legal developments and scholarly works have been considered where relevant and practicable.

2. Compatible conservation and management measures

2.1 INTRODUCTION

Since the *Fishing & Conservation Convention* (1958) was concluded, the duty of international cooperation has become paramount for the adoption and implementation of CMMs aimed at assuring the biological integrity of fish stocks across their full distribution and migratory range. However, the institution of EEZ under the Convention, and the resulting regime for straddling stocks in article 63(2) LOSC and for stocks of highly migratory species in article 64 LOSC, has proven insufficient to ensure the effective conservation and management of these stocks in keeping with the ecosystem approach. To achieve the long-term conservation and sustainable use of these stocks, the Agreement, by virtue of the compatibility rule, reconciles the jurisdictional divide between sea areas under national jurisdiction and the high seas. This rule has been carefully formulated to preserve the balance of interests established by the Convention, ensuring that it neither curtails the freedom of fishing on high seas enjoyed by all States nor infringes upon the exercise of sovereign rights by coastal States within their EEZs. Nevertheless, the substantive porosity of the rule has allowed earlier controversies surrounding the interpretation of articles 63(2) and 64 LOSC to be revived—this time within the context of the Agreement—depending on how States have previously construed these provisions under the Convention. This chapter offers a critical analysis of the arguments predicated on competing interpretations, assessed in light of the need for interpretative consistency between the two treaties.

2.2 THE ORIGINS IN THE 1958 UNITED NATIONS FISHING & CONSERVATION CONVENTION

A common criticism levelled at the *Fishing & Conservation Convention* (1958) has been that it lacks specific substantive rules.[1] Addressing scientific aspects of marine living resources in an international legal instrument proved a formidable task, especially when at the time there was no uniform practice or generally accepted rules among States to inform regulatory regimes. In fact, Commissioner Jean-Pierre Adrien François (The Netherlands) in assuming the role of Special Rapporteur on the topic had voiced the widespread hesitancy within the Commission as to the final result of its undertaking in relation to fisheries.[2] At his prompting, UNGA convened the International Technical Conference on the Conservation of the Living Resources of the Sea ('1955 *Rome Conference*') to provide practical insights into conservation matters.[3] Among the directions outlined in the conference's report was the view that any prospective fisheries convention would need to 'clearly specify the kinds or types of measures, which may be used in order to achieve their objectives' and to 'have clear rules regarding the rights and duties of member States'.[4] Yet neither the Commission nor the subsequent UNCLOS/I met these expectations, and ultimately the convention did not contain definitive rules on CMMs or the exclusive allocation of fish stocks. A perusal of its text reveals that, despite the aforementioned technical advice, the only substantive rule is the one referring to the coastal State's special interest in high seas areas beyond its territorial waters. The doctrine of abstention, which previously featured mainly in bilateral arrangements, came under consideration by ILC but was not included in any of its draft articles. Nor was it implied in the provisions of the convention, although it was extensively discussed at UNCLOS/I, and it has since gradually fallen into disuse.[5]

---

1 *e.g.*, Oda, 'The 1958 Geneva Convention on the Fisheries: Its immaturities', *Die Friedens-Warte* (1959) 317.

2 ILC 63rd meeting (7/7/1950) [1950*YbILC*(I) 178].

3 Following the eponymous A/Res.900(IX) (14/12/1954), the conference was held in Rome 18/4 to 10/5/1955.

4 A/Conf.10/6 'Report of the International Technical Conference on the Conservation of the Living Resources of the Sea' (10/5/1955), ¶76.

5 Scheiber, 'Origins of the Abstention Doctrine in Ocean Law: Japanese–US relations and the Pacific fisheries, 1937–1958', ELQ (1989) 23.

2.2.1 An Early Notion of Compatibility Disputes in International Fisheries Law

Yet the convention marked a turning point in fisheries law by institutionalising cooperation as a normative process. This was clearly born out of necessity, given the protracted disputes simmering in the background. The most prominent of these, at the time, were disputes involving Japan on one side, and the USSR in the northwest Pacific Ocean, China in the Yellow Sea, and South Korea over the *Rhee Line* maritime boundary around the peninsula, on the other. The disputes between Chile, Ecuador, Peru and the United States in the south Pacific Ocean, as well as those in the Atlantic Ocean involving the United Kingdom—with Argentina in the south and with Iceland in the north—were equally acrimonious. For this reason, the Commission's primary aim was to fill the regulatory void in fisheries by establishing a formal and binding framework to address the absence of substantive rules, thereby securing recognition by States of a legal foundation upon which CMMs could be agreed. Unlike the other *Geneva Conventions*, which primarily codified and restated substantive rules, this convention has been regarded as 'a moral code'. It enunciates declaratory principles alongside robust operational procedures, thereby providing a foundational basis for the development of regulatory regimes rather than constituting a comprehensive treaty on such matters in its own right.[6] This law-making conceptual basis was distinctly reflected in the structure of most of its articles and, as other commentators have cogently observed, it is the only instrument adopted in 1958 that was drafted in terms of provisions recognizing a need—namely, the necessity of conservation—and subsequently establishing a process to address that need.[7]

The 1955 *Rome Conference* had observed that the regulatory challenges posed by fish stocks traversing multiple jurisdictions, including scientific and technical issues, could only be effectively addressed if the entire area were subject to a common system of conservation. This, in turn, necessitated that CMMs be adopted by agreement among the States concerned. Disagreements were anticipated to arise primarily over the spatial scope of such measures and the need to prevent those adopted by one State—or through agreement among the States concerned—from being nullified by the refusal of others to comply.[8]

---

6 Koers, *International Regulation of Marine Fisheries* (1973) 118.

7 *see* Jessup, 'The Geneva Conference on the Law of the Sea; A study in international law-making', AJIL (1958) 732; Dhokalia, *The Codification of Public International Law* (1970) 305; and Goldie, 'The Management of Ocean Resources: Regimes for structuring the maritime environment', in Black–Falk (Eds) *The Future of the International Legal Order* (1972) 212.

8 A/Conf.10/6 (1955), *supra cit.* (*fn*4), ¶¶78–80.

The subject matter of disputes arising from the application and interpretation of articles 6, 7 and 8 FCC (see Table 4.3), in fact, reflects an early form of compatibility disputes, as they principally concern the validity and bindingness of measures affecting, in essence, sea areas under coastal jurisdiction and those adjacent and beyond, on the high seas. In a more assertive tone—although expressed through the resolutions on 'Co-operation in Conservation Measures' and on 'Special Situations relating to Coastal Fisheries'—UNCLOS/I recommended that, where fisheries occur in both sea areas, coastal States should cooperate with the relevant international conservation organizations responsible for adopting and applying conservation measures in the adjacent high seas, and implement and enforce, to the fullest extent practicable, the necessary conservation measures within sea areas under their national jurisdiction.[9] As noted later by the former UNCLOS/III President Koh, the precursory elements of the ecosystem approach can be traced back to the *Fishing & Conservation Convention* (1958), observing that the treaty promoted multilateralism in the implementation of regulatory schemes spanning the entire fishery.[10] As the first international instrument to apply this approach to fisheries, the convention, in the author's view, represents the point at which compatibility disputes began to emerge as a distinct category of fishery disputes in international law.

2.2.2 The Concept of Special Interest

The most apparent, yet unclear, substantive concept of the convention is in reference to the coastal State's special interest in maintaining the productivity of living resources in high seas areas beyond its territorial sea.[11] This concept was first proclaimed by the United States in response to the urgent need for protecting coastal fishery resources from destructive exploitation, emphasizing the importance of 'having due regard to conditions peculiar to each region and situation, and to the special rights and equities of the coastal State and of any other State which may have established a legitimate interest therein'.[12] Nonetheless, its recognition under the convention was minimal, deliberately

---

[9] A/Conf.13/L.56 'Resolutions adopted by the United Nations Conference on the Law of the Sea' (25/4/1958) [*Geneva Off.Records* II:144].

[10] Koh, 'The Origins of the 1982 Convention on the Law of the Sea', MLR (1987) 13.

[11] Oda, 'International Law of the Resources of the Sea', *Recueil des Cours* (1969) 417.

[12] United States, 'Proclamation by the President with respect to Coastal Fisheries in Certain Areas of the High Seas' (28/9/1945) [(1946) AJIL *Suppl.* 29(1):46–47]; and *see* Kraska, 'North America', in Kraska–Ryou-Ellison (Eds) *Cultural Influences on the Law of the Sea* (2025) 252–257.

avoiding the assignment of priority in fishing matters to the coastal State,[13] and was instead distinguished from the concept of preferential rights as envisaged in the resolution on 'Special Situations relating to Coastal Fisheries'. The Commission made it clear that the special nature of a coastal State's interest should be interpreted and applied functionally; specifically, that such interest arises in each case solely from the State's geographic position relative to the regulation of the fishery in question. The substantiation of this interest did not take precedence over the rights of DWFS with respect to high seas fishing activities.[14]

Admittedly, the vague legal nature of the concept—coupled with its undefined spatial scope—gave rise to serious questions about its implications for the progressive development of a new category of rights for coastal States. In particular, whether the special interest not only affirmed the coastal State's right to participate on an equal footing with other States in regulating fish stocks that span both areas under coastal jurisdiction and the high seas but also extended to a unilateral right to impose national measures on DWFS operating in the latter areas.[15] Underlying these questions was the looming risk of doctrinal abuse, whereby coastal States might engage in auto-interpretation of their newly asserted rights, to the detriment of other States fishing in areas of special interest. To safeguard the rights of all States concerned in such areas, the Commission—as will be discussed below—embedded procedural clauses within the relevant substantive articles, subjecting any disagreement over the adoption and application of regulatory measures to compulsory dispute settlement procedures.

In a high seas area adjacent to the territorial sea

The customary freedom of all States to allow their nationals to fish on the high seas[16] is qualified, inter alia, by the interests and rights of coastal States as recognized in the convention.[17] The provisions that acknowledge a distinct interest for coastal States—and on that basis confer additional rights compared to other States—are those referring to a 'special interest'. Article 6(1) FCC affirms, and accordingly confers upon the coastal State, 'a special interest in the maintenance of the productivity of the living resources in any area of the

13 As reflected in the fruitless initiatives aimed at establishing the priority of coastal State interests in conservation disputes with DWFS; e.g. in A/Conf.13/C.3/L.21 'Costa Rica, Mexico, Peru, United Arab Republic and Chile: Proposal (article 50)' (25/3/1958) [*Geneva Off.Records* V:140].

14 A/CN.4/104 (1956), *cit.* Chapter 1 (*fn*39), [1956*YbILC*(II) 288(¶14)].

15 *e.g.*, Sorensen, 'The Law of the Sea', *International Conciliation* (1958) 218.

16 HSC 2.

17 FCC 1(1).

high seas adjacent to its territorial sea'. States whose nationals are engaged in such areas are discouraged from adopting conservation measures that conflict with those implemented by the coastal State. The States concerned are required to enter into negotiations with a view to reaching agreement on the necessary measures in the relevant area. Where negotiations fail to result in an agreement, article 7 FCC permits the coastal State to prescribe unilateral conservation measures. However, this postulated 'special' right to act unilaterally in high seas areas adjacent to its territorial sea is, in substance, a right exercised at the sufferance of DWFS. At best, it may be characterized as a right of tolerance, which may furthermore be procedurally suppressed.

A useful example can be drawn from a situation where a disagreement arises between a coastal State and another State, whose nationals fish in high seas areas adjacent to the former's territorial sea, concerning conservation measures for a stock occurring within both jurisdictions. Article 6(5) FCC places both the DWFS and coastal State concerned under an obligation to engage in negotiations for a period of 12 months before either of them may resort to the dispute settlement procedures of the convention. Additionally, article 7(1) FCC permits the coastal State to prescribe and apply unilateral measures if the disagreement remains unresolved 6 months after the commencement of negotiations. The validity and bindingness of such measures on DWFS are, nevertheless, conditioned upon the fulfilment of three requirements: *in seriatim*, that the measures are necessitated by the 'need for urgent application ... in the light of the existing knowledge', are 'based on appropriate scientific findings', and 'do not discriminate in form or in fact against foreign fishermen'.[18]

At this critical juncture in the course of a disagreement, the coastal State's special right is subject to its first instance of procedural suppression; namely, the quasi-approval of DWFS by not objecting to the unilateral measures. Insofar as such approval is implied, the conservation measures cease to be unilateral and are, in essence, perceived as consented to, or indirectly agreed upon, *sensu lato*. The second instance of procedural suppression arises when the measures adopted by the coastal State are not accepted by DWFS. On that occasion, any of the States concerned may initiate the dispute settlement procedure under article 9(1) FCC.[19] Procedurally, an objection to unilateral measures triggers a reduction of the negotiation period from 12 to 6 months. Although the convention provides that the implicated measures shall remain provisional *sub judice*, these may be suspended if not justified by a *prima facie* need for urgent application.[20] Essentially, the coastal State's right of unilateral

[18] FCC 7(2).

[19] FCC 7(4).

[20] *idem*; the reference to the Special Commission is construed to apply to any other *ad hoc* or standing adjudicative forum under articles 9 and 10 FCC.

prescription is contingent upon either the consent of DWFS or a binding decision rendered through the third-party compulsory dispute settlement procedure. The only period during which a coastal State may apply such measures without restriction is the interim between the expiry of the 6 month threshold for unilateral prescription and the issuance of a preliminary decision on their validity.[21] In the case of compulsory arbitration by the Special Commission under article 9 FCC, this interval would be less than 5 months, or at most, just under 8 months from the time of its appointment.[22]

While the appointment of the Special Commission can take up to 6 months from the time a request for settlement is made, the actual duration is typically shorter, given that provisional measures are prescribed early in the process. In view of the brevity of this period—especially when contrasted with the long-term objectives of conservation—it cannot be gainsaid that the concept of 'special interest' falls considerably short of constituting a meaningful legal right capable of evidencing a predominant role for coastal States in adopting unilateral measures. From the scholarly viewpoint of the time, the concept was described as 'a product of collective imagination'.[23] This frustration was especially pronounced among coastal States—most vocally those in the Latin American region—which viewed the concept of special interest in high seas areas adjacent to the territorial sea as so abstrusely framed and encumbered with conditions that it effectively nullified its impact, if not rendered it entirely illusory.[24]

In high seas areas not adjacent to the territorial sea

The convention likewise envisages the concept of special interest as one that may be invoked by a coastal State in the conservation of fisheries located in high seas areas not adjacent to its national jurisdiction, including those where its nationals may not be actively engaged in fishing for the stocks concerned. In this context, the concept provoked considerable reticence on the part of DWFS, with respect to the factual circumstances that might give rise to it, especially since, in theory, it could be broadly invoked on grounds such as

---

21 FCC 7(3).

22 FCC 9(5).

23 *see* Burke, 'Some Comments on the 1958 Conventions', *American Society of International Law Proceedings* (1959) 205 and Extavour, *The Exclusive Economic Zone* (1979) 114.

24 *see* the statement recorded by Peru in A/Conf.13/5 'Comments by Governments on the Draft Articles Concerning the Law of the Sea' (23/10/1957) [*Geneva Off.Records* I:98] and the remarks by the delegation of Costa Rica at the Third Committee 12th meeting (20/3/1958) [*Geneva Off.Records* V:29].

economic dependence, food security, and other national interests.[25] In connection with an actively fishing coastal State, the relevant circumstances giving rise to a special interest are more straightforward, particularly in relation to stocks of highly migratory species that are harvested at various stages of their biological cycle in high seas areas far beyond those adjacent to its territorial sea. In that case, the coastal State may request DWFS to adopt the necessary conservation measures. Failing agreement, the earlier outlined dispute settlement procedure can be initiated.[26]

2.3 ECOLOGICAL CONCERNS IN THE REGULATION OF FISHERIES UNDER THE 1982 UNITED NATIONS CONVENTION ON THE LAW OF THE SEA

The *Fishing & Conservation Convention* (1958) crafted a remarkable balance between the need to adopt CMMs for common interest fish stocks in high seas areas and the imperative of upholding the sovereign rights of States consistently with the other *Geneva Conventions* at the same time. Owing to the functional basis of the rules on special interest, coastal States exercising such specialized competence outside the scope of their territorial jurisdiction could not thereby acquire exclusive fishing rights on the high seas.[27] Nevertheless, this was not satisfactory for the coastal States, not least because universal agreement on extending the limits of territorial sea more than the maximum breadth of 3 nm as recognized by custom in most regions at the time remained elusive. A policy impulse, prompted by the failure of UNCLOS/II to adopt the 6 nm + 6 nm formula, was reflected in the national legislation,[28] and various declarations advancing extraterritorial claims by several coastal States

[25] García-Amador, *The Exploitation and Conservation of the Resources of the Sea* (1959) 182–183.

[26] FCC 8.

[27] Verzijl, 'The United Nations Conference on the Law of the Sea, Geneva, 1958', *Nederlands Tijdschrift* (1959) 125 and Bowett, *The Law of the Sea* (1967) 11.

[28] *e.g.*, under the Brazilian decree No.68/459 *Regulamenta a Pesca, Tendo em Vista o Aproveitamento Racional e a Conservação dos Recursos Vivos do Mar Territorial Brasileiro* [(1971) 10ILM1224] in which the concept of special interest was connected to the exercise of territorial jurisdiction, and under No.189/1972 *Regulations Concerning the Fishery Limits Off Iceland* [(1972) 11ILM1112] whereby Iceland extended its fisheries limits from 12 to 50 nm.

across different continents and levels of economic development.[29] Most of these States, having promulgated national jurisdiction extending up to 200 nm, entered the UNCLOS/III negotiations with the expectation of legitimizing their prior claims.[30] While the very need to conserve marine living resources was the *raison d'être* of the demand by coastal States to extend further their authority over high seas areas by means of EFZ or otherwise broadly defined economic zone,[31] its *sui generis* character, as a functional zone situated beyond the territorial sea and over the high seas, was largely shaped during the negotiations by political causes rather than ecological considerations. Simply put, the 'nationalization' of the oceans could not solve the problem of conserving and managing fisheries without effective cooperation.[32] The institution of EEZ within the jurisdictional sea space—premised on the implicit assumption that all coastal States are equally willing and capable of effectively regulating living marine resources within the zone[33]—combined with the absence of substantive rules to promote an ecosystem approach and ensure cooperation, has been consistently highlighted as a major weakness of the Convention.[34]

On the matter of fishery regulation, the most telling example of the Convention's ecological deficit is the unqualified reliance on the conceptual objective of MSY, which is found at the core of conservation measures aiming at input/output controls and restrictions in fisheries management both on the high seas and in EEZs. This objective offers the sole technical reference in the Convention for guiding the establishment of TAC limits at harvesting levels that maintain a stock's maximum regenerative capacity in terms of biomass,

---

29 *e.g.*, the *Declaration of Santo Domingo*, adopted by the 'Specialized Conference of Caribbean Countries Concerning the Problems of the Sea', held 31/5 to 9/6/1972 in Santo Domingo de Guzmán [(1972) 66AJIL918]; *qv.* Anderson, *The Law of the Sea in the Caribbean* (2022) 21*ff* and the *Conclusions in the General Report of the African States* adopted by the African States in the 'Regional Seminar on the Law of the Sea' held on 20–30/6/1972 in Yaoundé [(1973) 12ILM210]; *qv.* Rembe, *Africa and the International Law of the Sea* (1980) 116*ff.*

30 Clingan, 'Emerging Law of the Sea: The economic zone dilemma', SDLR (1977) 536–543.

31 *see* the insights offered by Tommy Koh in 'The Third United Nations Conference on the Law of the Sea: What was accomplished?', LCP (1983) 6 and 'The Exclusive Economic Zone', MLR (1988) 26–27.

32 Ulfstein, '200 Mile Zone and Fisheries Management', NJIL (1983) 9.

33 Pardo, 'The Convention on the Law of the Sea: A preliminary appraisal', SDLR (1983) 498.

34 *see* Von Zharen, 'An Ecopolicy Perspective for Sustaining Living Marine Species', ODIL (1999) 15*ff*; Kirk, 'Maritime Zones and the Ecosystem Approach: A mismatch?', RECIEL (1999) 67; and Boyle–Redgwell, *International Law and the Environment* (2021) 725.

although it has faced severe criticism as a regulatory concept. This criticism arises from scientific concerns, including its neglect of environmental factors that influence reproductive capacity, as well as its failure to adequately address the natural variability of fish stocks. It also encompasses wider fisheries policy considerations, such as the omission of other guiding objectives, particularly the optimum utilization of resources in economic and social terms.[35] Although the MSY concept had been regarded as outdated by parts of the scientific community since the time of UNCLOS/III, it was nonetheless incorporated into the Convention. While it remains a debatable regulatory concept, often linked to significant failures that have contributed to the decline or even collapse of numerous fish stocks, it continues to be the primary reference on which the international community relies in pursuing SDG 14(4).[36] Moreover, with regard to fish stocks, the Convention has proven inadequate in overcoming the politically motivated jurisdictional division of ocean space between EEZ and the high seas, thereby failing to ensure effective regulation across their full distribution or migratory range. Apart from calling upon States to cooperate in the conservation and management of the stocks, it lacks substantive provisions to harmonize CMMs adopted in the respective sea areas.[37] Consequently, relevant provisions of substance with a bearing on the conservation and management of the stocks are to be found in two different parts of the Convention. These are mainly located in Part V, where articles 61–73 LOSC govern fisheries in EEZ, and in section 2 Part VII, where articles 116–120 LOSC apply to high seas fisheries.

With respect to the regime applicable on high seas fisheries, the Convention reiterates the freedom of fishing being qualified by reference to certain conditions. These include the treaty obligations of States and the rights, duties, and interests of coastal States in relation to fish stocks, as well as other relevant provisions of the Convention.[38] Furthermore, without departing from customary practice, it reaffirms the principles of nationality and non-discrimination

---

[35] Caddy–Griffiths, *Living Marine Resources and their Sustainable Development* (1995) 63 & 195, and Sovacool, 'A Game of Cat and Fish: How to restore the balance in sustainable fisheries management', ODIL (2009) 97.

[36] *see* Hey, 'The Persistence of a Concept: Maximum sustainable yield', in Freestone (Ed.) *The 1982 Law of the Sea Convention at 30* (2013) 89; Kesteven, 'MSY Revisited: A realistic approach to fisheries management and administration', *Marine Policy* (1997) 73; and Huck, *Sustainable Development Goals* (2022) 520*ff.*

[37] Nandan, 'An Introduction to the 1982 United Nations Convention on the Law of the Sea', in Vidas–Østreng (Eds) *Order for the Oceans at the Turn of the Century* (1999) 12.

[38] LOSC 116.

by holding each State responsible for prescribing, with respect to those under its exclusive flag jurisdiction, the necessary measures for the conservation of fish stocks on the high seas.[39] States, whose nationals exploit identical, dependent or associated stocks, are obligated to enter into negotiations with a view to adopting CMMs.[40] This constitutes a *de contrahendo* commitment to cooperate,[41] either directly or through the appropriate RFMO/A, that legally binds them to reach a definitive agreement on the necessary measures. The adoption of conservation measures on the high seas, and in particular the setting of TAC limits, the only substantive measure explicitly mentioned, requires consideration of two key criteria. First, in taking advantage of the best scientific evidence and information available, States are required to ensure the maintenance or restoration of harvested stocks at levels that produce MSY, as qualified, inter alia, by relevant environmental and economic factors. However, the Convention contains no provision for balancing these factors, an omission that highlights the need for conservation measures to be agreed upon through negotiation.[42] Second, States are also required to ensure the maintenance or restoration of any associated or dependent stocks at levels above those at which their reproduction may become seriously threatened.[43] The apparent regulatory imprecision in these criteria—regarding both their purpose and the expected conduct of States in fulfilling them—was intentionally drafted, reflecting the difficulty of reaching consensus during UNCLOS/III on specific conservation objectives other than MSY. Hence, the Convention offers no instructive guidance on how the duty to cooperate should be carried out to reach agreement on conservation measures aimed at fulfilling this objective, nor does it specify the corresponding management measures, if any, to be taken.[44] As pointed out elsewhere, the high seas fisheries provisions were drafted as part of a procedural tactic, with the expectation that open substantive questions, especially those concerning environmentally sustainable harvesting, would be resolved later through diplomatic negotiations or international adjudication.[45]

---

39 LOSC 117 & 119(3).

40 LOSC 118.

41 On this kind of obligation *see Tacna–Arica Question* (Award) 1925, 929–930; *qv*. McNair, *The Law of Treaties* (1961) 27–30.

42 Takei, *Filling Regulatory Gaps in High Seas Fisheries* (2013) 71–74.

43 LOSC 119.

44 Oda, 'Sharing of Ocean Resources – Unresolved Issues in the Law of the Sea', *New York Journal of International and Comparative Law* (1981) 10.

45 Stephens, 'The Limits of International Adjudication in International Environmental Law: Another perspective on the Southern Bluefin Tuna Case', IJMCL (2004) 192.

Regarding the EEZ regime, the coastal State is accorded sovereign rights for the purpose of exploring and exploiting fisheries in the zone, while at the same time bearing responsibility for their conservation and management.[46] As these are economic and resource-oriented rights with a clearly defined purpose, the coastal State's exercise of corresponding jurisdiction is functional in nature. In doing so, coastal States are subject to one general and two specific obligations. To begin with the general obligation, coastal States 'shall have due regard to the rights and duties of other States and shall act in a manner compatible with the provisions of the Convention'.[47] As noted by ITLOS, this provision has clear application to fisheries of common interest.[48] Regarding the more specific obligations, coastal States are bound to promote both the aim of conservation and the objective of optimum utilization. With respect to conservation, the coastal State—taking into account the best scientific evidence available—must ensure, through appropriate CMMs, that the maintenance of living resources in EEZ is not endangered by overexploitation. Among these measures, the establishment of TACs must be aimed at maintaining or restoring harvested stock populations to levels that produce MSY.[49] The second specific obligation relates to the objective of optimum utilization, which must be pursued without prejudice to the aim of conservation. To promote the optimum utilization of fisheries in EEZ, the coastal State is required to determine its harvesting capacity in relation to the national TAC quotas and, in the event of a surplus, grant access to other States for fishing the stock.[50]

The provisions that directly address straddling and highly migratory fish stocks are found in articles 63(2) and 64 LOSC, respectively, both situated within Part V, titled 'Exclusive Economic Zone', of the Convention. They essentially defer the regulation of fish stocks to cooperation, without providing States with guidance on the substantive content, timing, or necessity of CMMs, which are to be adopted by agreement. This lack of direction is attributed to unresolved differences over the balance of interests in such stocks between DWFS and coastal States. As a result, these provisions are considered among the most contentious in the Convention, underscoring the exceptional importance of the dispute settlement mechanism in their ongoing normative

46 LOSC 56(1)(a).

47 LOSC 56(2), which importantly also forms part of customary international law as per the pronouncements in *Alleged Violations of Sovereign Rights and Maritime Spaces in the Caribbean Sea* (Merits) 2022, §57 & §§215–216.

48 *SRFC Request* (Advisory Opinion) 2015, §216.

49 LOSC 61.

50 LOSC 62.

development.[51] In the *SRFC Request*, ITLOS elucidated significant aspects of articles 63 and 64 LOSC when interpreted in light of sustainable management responsibilities.[52] Given the jurisdictional scope it defined for itself in the instant case, the Tribunal focused on the rights and obligations of the coastal State in their EEZ. Correspondingly, articles 63(2) and 64(1) LOSC, although directly applicable in general, were excluded to the extent that they concern fishing in sea areas adjacent to and beyond the zone. As a result, issues relating to the conservation and management of fish stocks while on the high seas were not addressed. Despite these caveats, the Tribunal emphasized that the principle of cooperation reflected in both articles is well established.[53]

2.3.1 Straddling Fish Stocks

The applicable regime for straddling stocks is set out in article 63(2) LOSC, which provides:

> Where the same stock or stocks of associated species occur both within the exclusive economic zone and in an area beyond and adjacent to the zone, the coastal State and the States fishing for such stocks in the adjacent area shall seek, either directly or through appropriate subregional or regional organizations, to agree upon the measures necessary for the conservation of these stocks in the adjacent area.

It is apparent from a plain reading that the provision lacks in offering any specific substantive guidance to address problems regarding the conservation and management of these stocks. Nonetheless, it articulates two fundamental stipulations that define the scope of the duty of cooperation between DWFS and coastal States.

The first stipulation concerns the spatial scope of cooperation, as it confines the subject matter of negotiations on applicable conservation measures to 'an

---

51 Borg, 'The Influence of International Case Law on Aspects of International Law Relating to the Conservation of Living Marine Resources Beyond National Jurisdiction', YbIEL (2012) 50.

52 The exact question on this matter was posed in the terms of 'What are the rights and obligations of the coastal State in ensuring the sustainable management of shared stocks and stocks of common interest, especially the small pelagic species and tuna?' as contained in the 'Resolution of the Conference of Ministers of the SRFC' 14th Extraordinary Session (25–29/3/2013), and submitted to the Tribunal by the Permanent Secretary of the SRFC in the 'Request for Advisory Opinion' (27/5/2013); for a general commentary *see* Serdy, 'Managing Transboundary Fish Stocks for Sustainability', in Øystein (Ed.) *The Development of the Law of the Sea Convention* (2021) 123*ff*.

53 *SRFC Request* (Advisory Opinion) 2015, §§179–182, 200–204 & 213.

area beyond and adjacent' to EEZ. Certain approaches have interpreted this as an expansion of the coastal State's interest, first, by excluding EEZ itself from the scope of negotiations and second, by asserting the coastal State's primary role over DWFS in the adjacent high seas areas. In support of this interpretation, it has been argued that the freedom of high seas fishing is qualified by article 116(b) LOSC, in light of the rights, duties, and interests of the coastal State as provided in articles 63(2) and 64 LOSC. This view has been conveyed by the declaration of Uruguay upon signing the Convention, in which it asserted that DWFS 'are duty bound to agree with the coastal State' on the measures necessary for the conservation of harvested, associated, or dependent stocks, both in EEZ and in the high seas areas.[54] The declaration propounds a 'special interest' in the conservation and management of straddling stocks in high seas areas adjacent to, and beyond, the EEZ in accentuating the sovereign nature of the coastal rights in the zone. This assertion goes significantly further than contemporaneous declarations by other coastal States, which, on this precise point, employed more restrained language, stating merely that DWFS 'are duty bound to enter into arrangements with the coastal State' regarding the necessary conservation measures.[55] Even so, a commensurate obligation rests with the coastal State. In exercising its sovereign rights for the conservation and management of a stock within EEZ, the coastal State is, as previously noted, subject to the general obligation to act in a manner compatible with the provisions of the Convention and with due regard for the rights and duties of other States.

Based on this counterpoint, scholarly commentary tends to moderate the arguments advocating a special interest, by suggesting that such an interest may be implied to the extent that CMMs adopted on the high seas must be consistent with those prescribed by the coastal States within its EEZ. Although this reading does not contradict the procedural requirement of cooperation, the substantive concept of special interest—even in its most attenuated version—remains at considerable variance with the wording of article 63(2) LOSC.[56] Other commentators emphasize the integrity of the obligations in article 116(b) LOSC, concluding that 'coastal States do not have a special right' concerning the exploitation of straddling stocks on the high seas and may engage in such exploitation only under the general freedom of fishing to which all

---

[54] On 23/2/1983 [CN.7.1983:XXI-6], F.

[55] *e.g.* Cabo Verde on 23/2/1983 [CN.7.1983:XXI-6], VII.

[56] *see* the view held by Barbara Kwiatkowska in *The 200 Mile Exclusive Economic Zone in the New Law of the Sea* (1989) 79–80 and 'Creeping Jurisdiction Beyond 200 Miles in the Light of the 1982 Law of the Sea Convention and State Practice', ODIL (1991) 169.

States are entitled.[57] The latter aligns more closely with a balanced approach to the respective rights by highlighting the duty of cooperation in the adoption of high seas measures agreed upon by the States concerned. While acknowledging that article 116(b) LOSC presents interpretative challenges, owing to its cross-reference to articles 63(2) and 64 LOSC, the duties and requirements imposed on coastal States do not serve to restrict the right to engage in high seas fishing exclusively to the coastal State.[58]

Not far from the arguments above lies the second stipulation, which addresses the degree of engagement expected in terms of cooperation. The specific duty, being phrased through the idiom of 'shall seek … to agree upon the measures', suggests that the commitment to negotiate with respect to straddling stocks constitutes a *pactum de negotiando*.[59] While the normative content of the duty requires States to engage in meaningful negotiations and to make a genuine effort *bona fide* to agree on CMMs, it does not entail a binding commitment to reach a final agreement.[60] A genuine disagreement may indeed persist regarding the objective determination of the 'measures necessary' for the conservation of the stocks. While the term 'necessary' is conventionally used in treaty language to mean 'indispensable, requisite, or essential', in the context of provisions aimed at conserving marine living resources—importantly, as part of the evolving obligations to protect and preserve the marine environment—it is broadly interpreted to encompass a variety of measures that contribute to achieving these objectives.[61] It may be questioned whether the existence of a *pactum de negotiando* within the Convention precludes a State from resorting to dispute settlement procedures when agreement proves unattainable, or the parties otherwise reach an impasse in negotiations.

---

[57] Hey, *The Regime for the Exploitation of Transboundary Marine Fisheries Resources* (1989) 53–54, who was importantly followed on this point by the ILA Report 1992, with Rainer Lagoni as Rapporteur of the Committee on the *Principles Applicable to Living Resources Occurring both within and without the Exclusive Economic Zone or in Zones of Overlapping Claims* (65th Session, held in Cairo on 20–26/4/1992), 254–277.

[58] Oda, 'Fisheries under the United Nations Convention on the Law of the Sea', AJIL (1983) 749–751.

[59] 1993 *Virginia Commentary* II ¶63.12(646).

[60] On the essence of the pactum and the distinction between the quality of the legal engagement to negotiate from that of agreeing *see Railway Traffic between Lithuania and Poland* (Advisory Opinion) 1931, p.116 and *Greece–Germany Arbitration* (Decision) 1972, pp.42–60.

[61] *Climate Change and International Law* (Advisory Opinion) 2024, §§203, 387 & 419–424.

As a starting point in addressing this question, the following general jurisprudential authorities are instructive. In the *Affaire du Lac Lanoux*, the arbitral tribunal, recognizing the nuance inherent in assessing the extent to which a State conducts negotiations in good faith, aptly observed that:

> d'après les règles de la bonne foi, l'obligation de prendre en considération les différents intérêts en présence, de chercher à leur donner toutes les satisfactions compatibles avec la poursuite de ses propres intérêts et de montrer qu' il a, à ce sujet, un souci réel de concilier les intérêts.[62]

The commitment to reconciling interests as a matter of genuine concern was further affirmed by ICJ in the *North Sea Continental Shelf Cases*, where the Court considered the parties to be under an obligation to enter into negotiations with a view to reaching an agreement, rather than treating such engagement as a mere formality. This obligation requires States to conduct themselves in a manner conducive to meaningful negotiations, an expectation that is not met when either party rigidly adheres to its own position without demonstrating any willingness to consider adjustments.[63] In the context of fisheries, the essence of this dictum finds expression in a subsequent pronouncement by ICJ in the *Fisheries Jurisdiction Cases*. As at that time, UNCLOS/III was only beginning to grapple with the prospective regime of economic rights in specific sea areas, there existed no formally accepted legal grounding—apart from the customary 6 nm + 6 nm formula, which UNCLOS/II had not adopted—for a coastal State to extend its exclusive fishing jurisdiction or to unilaterally establish a 200 nm special fishery zone.[64] The juridical issues were therefore framed largely in terms of whether, and to what extent, other States should enjoy fishing rights over marine living resources within such zones. The Court did not adjudicate the parties' conflicting claims to substantive fishing rights per se but instead identified relevant factors to be considered in negotiating an equitable resolution of their differences with respect to the exercise of such rights in high seas areas to which both parties had a right to access. It nevertheless observed that, even where a coastal State is entitled to preferential rights, their exercise is not consistent with the total exclusion of other States from

[62] *Affaire du Lac Lanoux* (Award) 1957, §22.

[63] *North Sea Continental Shelf Cases* (Judgment) 1969, §85.

[64] *see* Churchill, 'The Fisheries Jurisdiction Cases: The contribution of the International Court of Justice to the debate on coastal States' fisheries rights', ICLQ (1975) 86*ff* and Fleischer, 'The Right to a 200-Mile Exclusive Economic Zone or a Special Fishery Zone', SDLR (1977) 572–579.

fishing on the high seas. Nor is the coastal State free to determine unilaterally, and according to its own uncontrolled discretion, the scope of those rights.[65]

Without conflating the concept of 'preferential rights' with that of 'special rights', the logic underpinning the above dictum suggests that, in the present context, disagreements concerning the clarification of rights in fisheries of common interest cannot remain unresolved merely owing to their concurrent exercise by the respective States. In the same vein, and in light of the *pactum de negotiando* spelled out in the provision governing straddling stocks, where the States fail to reach agreement on CMMs despite negotiating in good faith, they are obliged to engage with the available third-party dispute resolution process. This proposition is notably espoused by leading academic authorities who, while advocating an extension of coastal jurisdiction derived from the exercise of 'special interest', nonetheless maintain that its application should be subject to the application of compulsory dispute settlement procedures in the event of disagreement with DWFS.[66]

2.3.2 Fish Stocks of Highly Migratory Species

The negotiations during UNCLOS/III also revealed divergent views among States regarding whether highly migratory fish stocks—particularly tuna and tuna-like species[67]—should be included in a special regime conceding sovereign rights to coastal States within EEZ, or whether they should be regulated on the basis of agreement and legal parity with DWFS. The compromise reached is concealed in article 64 LOSC, which reads as follows:

> 1. The coastal State and other States whose nationals fish in the region for the highly migratory species listed in Annex I shall cooperate directly or through appropriate international organizations with a view to ensuring conservation and promoting the objective of optimum utilization of such species throughout the region, both within and beyond the exclusive economic zone. In regions for which no appropriate international organization exists, the coastal State and other States whose nationals harvest these species in the region shall cooperate to establish such an organization and participate in its work.

---

65 *Fisheries Jurisdiction Cases* (Merits) 1974, §§27, 54 & 75.

66 *see* Miles–Burke, 'Pressures on the United Nations Convention on the Law of the Sea of 1982 Arising from New Fisheries Conflicts: The problem of straddling stocks', ODIL (1989) 352–353 and Saunders, 'Jurisdiction and Principle in the Implementation of the Law of the Sea: The case of straddling stocks', in Carmody–Rhodes–Iwasawa (Eds) *Trilateral Perspectives on International Legal Issues* (2003) 384.

67 Munro, 'Extended Jurisdiction and the Management of Pacific Highly Migratory Species', ODIL (1990) 289.

2. The provisions of paragraph 1 apply in addition to the other provisions of this Part.

At first glance, the controversy over the applicable regime for these species may seem less intense than that surrounding straddling stocks. In line with the high seas fishing regime, States are *de contrahendo* obligated to cooperate. This obligation is of a due diligence nature, requiring States to act in good faith by engaging in meaningful negotiations and making a genuine, substantive effort to reach agreement either directly or through the appropriate RFMO/A.[68] In contrast to the regime governing straddling stocks, the duty to cooperate in respect of highly migratory species is directed towards the adoption of effective measures necessary to ensure their conservation and to promote the objective of optimum utilization throughout their migratory range, both within and beyond EEZ. Yet uncertainties subsisted in the Convention concerning the material scope of the coastal State's sovereign rights over these species, especially with respect to exclusive fishing rights within EEZ and the unilateral adoption of in-zone CMMs. Perspectives on these issues have evolved over time, often in relation to certain species or regions, and in some instances, previously held positions were subsequently recanted following the Convention's entry into force.[69]

The provision made in paragraph (1) reflects an ecological aspiration to regulate these species across their entire migratory range, within a framework of cooperative governance. This objective served as the principal impetus for a considerable portion of the international community in advancing towards UNCLOS/III, as heralded in the *Moscow Declaration on Principles of Rational Exploitation of the Living Resources of the Seas and Oceans in the Common Interests of All Peoples of the World* (1972)[70] and the *African Declaration on the Issues of the Law of the Sea* (1973).[71] The duty of cooperation was subsequently elaborated in numerous national preparatory proposals,

68 *SRFC Request* (Advisory Opinion) 2015, §210.

69 A clear example is found in the practice of the United States, whose national policy demonstrates a balance between distant-water fishing and coastal interests; *see* the enactment of the *Magnuson–Stevens Fishery Conservation and Management Act* [94th Congress Public Law, 265 (13/4/1976)]; *qv.* Carr, 'Transformations in the Law Governing Highly Migratory Species: 1970 to the present', in Caron–Scheiber (Eds) *Bringing New Law to Ocean Waters* (2004) 55, and its revision by the *Magnuson–Stevens Fishery Conservation and Management Reauthorization Act* (2006) [109th Congress Public Law, 479 (12/1/2007)].

70 A/AC.138/85 (17/8/1972) [(1973) 12ILM214], circulated at the request of Bulgaria, Czechoslovakia, Hungary, Poland and the USSR.

71 A/AC.138/89 (2/7/1973) [(1973) 12ILM210].

with Japan taking the lead early on in efforts by DWFS to secure a jurisdictional regime for highly migratory species that was free from any connotation of 'preferential' or otherwise 'special' interests, and aligned as closely as possible with the precepts of collective governance,[72] now forming the core normative basis of article 64 LOSC. Nevertheless, the provision leads to two successive and closely linked questions. One question is whether the legal import of the obligation to ensure the aim of conservation and promote the objective of optimum utilization applies not only to the high seas but also within EEZ, and to what extent this potentially qualifies the sovereign rights of the coastal State.[73] This inevitably hinges on a second question concerning the interaction between CMMs adopted for the high seas by all States concerned and those prescribed by the coastal State within its EEZ. When these questions are generally considered in the context of the EEZ regime, the spectrum of academic answers tends to polarize; on one end, some argue that highly migratory stocks are exempt from the coastal State's exclusive jurisdiction in EEZ, while on the other, some counter that the coastal State's sovereign rights hold sway both within and beyond the zone.[74]

The prevailing view among coastal States is that these species fall entirely within their exclusive jurisdiction in EEZ, as article 64 LOSC is not intended to infringe upon their sovereign rights to explore, exploit, conserve, and manage marine living resources in the zone in accordance with article 56 LOSC. Adherents to this view interpret the addition of paragraph (2) to article 64 LOSC as a provision that preserves the sovereign nature of coastal State rights over these species.[75] Moreover, several coastal States have asserted a superior position over DWFS by reviving the concept of special interest in maintaining the productivity of fish stocks in high seas areas adjacent to and beyond the zone. This stance becomes evident from the early stages of UNCLOS/III in national proposals submitted by States from the Latin American region,[76] and it has

[72] A/AC.138/SC.II/L.12 'Regime of fisheries on the high seas' (1972) [(1973) 12ILM25], and *see* further Kawasaki, 'The 200-Mile Regime and the Management of Transboundary and High Seas Stocks', *Ocean Management* (1984) 19.

[73] Lugten, *A Review of Measures Taken by Regional Marine Fishery Bodies to Address Contemporary Fishery Issues* (1999) 10.

[74] *compare* Kelly, 'Law of the Sea: The jurisdictional dispute over highly migratory species of tuna', *Columbia Journal of Transnational Law* (1988) 475 *with* de Klemm, 'Migratory Species in International Law', *Natural Resources Journal* (1989) 941–944.

[75] Burke, *The New International Law of Fisheries: UNCLOS 1982 and Beyond* (1994) 218–219.

[76] *eg.* A/AC.138/SC.II/L.27/Corr.2 'Draft articles for inclusion in a convention on the law of the sea, submitted by the delegations of Ecuador, Panama and Peru'

been fervently rehearsed since the adoption of the Convention.[77] For example, Argentina, Chile, and Peru have, at various times, enacted legislation prescribing CMMs with a scope extending into high seas areas.[78] Ecuador has spelt out the contemporary understanding of this position in its interpretative declaration upon accession to the Convention, wherein it affirmed the full validity of the *Santiago Declaration* (1952),[79] and asseverated exclusive sovereignty and jurisdiction over the sea adjacent to its coast 'up to a minimum distance of 200 nm' as a norm of its maritime policy, aimed at securing adequate livelihood conditions and providing its population with the means for economic development. Accordingly, it declared—without distinguishing between straddling and highly migratory fish stocks—that, in accordance with the relevant provisions of the Convention:

> [W]hen the same or associated fish stocks are found both within the Ecuadorian 200-mile zone and in a maritime area adjacent to the said zone, the States whose nationals fish for those species in the area adjacent to the Ecuadorian zone must agree with Ecuador the measures necessary to conserve and protect them, as well as to promote their optimum utilization. In the absence of such agreement, Ecuador reserves to itself the exercise of its rights under article 116 and other provisions of the Convention, as well as all other relevant rules of international law.[80]

This declaration has been challenged by other parties as a disguised reservation that undermines the legal effect of the Convention's provisions, in violation of article 309 LOSC. The European Union, in particular,[81] along with several of its member States individually, has filed communications either expressly opposing the declaration,[82] or stressing that it is 'unclear in important respects'[83] and contesting specific elements, such as its 'interpretation of

(13/7/1973) [(1973) 12ILM1229], article 21.

77 *see* the views expressed in the collection of chapters by Orrego-Vicuña (Ed.) *The Exclusive Economic Zone, A Latin American Perspective* (1984), and more recently in Valencia-Rodríguez, 'The Contributions of Latin America to the Implementation of the UNCLOS', in Del Castillo (Ed.) *Law of the Sea, From Grotius to the International Tribunal for the Law of the Sea* (2015) 144*ff.*

78 *see* García-Amador, 'The Latin American Contribution to the Development of the Law of the Sea', AJIL (1974) 33 and Paolillo, 'The Exclusive Economic Zone in Latin American Practice and Legislation', ODIL (1995) 105.

79 *Joint Declaration concerning Fishing Problems in the South Pacific* and *Declaration on the Maritime Zone* by Chile, Ecuador and Peru on 18/8/1952 [1006UNTS318–326].

80 On 24/10/2012 [CN.595.2012:XXI-6].

81 On 30/10/2013 [CN.862.2013:XXI-6].

82 *e.g.* Spain on 18/10/2013 [CN.839.2013:XXI-6].

83 *e.g.* Ireland on 22/10/2013 [CN.854.2013:XXI-6].

the rights of coastal States in the EEZ'.[84] The Convention is considered to be in force *inter partes* on the basis of clarifications purportedly provided by Ecuador, which stated that it does not intend to exclude or modify the legal effect of any provision;[85] yet, it has not withdrawn or formally revised its declaration. This disagreement was muted in the context of the ambitious *Trade Agreement between the European Union and its Member States, of the one part, and Colombia and Peru, of the other part* (2012), which promoted significant tuna-canning sector and related market-access interests. Ecuador, after discontinuing earlier negotiations, eventually joined the agreement through the *Protocol of Accession to the Trade Agreement between the European Union and its Member States, of the one part, and Colombia and Peru, of the other part* (2016).

2.4 THE RULE OF COMPATIBLE CONSERVATION AND MANAGEMENT MEASURES IN THE 1995 UNITED NATIONS FISH STOCKS AGREEMENT

The fisheries provisions were agreed at the midpoint of UNCLOS/III, during a pivotal phase of the negotiations conducted under the package deal principle at the Third Session (17/3 to 9/5/1975, Geneva). At that juncture, each Chairman of the three main Committees prepared a tentative draft text reflecting the progress of discussions on the subjects within their respective mandates. These were consolidated into the Informal Single Negotiating Text ('ISNT'). The articles concerning fish stocks were agreed upon during the negotiations over the ISNT,[86] and their formulation remained essentially unchanged—aside from drafting refinements—after the adoption of the Revised Single Negotiating Text ('RSNT') at the Fourth Session (15/3 to 7/5/1976, New York).[87] The early finalization of these articles was not, however, indicative of conceptual clarity and precision in their provisions. As part of the broader package deal, the negotiations on fisheries were mostly influenced by concerns over the remaining extent of the high seas once EEZs had been established, rather than by efforts to design a coherent regulatory regime applicable across jurisdictional sea areas. Unlike the high seas fisheries provisions, which were substantially inherited from the *Fishing & Conservation Convention* (1958), those

84 *e.g.* The Netherlands on 25/10/2013 [CN.857.2013:XXI-6].

85 *see* the statement contained in the declaration made by Greece on 31/10/2013 [CN.861.2013:XXI-6].

86 A/Conf.62/WP.8/Part-II 'Informal Single Negotiating Text (Part II)' (7/5/1975) [*UNCLOS/III Off.Records* IV:152], articles 52–53.

87 A/Conf.62/WP.8/Rev.1/Part-II 'Revised Single Negotiating Text (Part II)' (6/5/1976) [*UNCLOS/III Off.Records* V:151], articles 52–53.

concerning straddling and highly migratory fish stocks arose out of carefully balanced trade-offs among competing State interests during the negotiation process. As expected, reconciling the rights and obligations that inform the exercise of regulatory authority by States over these stocks has proved to be one of the most complex aspects of the Convention's implementation.[88]

More specifically, the Convention has often been regarded as so imprecise—if not internally inconsistent—that it effectively leaves the jurisdictional regime of the fish stocks practically unsettled.[89] Although that is one—and a by no means unjustified—perspective on the Convention's contribution, another would lay the emphasis on the legal regime's deliberate flexibility as a purposeful outcome of progressive development, which ultimately had to be accommodated within, and accepted as part of, the package deal. Recognizing that extended jurisdiction zones were bound to materialize through the developing yet fragmented practice among States, the Convention sought to channel normative pressures arising from the growing wave of unilateral claims by the majority of coastal States, which might otherwise have driven an unchecked expansion of coastal jurisdiction seaward. By the end of 1977, while the UNCLOS/III negotiations were still underway, most of the world's significant fishing grounds had already been encompassed within EFZs unilaterally declared by coastal States.[90] Although the Convention formally instituted the regime of EEZ by vesting coastal States with sovereign rights over marine natural resources of the zone, along with functional jurisdiction, it concurrently introduced procedural safeguards to prevent the zone's gradual transformation into a *de facto* 200 nm territorial sea.[91] By so doing, it clearly signalled that the regulation of fish stocks necessitates cooperation through negotiation, an obligation to which both DWFS and coastal States are deferred through the Convention's procedural architecture.[92]

---

[88] As later noted by the *Fish Stocks Conference* in A/Conf.164/INF/5 'Background paper (Prepared by the Secretariat)' (8/7/1993), 8 & 17.

[89] *see* Pontecorvo, 'The Impact of the Law of the Sea Treaty on the Organization of World Fisheries: Some preliminary observations on production', in Brown–Churchill (Eds) *The UN Convention on the Law of the Sea: Impact and implementation* (1987) 138; Kwiatkowska, 'The High Seas Fisheries Regime: At a point of no return?', IJMCL (1993) 331; and Bjørndal *et al.*, 'The Management of High Seas Fisheries', *Annals of Operational Research* (2000) 185.

[90] Hudson, 'Fishery and Economic Zones as Customary International Law', SDLR (1979) 661 and Copes, 'The Impact of UNCLOS/III on Management of the World's Fisheries', *Marine Policy* (1981) 217–218.

[91] Sellars, *A 'Constitution for the Oceans'* (2025) 326–327.

[92] *see* Stevenson–Oxman, 'The Future of the United Nations Convention on the Law of the Sea', AJIL (1994) 497–498 and Charney, 'Entry into Force of the 1982

Thus, the *Fish Stocks Conference* was principally mandated to develop a coherent follow-up regime for the effective implementation of the Convention's provisions on the conservation and management of straddling and highly migratory fish stocks. While, prior to these multilateral negotiations, both sides had already come to the realization that CMMs must extend across the distribution or migration range of the stocks—requiring consistency between national measures adopted by the coastal State within their EEZs and international measures collectively adopted by all interested States for the high seas—this ecological prerequisite initially appeared innocuous, ostensibly reiterating the obvious. Coastal States, while anxious at first to secure their legitimate interests in the adoption of international measures, soon advanced their position to contend that in-zone national measures should serve as the basis for the entire regulatory regime. Perceived as yet another manifestation of the creeping jurisdiction syndrome, DWFS remained adamant in asserting their uncontested authority to adopt international measures, thereby obliquely enabling them to influence the regulatory regime within EEZ. Unsurprisingly, consensus could not be reached on the formulation of a clear-cut rule.[93]

Instead, the Agreement set out in article 7 FSA a rule, which in short, envisages that CMMs undertaken within the respective jurisdictional sea areas shall be compatible. Owing to the interpretative complexities it presents, the rule has become one of the most ambiguous concepts in the regulation of fisheries.[94] The phrasing of the *chapeau* in article 7(2) FSA eschews a direct engagement with the spatial scope of the rule and rather proceeds to stipulate that the relevant measures 'shall be compatible in order to ensure' the conservation and management of the stocks 'in their entirety', and accordingly it places

---

Convention on the Law of the Sea', VJIL (1995) 402.

[93] *see* Kaitala–Munro, 'The Management of High Seas Fisheries', *Marine Resource Economics* (1993) 318; Orrego-Vicuña, 'Towards an Effective Management of High Seas Fisheries and the Settlement of the Pending Issues of the Law of the Sea: The view of developing countries ten years after the signature of the Law of the Sea Convention', in Miles–Treves (Eds) *The Law of the Sea* (1993) 420–424; and Townsend-Gault, 'The "Territorialisation" of the Exclusive Economic Zone: A requiem for the remnants of the freedom of the seas?', in Schofield–Lee–Kwon (Eds) *The Limits of Maritime Jurisdiction* (2014) 65*ff.*

[94] *see* Juda, 'The United Nations Convention on Straddling and Highly Migratory Stocks: Policy problems in implementation', in Nordquist (Ed.) *Implementing the 1982 Law of the Sea Convention* (1996) 165–167; Schram–Tahindro, 'Developments in Principles for the Adoption of Fisheries Conservation and Management Measures', in Hey (Ed.) *Developments in International Fisheries Law* (1999) 272–281; and Nelson, 'The Development of the Legal Regime of High Seas Fisheries', in Boyle–Freestone (Eds) *International Law and Sustainable Development* (1999) 130.

both DWFS and coastal States under the 'duty to cooperate for the purpose of achieving compatible measures in respect of such stocks'. While no definitive answer is provided as to whether in-zone national measures or those internationally adopted on the high seas should take precedence in informing regulatory efforts, article 7(2) FSA identifies, through a series of criteria, the factors relevant to determining the compatibility of CMMs.

2.4.1 The Ambiguity in the Rule of Compatible Conservation and Management Measures

The controversy surrounding the rule emanates from the specific legal meaning ascribed to the term 'compatibility' and its implications for the respective obligations imposed on DWFS and coastal States in agreeing CMMs. It is undisputed that compatibility, as a regulatory concept, introduces a new—and, in principle, shared—obligation for both DWFS and coastal States. In contrast, the congruity of the fishing interests underlying the rule remains a matter of sharp contention. While the crux of this debate clearly arises from the jurisdictional differentiation between the legal regimes for straddling stocks and highly migratory species under the Convention, the compatibility rule carries significant potential for controversy with regard to its interpretation and application. In this regard, the substantive formulation of the rule warrants close analysis.

To begin with, it should be noted that *litt.*(a)–(b) of article 7(1) FSA closely recite articles 63(2) and 64 LOSC, respectively. This logically implies that any interpretation of the rule must be undertaken with great care to preserve the jurisdictional balance envisaged in these provisions of the Convention, and to avoid, within the context of the Agreement, tilting that balance either in favour of or against any of the fishing interests involved.[95] This is regarded as the prevailing approach to interpreting compatibility, as it avoids encroaching upon the rights established for DWFS and coastal States under the two regulatory regimes.[96] Yet the re-introduction of this jurisdictional division consequently poses a challenge to the very premise of the compatibility concept, given that the rule was adopted to address the ecological deficit originally created by these same provisions. The text of the Agreement remains vague on this matter, notably in article 3(1) FSA, where it addresses the spatial scope of the compatibility rule by providing the following:

---

[95] Oude-Elferink, 'The Determination of Compatible Conservation and Management Measures for Straddling and Highly Migratory Fish Stocks', *Yearbook UN Law* (2001) 555–556.

[96] Borg, *Conservation on the High Seas* (2012) 135.

> Unless otherwise provided, this Agreement applies to the conservation and management of straddling fish stocks and highly migratory fish stocks beyond areas under national jurisdiction, except that articles 6 and 7 apply also to the conservation and management of such stocks within areas under national jurisdiction, subject to the different legal regimes that apply within areas under national jurisdiction and in areas beyond national jurisdiction as provided for in the Convention.

The uncertainty stemming from the circumlocutory language of the above provision is palpable. While the application of the Agreement is 'subject to the different legal regimes that apply within areas under national jurisdiction and in areas beyond national jurisdiction' under the Convention—with the reference to 'areas under national jurisdiction' to undoubtedly subsume EEZ[97]—the rules on precautionary approach and compatible CMMs apply by way of exception 'within areas under national jurisdiction'. The proviso containing the exception is thus critical in defining the spatial scope of the compatibility rule.

However, to elaborate further on the preceding point, article 63(2) LOSC—reproduced almost verbatim in article 7(1)(a) FSA—stipulates that States shall seek to agree upon the measures necessary for the conservation of these stocks 'in the adjacent high seas area'. The case of straddling stocks becomes even more complex in light of the general principle enshrined in article 5(a) FSA, which firmly stipulates that DWFS and coastal States 'shall, in giving effect to their duty to cooperate … adopt measures to ensure long-term sustainability of straddling and highly migratory fish stocks and promote the objective of their optimum utilization'. By setting out this common objective, the Agreement not only dispels the previous uncertainty arising from the *de negotiando* commitment of States to agree on measures but also contributes to unifying the two jurisdictional regimes. As another commentator has rightly observed on this specific aspect, the Agreement strives to overcome this division by promoting an ecosystem-based model for a biologically unified area composed of two distinct jurisdictional components.[98] Still, a close reading of the text reveals that the general principle appears to oppose article 7(1)(a) FSA, which recites article 63(2) LOSC. Unless the scope of the latter provision is narrowly construed to fit in with the general principle, the contradiction arising from a literal reading of the two articles appears unavoidable. To resolve this, a purposive interpretation of the general principle in article 5(a) FSA has been proposed; one

---

[97] Tsamenyi–Hanich, 'Fisheries Jurisdiction under the Law of the Sea Convention: Rights and obligations in maritime zones under the sovereignty of coastal States', IJMCL (2012) 789.

[98] Orrego-Vicuña, 'The International Law of High Seas Fisheries: From freedom of fishing to sustainable use', in Stokke (Ed.) *Governing High Seas Fisheries* (2001) 38–40.

that seeks to collapse the division between the two jurisdictional regimes and to unify them for the purposes of conservation and management.[99]

Building upon the foregoing analysis, the debate over the spatial scope of the compatibility rule with regard to highly migratory species does not appear to have diminished. The provision in article 7(1)(b) FSA, by reciting article 64(1) LOSC, places States under the obligation to ensure conservation and promote the objective of optimum utilization of such species 'throughout the region, both within and beyond the exclusive economic zone'. This reading finds support in the wording of the exception in article 3(1) FSA and is further corroborated by the general principle in article 5(a) FSA. Even so, there have been attempts by coastal States to interpret article 7(1) FSA broadly, as a basis for asserting 'priority interests' over DWFS on the high seas.[100] The most explicit representation of this interpretation is found in the declaration filed by Argentina upon ratifying the Convention on 1/12/1995. This is quoted at length below, as it demonstrates a coastal State's immediate response intended to preempt the legal effect of the compatibility rule. Although Argentina signed the Convention on 5/10/1984, its instrument of ratification was submitted only several months after the adoption of the Agreement. The timing of this act carries its own significance, as it was accompanied by a declaration stating that:

> [It] accepts the provisions on the conservation and management of the living resources of the high seas, but considers that they are insufficient, particularly the provisions relating to straddling fish stocks or highly migratory fish stocks [...] [B]earing in mind its priority interest in conserving the resources of its [EEZ] and the area of the high seas adjacent thereto, considers that, in accordance with the provisions of the Convention, where the same stock or stocks of associated species occur both within the [EEZ] and in the area of the high seas adjacent thereto, the Argentine Republic, as the coastal State, and other States fishing for such stocks in the area adjacent to its [EEZ] should agree upon the measures necessary for the conservation of those stocks or stocks of associated species in the high seas. Independently of this, [its] understanding [is] that in order to comply with the obligation laid down in the Convention concerning the conservation of the living resources in its [EEZ] and the area adjacent thereto, it is authorized to adopt, in accordance with international law, all the measures it may deem necessary for the purpose.[101]

---

[99] Tahindro, 'Conservation and Management of Transboundary Fish Stocks: Comments in light of the adoption of the 1995 Agreement for the Conservation and Management of Straddling Fish Stocks and Highly Migratory Fish Stocks', ODIL (1997) 14–21.

[100] Morin, '"Creeping Jurisdiction" by the Small Islands of the Pacific Ocean in the Context of Management of the Tuna Fisheries', IJMCL (2015) 481–483.

[101] On 1/12/1995 [CN.425.1995:XXI-6], C.

This understanding signifies a regression to views expressed in other declarations by coastal States shortly after the adoption of the Convention,[102] wherein the provisions concerning fish stocks were interpreted as either reserving to coastal States the right to adopt laws and regulations to ensure the conservation of stocks in adjacent high seas areas or asserting that DWFS are obligated to agree with the coastal State on any such necessary conservation measures.[103]

Where the above interpretation of the Convention has been carried over to the Agreement, the resulting disagreement over the spatial scope of the rule remains unabated. This is reflected in declarations such as the one filed by Uruguay, claiming that the effectiveness of CMMs under the Agreement '[w]ill depend on whether the measures that are applied in areas beyond national jurisdiction take duly into account and are compatible with, those adopted by the relevant coastal States with respect to the same stocks in areas under their national jurisdiction'.[104] The opposing interpretation is reflected in the declaration upon ratification of the Convention by The Netherlands, in which it is categorically stated that the Convention confers no exclusive jurisdiction on coastal States beyond EEZ. Accordingly, the conservation and management of straddling and highly migratory fish stocks should, in line with articles 63(2) and 64 LOSC, take place on the basis of cooperation.[105]

2.4.2 Interpretations Favouring the Extension of Coastal Interests into High Seas Areas

Interpretations favouring the extension of coastal interests construe the wording of the compatibility rule as granting a degree of 'priority' to in-zone national measures, thereby enabling coastal States to exercise a 'considerable' or even 'decisive' influence over the conservation of the fish stock as a whole, including in the adjacent high seas areas.[106] For the proponents of this reading,

102 Vignes, 'Les Déclarations faites par les États Signataires de la Convention sur le Droit de la Mer sur la Base de l'Article 310 de cette Convention', *Annuaire Français* (1983) 731.

103 *e.g.* the declaration upon signature made by São Tomé and Príncipe on 3/8/1983 [CN.212.1983:XII].

104 Declaration upon signature on 18/3/1996 [CN.19.1996:XXI-7], I(¶¶2–3) as confirmed subsequently upon ratification on 10/9/1999 [CN.804.1999:XXI-7].

105 On 28/6/1996 [CN.214.1996:XXI-6], V.

106 *see*, among others, Orrego-Vicuña, *The Changing International Law of High Seas Fisheries* (1999) 63–76 and 183–194; Burke, 'Compatibility and Precaution in the 1995 Straddling Stock Agreement', in Scheiber (Ed.) *Law of the Sea* (2000) 111; Molenaar, 'Addressing Regulatory Gaps in High Seas Fisheries', IJMCL (2005) 533; Van Dyke, 'Allocating Fish Across Jurisdictions', in Ndiaye–Wolfrum

the primacy of coastal interests is normatively sustained by article 7(2)(a) FSA, which stipulates that compatible measures shall take into account the in-zone national measures that are adopted and applied in accordance with article 61 LOSC and make sure that the effectiveness of such measures is not undermined by international measures adopted for the high seas. This is followed by stressing article 7(2)(d) FSA, which further provides that account must be taken of the geographical particularities of the region concerned, including the extent to which the stocks occur and are fished in areas under national jurisdiction. These criteria are interpreted as giving primacy to in-zone national measures, positioning them as the predominant components of the regulatory regime. Yet it remains unclear whether this precedence is founded in a concrete legal principle or merely reflects a form of temporal, or otherwise substantively distinct, priority over DWFS. For example, it has been argued that the conservation principles in both the Convention and the Agreement should translate into a diverse category of norms, not exclusively legal in nature, which ought to be given weight in the application of allocation criteria. These may include traditional fishing practices, zonal attachment, resource dependency, as well as privileges arising from incentive or reward schemes.[107]

Indicative of State practice lacking a principled legal approach to the interpretation of article 7 FSA is the *Joint Declaration of Lima* (2010), issued by several Latin American States and communicated to the resumed *FSA Review Conference* that same year. Therein it was contended that articles 5 (General principles), 6 (Application of the precautionary approach), and 7 (Compatibility of conservation and management measures) FSA should be interpreted in a manner that does not conflict with the sovereign rights exercised within EEZ. Furthermore, the compatibility rule should acknowledge in clear terms the necessity to refrain from applying high seas measures that could undermine the effectiveness of those applied by the coastal State.[108] The very act of issuing a joint declaration that openly calls for the revision of the compatibility rule

---

(Eds) *Law of the Sea, Environmental Law, and Settlement of Disputes* (2007) 825–826; and Henriksen–Hoel, 'Determining Allocation: From paper to practice in the distribution of fishing rights between countries', ODIL (2011) 71–75 followed by Goodman, *Coastal State Jurisdiction over Living Resources in the Exclusive Economic Zone* (2022) 39.

107 *see* Hoel–Kvalvik, 'The Allocation of Scarce Natural Resources: The case of fisheries', *Marine Policy* (2006) 350 and Van Dyke, 'Allocating Fish Across Jurisdictions', in Allen–Joseph–Squires (Eds) *Conservation and Management of Transnational Tuna Fisheries* (2010) 172–175.

108 A/Conf.210/2010/6 'Note verbale from the Permanent Missions of Chile, Colombia, Costa Rica, Ecuador, Mexico and Peru to the United Nations addressed to the Secretariat' (20/5/2010), Annex (¶3(1)).

reflects underlying uncertainty about the legal basis for extending coastal State jurisdiction into high seas areas through doctrinal interpretation. Moreover, the undersigning States do not share a unified interpretation of the scope of sovereign rights in EEZ. For example, Peru has maintained an understanding that essentially territorialize the sea area lying within 200 nm of its coast in considering it 'el dominio marítimo del Estado' in which 'el Estado ejerce soberanía y jurisdicción'.[109] This longstanding and deeply held claim of territorial sovereignty is the main reason for Peru not becoming party to the Convention.

Among the other coastal States that signed the declaration, Chile's position is the most illustrative—in light of its *el mar presencial* narrative—given its pivotal role in the development of the EEZ regime, to which it has been 'unalterably bound'[110] since becoming the first State to formally declare it a legal concept in July 1947. This conveys a broad doctrinal frame encompassing geopolitical, economic, and strategic aspirations, infused with a strong sense of national identity. It asserts effective and continuous influence over a vast expanse of the high seas in the southeastern Pacific Ocean by conducting economic and scientific activities that contribute to its national development, while simultaneously countering foreign activities that could, directly or indirectly, affect Chile's maritime heritage and national security. Although Chile avows that this position does not amount to a claim of sovereign or exclusive coastal State rights, it has acknowledged that the doctrine may incur jurisdictional implications in cases of ineffective international cooperation, particularly when the coastal State prescribes CMMs for high seas areas in the absence of an agreed solution with DWFS. Evidently, it has been advanced to justify unilateral action and has, as such, been fittingly described as a new instalment to the *Santiago Declaration* (1952), reflecting the seaward-looking posture that Chile and other States have historically assumed in consistently expanding towards, rather than retreating from, the concept of territorial sovereignty within and beyond the zone.[111] It has thus strongly influenced Chile's

---

109 *Constitución Política del Perú*, promulgated on 29/12/1993, as subsequently amended, article 54.

110 As described by the Chilean representative Fernando Zegers during the UNCLOS/III Plenary 164th meeting (1/4/1982) [*UNCLOS/III Off.Records* XVI:68(¶128)].

111 *compare* the views held by Francisco Orrego-Vicuña in 'The "Presential Sea": Defining coastal States' special interest in high seas fisheries and other activities', GYIL (1993) 264 and 'Toward an Effective Management of High Seas Fisheries and the Settlement of the Pending Issues of the Law of the Sea', ODIL (1993) 88 *with* those of Clingan, 'Mar Presencial: Deja vu all over again? A response to Francisco Orrego Vicuña', ODIL (1993) 94; Dalton, 'The Chilean

approach to the conservation and management of fish stocks, which has been promoted at the political level and implemented through legal measures at the national level.[112]

At the international level, and within the immediate context of fisheries, the development of this narrative is best understood retrospectively by reference to three dates, albeit presented in a non-chronological order. First, upon acceding to the Agreement in 2016, Chile filed a declaration in which it emphasized the importance of the principles of ecosystem and precautionary approach for the conservation and sustainable use of fish stocks. Nonetheless, it affirmed that the provisions of the Agreement concerning sovereign rights, jurisdiction, and competence of coastal States within EEZ, as well as those relating to the high seas, must be interpreted and implemented in accordance with the Convention.[113] Second, in 1982, at the time of signing the Convention, Chile issued a statement affirming that the *sui generis* legal character of the zone is 'bound to the territorial sovereignty and actual territory' of the coastal State.[114] The third date is situated between the two previously noted dates: in 1997—i.e. two years after the adoption of the Agreement—when Chile decided to ratify the Convention. Upon ratification, it filed a declaration confirming the statement previously made upon signature. As to the conservation of marine living resources in EEZ and adjacent high seas areas, Chile recorded its understanding that, in the absence of agreement on the measures necessary for the conservation of fish stocks on the high seas, the coastal State 'reserves the right to exercise its rights under article 116 [LOSC] and other provisions of the Convention, and the other rights accorded to it under international law'.[115]

The narrative now appears to have been diluted in its legal weight, at least with respect to article 7 FSA, as evidenced by the submissions made to ITLOS during the *SRFC Request* proceedings. On this matter, Chile took the position that, in determining compatible CMMs, due account should be taken of any in-zone measures previously adopted by the coastal State in areas under its national jurisdiction, to ensure that their effectiveness is not undermined. Likewise, when prescribing new in-zone national measures, the coastal State should exercise equal care to avoid undermining the effectiveness of international measures already adopted for the high seas. Finally, Chile observed that

---

Mar Presencial: A harmless concept or a dangerous precedent?', IJMCL (1993) 397–398; and Kibel, 'Alone at Sea: Chile's *presencial* ocean policy', *Journal of Environmental Law* (2000) 44.

112 *e.g. Ley General de Pesca y Acuicultura* No.19080 (6/9/1991), article 165; as consolidated by Decree No.430 (28/9/1991) and extended by No.598 (15/10/1999).

113 On 4/3/2016 [CN.78.2016:XXI-7].

114 On 23/2/1983 [CN.7.1983:XXI-6].

115 On 23/10/1997 [CN.380.1997:XXI-6], ¶5.

such measures should be agreed upon within a reasonable timeframe; otherwise, recourse to dispute settlement procedures may be warranted.[116]

2.4.3 Interpretations Maintaining a Balance of Interests between Coastal and Distant Water Fishing States

Neutral interpretations of the compatibility rule seek to balance the interests of DWFS and coastal States in the sustainable regulation of fish stocks, without assigning precedence to either national measures adopted by coastal States within their EEZs or international measures collectively adopted by all interested States for the high seas. Given the distribution and migration of fish stocks across jurisdictional sea areas, a cooperative approach is essential to foster compatible CMMs that are determined, adopted, and effectively implemented, whether at the regional, subregional, or national level, as appropriate. In practical terms, this may compel DWFS and coastal States to engage in science-based compromises in the interests of the long-term conservation and sustainable use of the stocks. High seas measures and in-zone measures must be compatible *inter se*, with both regulatory regimes regarded, in principle, as mutually adaptable. Depending on the factual circumstances and the available scientific evidence, the criteria set out in article 7(2) FSA are to be accorded only relative importance in achieving compatible CMMs and in determining, accordingly, their necessary spatial scope of application. The foundational commitment to adopt CMMs through direct and meaningful negotiations formed the basis of the compromise that enabled consensus on the compatibility rule. In the event of disagreement, any measures must be qualified considering the factual circumstances and scientific evidence specific to the stock in question and must be guided by the principles of an ecosystem and precautionary approach. This marks a significant progressive development in the law of conservation and management, as it has constrained the reliance on unconditional claims and self-judging arguments regarding coastal States' special interests in high seas areas.[117]

---

[116] *SRFC Request* (Advisory Opinion) 2015, 'Written Statement of the Republic of Chile' (29/11/2013), 356 and the Statement recorded in ITLOS/PV.14/C21/2/Rev.l 'Verbatim Records – Minutes of Public Sittings' (2–5/9/2014), 1407.

[117] *see*, among others, Anderson, 'The Straddling Stocks Agreement of 1995: An initial assessment', ICLQ (1996) 470 & 475; Jacobson, 'Conserving and Managing Living Marine Resources: The second story', in Nordquist (Ed.) *Implementing the 1982 Law of the Sea Convention* (1996) 28–29; Hayashi, 'The 1995 UN Fish Stocks Agreement and the Law of the Sea', in Vidas–Østreng (Eds) *Order for the Oceans at the Turn of the Century* (1999) 51; Oude-Elferink, *The Impact of Article 7(2) of the Fish Stocks Agreement on the Formulation of Conservation*

This interpretation faithfully reflects the reasons for convening the *Fish Stocks Conference*, as these were stressed to the national delegations by Chairman Nandan at the outset of the negotiations, when he stated that:

> It is clear from the mandate that this Conference is not about the extension of national jurisdiction or the abridgement of the right of States to fish in the high seas in accordance with the Convention. Nor is it a Conference for intrusion on, or the derogation of, the sovereign rights of coastal States in their exclusive economic zones. It is a Conference, however, to resolve the festering problems of high seas fishing in order to give full and faithful effect to the very delicately balanced provisions of the Convention.[118]

Steered by this diplomatic mindset, the pursuit of compatible CMMs was framed since the initial phase of the conference around the idea of establishing international standards for the conservation of stocks on the high seas, which could also serve as recommended national minimums for implementation by the coastal State within its EEZ.[119] In addition, the guiding tenet of the negotiations on the compatibility rule disallowed any interpretation harbouring covert ambitions of creeping jurisdiction, a problematic tendency that was clearly identified and deliberately addressed in order to be counteracted.[120] For that reason, the conference aimed to impose common obligations focused on ensuring effective conservation, rather than creating additional rights for either category of States. Consistent with this approach, it rejected any proposal that conflicted with the Convention, including those that sought to assert a special

& Management Measures for Straddling & Highly Migratory Fish Stocks (1999) 4–5*ff*, followed by Tanaka, *The International Law of the Sea* (2023) 328–329 and Murphy, 'Straddling and Highly Migratory Fish Stocks Compatibility Requirement and Scope of Rights of High Seas Fishing Nations', in Kunoy–Heidar–Yiallourides (Eds) *International Fisheries Law* (2024) 228–235; Hanich–Ota, 'Moving Beyond Rights-Based Management: A transparent approach to distributing the conservation burden and benefit in tuna fisheries', IJMCL (2013) 140–141; Harrison, *Saving the Oceans through Law* (2017) 176; and Kawano, 'Implementation of the Rules of the UNCLOS through Universal and Regional Organizations', in Ribeiro–Bastos–Henriksen (Eds) *Global Challenges and the Law of the Sea* (2020) 15–20.

118 A/Conf.164/7 'Statement made by the Chairman of the Conference at the opening of the Organizational Session, held on 19 April 1993' (4/5/1993), 2.

119 A/Conf.164/10 'A guide to the issues before the Conference (Prepared by the Chairman)' (24/6/1993), 10.

120 A/Conf.164/INF/2 'Report of the FAO technical consultation on high seas fishing' (14/5/1993), in 'High Seas Fisheries Management: New concepts and techniques', ¶43.

interest for coastal States.[121] The balance of fishing interests is substantively maintained through the compatibility criteria contained in *litt.*(b)–(c) of article 7(2) FSA, which require coastal States to take into account previously agreed measures for the high seas in accordance with the Convention. Moreover, *litt.* (e)–(f) of article 7(2) FSA, being addressed equally to all States, recognize the factor of respective fishing dependency on the stocks and impose on both categories of States the obligation to ensure that CMMs do not ultimately result in harmful impacts on living marine resources as a whole.

Turning to paragraph (1), neutral interpretations maintain that nothing in the text of article 7 FSA designates this provision as the overarching context within which the compatibility criteria must be read. Rather, the criteria set out in paragraph (2) and article 7 FSA in its entirety are to be applied in light of the object and purpose of the Agreement. In any case, this follows from the general rule of treaty interpretation.[122] In addition, care should be taken to ensure the most integrative construction of all relevant provisions by interpreting the treaty text as a whole, with the aim of eliminating or reducing ambiguities wherever possible.[123] The context includes, but is not confined to, the article or section of the treaty in which a clause is found.[124] The object and purpose of the Agreement are definitively set out in the second preambular recital and article 2 FSA, which requires States 'to ensure the long-term conservation and sustainable use of straddling fish stocks and highly migratory fish stocks through effective implementation of the relevant provisions of the Convention'. It follows, then, that while the interpretation of paragraph (2) should not negate the relevance of paragraph (1) of article 7 FSA, this does not confer upon the latter any interpretative contextual precedence. Addressing this point directly, an expert commentator—who served as Secretary of the *Fish Stocks Conference* during its final two sessions—has insightfully observed that the phrase 'without prejudice to the sovereign rights of coastal States',

---

[121] *see* Joyner–Von Gustedt, 'The Turbot War of 1995: Lessons for the law of the sea', IJMCL (1996) 453; Fortier, 'From Confrontation to Cooperation on the High Seas: Recent developments in international law concerning the conservation of marine resources', in Nisuke–McWhinney–Wolfrum (Eds) *Liber Amicorum Judge Shigeru Oda* (2002) 1385–1386; and Franckx, '200-Mile Limit: Between creeping jurisdiction and creeping common heritage?', *George Washington International Law Review* (2007) 494.

[122] VCLT 31(1).

[123] *Competence of the ILO in regard to International Regulation of the Conditions of the Labour of Persons Employed in Agriculture* (Advisory Opinion) 1922, pp. 23 & 39.

[124] Fitzmaurice, 'The Practical Working of the Law of Treaties', in Evans (Ed.) *International Law* (2024) 159.

as it appears in paragraph (1), does not extend to the situations governed by paragraph (2) of article 7 FSA, since the Convention itself provides no such basis beyond EEZ.[125]

Stated differently, even in the case of straddling stocks—whose legal regime, compared with that of highly migratory species, presents considerably more complex questions regarding the negotiation of CMMs—the coastal State is not granted any special rights beyond its EEZ under the Convention. Other than those arising from the duty to cooperate in agreeing on the necessary measures for the high seas, the coastal State has exactly the same rights as all other States concerned. In point of fact, this position not only finds support in scholarly analysis concerning both the Convention and the Agreement,[126] but has also been affirmed in commentary contemporaneous with the final negotiations on the fisheries provisions at UNCLOS/III, especially in reference to article 63(2) ICNT.[127] The inclusion of the *pacta de negotiando* and *de contrahendo* obligations in articles 63(2) and 64 LOSC, respectively, is regarded as an explicit expression of the Convention's refusal to confer any priority or otherwise decisive influence on the coastal States in high seas areas beyond and adjacent to its EEZ.[128] This is demonstrated by the failed efforts of coastal States to amend article 63(2) ICNT in order to grant themselves greater leverage over straddling stocks in adjacent high seas areas (see Box 2.1).[129] Markedly, the rejection of such proposals underscored a clear demarcation of the coastal State's functional rights, which—as part of the mini package deal for EEZ—were not to extend beyond the strict limit of 200 nm.[130] If anything, the declaratory rule in article 117 LOSC, containing a substantive obligation

---

[125] Hayashi, 'The Straddling and Highly Migratory Fish Stocks Agreement', in Hey (Ed.) *Developments in International Fisheries Law* (1999) 75–76.

[126] *see* Oda, *International Control of Sea Resources* (1989) xxi–xxii; Joyner–DeCola, 'Chile's Presential Sea Proposal Implications for Straddling Stocks and the International Law of Fisheries', ODIL (1993) 99–100 & 114; and Brown, 'United Nations Conference on Straddling Fish Stocks and Highly Migratory Fish Stocks: An analysis of international environmental law and the conference's final agreement', *Vermont Law Review* (1996) 575–576.

[127] A/Conf.62/WP.10 'Informal Composite Negotiating Text' (15/7/1977) [*UNCLOS/III Off.Records* VIII:1].

[128] de Mestral, 'Deux Récents Accords Bilatéraux en Matière de Pêche en 1977', *Canad. YbIL* (1977) 292.

[129] A/Conf.62/L.114 'Amendments to article 63, Submitted by Australia, Canada, Cape Verde, Iceland, the Philippines, São Tomé and Príncipe, Senegal and Sierra Leone' (13/4/1982) [*UNCLOS/III Off.Records* XVI:224]; and *see* further in Chapter 3 (*fn*246), associated main text.

[130] Treves, 'La Pêche en Haute Mer et l'Avenir de la Convention des Nations Unies sur le Droit de la Mer', *Annuaire Français* (1992) 889.

more potent to the requirement of undertaking negotiations under articles 63(2) and 64 LOSC,[131] restricts each State's jurisdiction over the conservation and management of the stocks on high seas to the fishing activities of its own nationals.

BOX 2.1 JOINT SUBMISSION BY AUSTRALIA, CANADA, CAPE VERDE, ICELAND, THE PHILIPPINES, SÃO TOMÉ AND PRÍNCIPE, SENEGAL, AND SIERRA LEONE AT UNCLOS/III TO AMEND ARTICLE 63 ICNT

A/Conf.62/L.114 (13 April 1982)*

Article 63, paragraph 2: amend to read as follows:

2. Where the same stock or stocks of associated species occur both within the exclusive economic zone and in an area beyond and adjacent to the zone, the coastal State and the States fishing for such stocks in the adjacent area shall, by mutual agreement, either directly or through appropriate subregional or regional organizations, adopt such measures as may be necessary for the conservation of these stocks in the adjacent area. (P) *In the event that agreement on such measures is not reached within a reasonable period, and proceedings are instituted before the appropriate tribunal pursuant to article 286, that tribunal shall determine the measures to be applied in the adjacent area for the conservation of these stocks. If definitive measures cannot be determined within a reasonable period, the tribunal, upon request of any of the interested States, shall determine provisional measures for that same area. In establishing definitive or provisional measures, the tribunal shall take into account those measures applied to the same stocks by the coastal State within its exclusive economic zone and the interests of other States fishing these stocks.* (P)

Note: * Incorporating document A/Conf.62/L.114/Corr.1 (14/4/1982). The symbol '(P)' has been used to highlight the textual disposition of clausal embeddedness.

Moreover, with respect to State practice on this point, it is noteworthy to consider Canada's narrative for extending coastal State jurisdiction into adjacent high seas areas, developed in line with its evolving fisheries policy and

131 Meseguer, 'Le Régime Juridique de l'Exploitation des Stocks Conjoints de Poissons au-delà des 200 Milles', *Annuaire Français* (1982) 898–899.

heightened during the *Fish Stocks Conference* negotiations amidst the escalation of its dispute with Spain.[132] In the aftermath of UNCLOS/II concluding without agreement on uniform legal rules for the territorial sea and fishing zones, Canada unilaterally proclaimed a 9 nm EFZ contiguous to its 3 nm territorial sea by enacting the Territorial Sea and Fishing Zones Act. This was later followed by the merging of the two sea areas into a 12 nm territorial sea and the establishment of additional extended jurisdiction zones for environmental purposes reaching up to 100 nm. Most of Canada's fishing-related claims were eventually consolidated into its 1977 proclamation of a 200 nm EFZ.[133] In the interim, Canada had already sought to protect its unilateral claims in adjacent high seas areas by filing a declaration concerning the compulsory jurisdiction of ICJ, whereby intended to exclude from such jurisdiction 'disputes arising out of or concerning jurisdiction or rights claimed or exercised by Canada in respect of the conservation, management or exploration of the living resources of the sea … in marine areas adjacent to [its] coast'.[134] Following the adoption of the Convention, Canada continued to entertain within its domestic political agenda the prospect of adopting a radical stance aimed at conserving straddling stocks beyond its EEZ in the northwest Atlantic Ocean. Indeed, as one author has remarked, a key reason for Canada's hesitation to ratify the Convention was the concern that such a policy decision might foreclose the argument for a 'functional' extension of unilateral fisheries jurisdiction across the full biological range of straddling stocks into the adjacent high seas.[135] The Canadian government's response to a technical report—commissioned by the parliamentary fisheries committee in 1990—indicates initial misgivings about adopting measures that would extend beyond its EEZ. The government

---

[132] *see* Gotlieb, 'The Canadian Contribution to the Concept of a Fishing Zone in International Law', *Canad. YbIL* (1964) 67*ff*; Morin, 'The Quiet Revolution: Canadian approaches to the law of the sea', in Zacklin (Ed.) *The Changing Law of the Sea* (1974) 24–26; Johnson, 'Canadian Foreign Policy and Fisheries', in Johnson–Zacher (Eds) *Canadian Foreign Policy and the Law of the Sea* (1977) 52*ff*; Fauteux, 'L'Initiative Juridique Canadienne sur la Pêche en Haute Mer', *Canad. YbIL* (1994) 33; and Applebaum, 'The UN Conference on Straddling Fish Stocks and Highly Migratory Fish Stocks: The current Canadian perspective', in Nordquist–Moore (Eds) *Entry into Force of the Law of the Sea Convention* (1995) 302–303.

[133] *Revised Statutes of Canada*, T-7 (1964); *Revised Statutes of Canada*, C-47 (1970); *Consolidated Regulations of Canada*, C-1548 (1977).

[134] On 7/4/1970 [724UNTS63]; *qv.* Macdonald, 'The New Canadian Declaration of Acceptance of the Compulsory Jurisdiction of the International Court of Justice', *Canad. YbIL* (1970) 34.

[135] McDorman, 'Will Canada Ratify the Law of the Sea Convention?', SDLR (1988) 557.

maintained that the recommendation for unilateral coastal measures 'va à l'encontre du droit international public de la mer'.[136] Nevertheless, it subsequently amended the Coastal Fisheries Protection Act, specifically the rules governing its application, to authorize the prescription of unilateral CMMs in high seas areas adjacent to Canada's EEZ. These amendments conferred extensive executive powers on national authorities, enabling them to board, inspect, arrest, and prosecute foreign vessels fishing in those areas,[137] thereby exceeding the enforcement powers available under the NAFO *Convention*. It is particularly noteworthy that Brian Tobin—then Minister for Fisheries and Oceans during the dispute with Spain, who is widely remembered for his oft-cited call '*I speak for those who have no voice, the fish*'—commented on the amendments, stating:

> [T]here is nothing in the books allowing to extend our authority beyond 200 nm, and this legislation will do just that. Some think it runs against international law … others … otherwise. [But] we cannot afford to wait for years while this is debated, because by that time, the fishery will be purely an academic thing.[138]

This arguably unveils the *arrière-pensée* behind the subsequently revised declaration under article 36(2) ICJ Statute, which exempted from the Court's compulsory jurisdiction 'disputes arising out of or concerning [CMMs] taken by Canada with respect to vessels fishing in the NAFO regulatory area, as defined in the [NAFO *Convention*], and the enforcement of such measures'.[139] This proved to be the crux of the matter in the *Fisheries Jurisdiction*, where the Court insinuated that Canada was aware of the questionable legality of its actions, noting that 'reservations from the Court's jurisdiction may be made by States for a variety of reasons; sometimes precisely because they feel vulnerable about the legality of their position or policy'.[140] In this respect, although the Court may be seen as having procedurally condoned Canada's activism on

---

[136] Canada–Ministère des Pêches et des Océans, 'Étude Indépendante sur L'État des Stocks de Morue du Nord: Réponse du Gouvernement du Canada' (7/5/1990).

[137] *Revised Statutes of Canada*, C-33 (1985), as amended for these matters on 12/5/1994 and 9/3/1995.

[138] *see* Kedziora, 'Gunboat Diplomacy in the Northwest Atlantic: The 1995 Canada–EU fishing dispute and the United Nations Agreement on Straddling and High Migratory Fish Stocks', *Northwestern Journal of International Law & Business* (1996) 1149 and Russell, *Who Rules the Waves?* (2010) 84*ff*.

[139] On 10/5/1994 [1776UNTS9], 2(d) modifying the declaration filed on 10/9/1985 [1406UNTS133]; *qv*. Lucchini, 'La Loi Canadienne du 12 Mai 1994: La logique extrême de la théorie du droit préférentiel de l'Etat côtier en haute mer au titre des stocks chevauchants', *Annuaire Français* (1994) 872–873.

[140] *Fisheries Jurisdiction* (Jurisdiction) 1998, §54.

the high seas,[141] its refusal to uphold jurisdiction on the merits may equally be interpreted as reflecting a form of doctrinal disillusionment with coastal States' aspirations to assert a 'special interest' in high seas areas.

Against a predetermined spatial reading of the rule, neutral interpretations construe compatibility as deriving from the concept of a stock's biological unity across its distribution or migration range.[142] To this end, the objective of compatibility—stipulating that CMMs adopted for the high seas and those applied within areas under national jurisdiction shall be compatible so as to ensure the conservation and management of fish stocks 'in their entirety', as stated in the *chapeaux* of article 7(2) FSA—embodies the core normative value of the rule. As has been pointed out, notably by the UN Office of Legal Affairs, Division for Ocean Affairs and the Law of the Sea ('DOALOS'), the inclusion of this spatial designation signals the application of ecosystem-based requirements, whereby regulatory regimes address the stock as an integrated whole rather than dividing it along jurisdictional sea boundaries.[143]

In fact, the only compatibility criterion that needs to be upheld when determining CMMs, among those listed in article 7(2), is that provided in *lit.*(f), which requires that 'such measures do not result in harmful impact on the living marine resources as a whole'. All other criteria, introduced with the phrasing 'take into account', denote considerations that may carry only circumstantial weight, depending on the specific factual context and scientific evidence available. In contrast, the use of the term 'ensure' reflects an obligation to achieve a defined outcome through compatible measures.[144] Article 3(1) FSA attests to the pre-eminence of the ecosystem criterion by plainly stipulating that both the precautionary approach and the compatibility rule apply to the conservation and management of such stocks in areas under national jurisdiction, a notable feature that is consistently highlighted in textbooks and

---

141 Oral, 'Protection of Vulnerable Marine Ecosystems in Areas Beyond National Jurisdiction: Can international law meet the challenge?', in Strati–Gavouneli–Skourtos (Eds) *Unresolved Issues and New Challenges to the Law of the Sea* (2006) 94.

142 As exemplarily put in the 'Position statement submitted by the European Economic Community', in A/Conf.164/L.8 'Letter from the Director-General Fisheries, Commission of the European Communities, addressed to the Chairman' (17/6/1993), ¶2.

143 *see* the consultations at the fifteenth round of ICSP as recorded in ICSP15/UNFSA/Inf.3 (22/6/2022), ¶30, which pursuant to A/Res.76/71 'Sustainable Fisheries' (9/12/2021) was convened on the topic of 'Implementation of an Ecosystem Approach to Fisheries Management'.

144 Oude-Elferink, *op. cit.* (*fn*95), 559–560.

general legal commentary.[145] Consequently, the effective implementation of the provisions on fish stocks also requires coastal States to adopt CMMs that are compatible for stocks occurring within their EEZs.[146]

On this particular point, the *SRFC Request* provides a definitive interpretation—especially regarding stocks of highly migratory species—in emphasizing the obligation of coastal States to cooperate with other concerned States, either directly or through the appropriate RFMO/As, with a view to ensuring conservation and promoting the objective of optimum utilization of stocks under their national jurisdiction. ITLOS observed that CMMs adopted for sea areas under national jurisdiction and those established for the high seas should be compatible throughout the region, both within and beyond EEZ. While recognizing the limits of its jurisdiction in the instant case, the Tribunal nonetheless stressed the importance of applying CMMs across the entire migration range of the species.[147] Judge Jin-Hyun Paik elaborated at greater length on the meaning and scope of cooperation as an obligation within the context of the compatibility rule. He pointedly observed that, while the Convention leaves 'unclear how this obligation is to be discharged', the Agreement is considerably more 'instructional in this regard'. To that end, he highlighted the procedural stipulation in article 7(4) FSA.[148]

2.4.4 The Special Case of High Seas Enclaves

In geographical terms, enclaves are especially prone to arise in areas with the natural characteristics of semi-enclosed seas.[149] The Convention defines such sea areas as 'a gulf, basin or sea surrounded by two or more States and connected to another sea or the ocean by a narrow outlet or consisting entirely or primarily of the territorial seas and exclusive economic zones of two or more coastal States'.[150] The extension of coastal jurisdiction to 200 nm has resulted in relatively small high seas pockets, surrounded by the outer limits of adjoining EEZs, for example, the fisheries known as the Donut Hole in the

---

145 *e.g.* Churchill–Lowe–Sander, *The Law of the Sea* (2022) 565 and Rothwell–Stephens, *The International Law of the Sea* (2023) 374.

146 Barnes, 'Fisheries and Marine Biodiversity', in Fitzmaurice *et al.* (Eds) *Research Handbook on International Environmental Law* (2010) 548.

147 *SRFC Request* (Advisory Opinion) 2015, §§213–215 & 219 *dispositif* (6).

148 *ibid.*, *Separate Opinion of Judge Paik*, §36.

149 Alexander, 'Regionalism and the Law of the Sea: The case of semi-enclosed seas', ODIL (1974) 151.

150 LOSC 122.

central Bering Sea and the Loophole in the Barents Sea.[151] The existence of these areas introduced an added layer of complexity, frequently precipitating disagreements over conservation and management practices or access to fishery resources.[152] In many cases, fish stocks within these regions were left in a precarious state of decline, if not on the verge of commercial collapse.[153]

The legal regime applicable to semi/enclosed seas is laid down by article 123 LOSC, which calls upon the bordering coastal States to 'cooperate with each other in the exercise of their rights and in the performance of their duties under [the] Convention'. In line with this, States 'shall endeavour, directly or through an appropriate regional organization to coordinate the management, conservation, exploration and exploitation of the living resources of the sea' and are encouraged to 'invite, as appropriate, other interested States or international organizations to cooperate with them in furtherance of the provisions of the [article]'.[154] While high seas enclaves are comparatively limited in size relative to surrounding EEZs, they nonetheless play a critical role in the effective management of fish stocks within adjacent national jurisdictions. Yet interpretations of the Convention's duty of international cooperation have shown considerable variation.[155] These interpretations reflect the article's hortatory and non-committal language—using terms such as 'should cooperate' and 'shall endeavour … to coordinate'—which, although normatively framed, leaves the specific content of the obligations indeterminate.[156] In terms of fishing interests, granting coastal States overall control of stocks in adjacent high seas enclaves could presumably offer a stronger basis for regulation. However, this was firmly opposed as a matter of principle during UNCLOS/III. As a

---

151 *see* respectively Burke, 'Fishing in the Bering Sea Donut: Straddling stocks and the new international law of fisheries', ELQ (1989) 285 and Stokke, 'The Loophole of the Barents Sea Fisheries Regime', in Stokke (Ed.) *Governing High Seas Fisheries* (2001) 273.

152 *see* Churchill–Ulfstein, *Marine Management in Disputed Areas: The case of the Barents Sea* (1992) 79–109 and Gudmundsson, 'Cod War on the High Seas: Norwegian–Icelandic dispute over "Loophole" fishing in the Barents Sea', NJIL (1995) 557.

153 Wespestad, 'The Status of Bering Sea Pollock and the Effect of the *Donut Hole* Fishery', *Fisheries* (1993) 18.

154 LOSC 123 (a) & (d).

155 Burke, *Fisheries Regulations under Extended Jurisdiction and International Law* (1992) 1.

156 *compare* Franckx–Benatar, 'The Duty to Co-Operate for States Bordering Enclosed or Semi-Enclosed Seas', *Chinese (Taiwan) Yearbook of International Law and Affairs* (2013) 69–73 *with* Lee–Kim, 'UNCLOS and the Obligation to Cooperate: International legal framework for semi-enclosed seas cooperation', in Zou (Ed.) *Maritime Cooperation in Semi-Enclosed Seas* (2019) 15–21.

result, article 123 LOSC deliberately avoids establishing any requirement or condition that would confer special or privileged treatment upon coastal States. Instead, it preserves a balance of jurisdictional functions grounded in qualified fishing interests and the overarching objective of effective conservation through cooperation.[157]

The Agreement reiterates the requirement of international cooperation by extending it to all States, while calling on them to take into account the natural characteristics of such areas and to act consistently with the provisions of the Convention.[158] The compatibility rule applies to fish stocks in semi/enclosed seas, with several criteria listed in article 7(2) FSA acquiring particular relevance in light of the geographical circumstances. This approach reflects the view that such areas do not warrant exceptional jurisdictional treatment but may nonetheless be recognized as presenting certain distinguishing features for practical purposes.[159] Of particular relevance to the present analysis, the Agreement separately addresses the situation of a high seas enclave that is entirely surrounded by an area under the national jurisdiction of a single State. This scenario is exemplified by the so-called Peanut Hole, located in the Sea of Okhotsk and encircled by the Russian EEZ. Article 16 FSA calls upon all States fishing in such enclaves to cooperate in establishing CMMs with respect to the stocks in the high seas areas. This provision establishes a distinct regulatory regime, reflecting the position advanced by Russia during the *Fish Stocks Conference*, where it made repeated submissions aimed at distinguishing this scenario from the regime governing semi/enclosed seas. Russia's efforts to gain recognition of a 'special interest'—which, in the absence of agreed CMMs, would have permitted the coastal State to unilaterally adopt provisional protection measures, including temporary moratoria—were ultimately unavailing.

The Agreement maintains a balanced approach, requiring States, in principle, to pay particular attention to the establishment of CMMs in accordance with article 7 FSA, while taking into account the natural characteristics of the sea area. Regarding the specific substantive criteria for the adoption of compatible CMMs on the high seas, such measures 'shall take into account the rights,

---

[157] *see* the extensive analysis of the negotiations in Symonides, 'The Legal Status of the Enclosed and Semi-Enclosed Seas', GYIL (1984) 315 and Miovski, 'Solutions in the Convention on the Law of the Sea to the Problem of Overfishing in the Central Bering Sea: Analysis of the convention, highlighting the provisions concerning fisheries and enclosed and semi-enclosed seas', SDLR (1989) 574.

[158] FSA 15.

[159] Lodge, 'The Fisheries Regimes of Enclosed and Semi-Enclosed Seas and High Seas Enclaves', in Hey (Ed.) *Developments in International Fisheries Law* (1999) 208.

duties and interests of the coastal State under the Convention', 'shall be based on the best scientific evidence available', and 'shall also take into account any conservation and management measures adopted and applied in respect of the same stocks in accordance with article 61 LOSC by the coastal State in the area under national jurisdiction'.[160] The compatibility rule, in the context of article 16 FSA, departs from the approach taken under article 7 FSA and may be understood as acknowledging that the coastal State's rights in this specific type of enclave differ in nature from those typically applicable under the general regime for fish stocks.[161] Still, this approach falls significantly short of the broader claims Russia repeatedly sought to secure during the negotiations. Having failed to elicit such recognition at the multilateral level, Russia subsequently confined its position to a bilateral arrangement with the United States concerning the straddling pollock stocks in the Sea of Okhotsk. Pursuant to that arrangement, the United States undertook to recognize that fishing for these straddling stocks in the central Sea of Okhotsk falls within the scope of Russia's rights, duties, and interests, and to observe Russia's CMMs, provided they are compatible across the distribution range of the stocks.[162] This typifies a 'bilateralist' form of regulatory arrangement, whereby straddling stocks are primarily managed by the coastal State(s), with DWFS receiving access to EEZ resources in return for curtailing or ceasing operations in the high seas enclave.[163]

2.5 ATTAINING INTERPETATIVE CONSISTENCY BETWEEN THE 1982 UNITED NATIONS CONVENTION ON THE LAW OF THE SEA AND THE 1995 UNITED NATIONS FISH STOCKS AGREEMENT

The precise nature of the relationship between the two treaties—and the resulting legal effects—extends far beyond a mere reading of the Agreement's nominal title, which presents it as an instrument for the 'implementation' of the Convention's provisions on the conservation and management of fish stocks. Given the range of issues likely to intensify, rather than alleviate,

[160] FSA 16(1).

[161] Oude-Elferink, 'The Sea of Okhotsk Peanut Hole, *De Facto* Extension of Coastal State Control', in Stokke (Ed.) *Governing High Seas Fisheries* (2001) 186.

[162] *Agreement between the Government of the United States of America and the Government of the Russian Federation on the Conservation of Straddling Fish Stocks in the Central Part of the Sea of Okhotsk* (1996), articles 1 & 2.

[163] Churchill, 'The Barents Sea Loophole Agreement: A 'coastal State' solution to a straddling stock problem', IJMCL (1999) 482.

ambiguities within the Convention's provisions, the possibility of incorporating interpretive guidelines into the Agreement had been suggested during the negotiations.[164] Although such guidelines were eventually not adopted, the treaty relationship between the two instruments is directly addressed in article 4 FSA, which states that 'nothing in [the] Agreement shall prejudice the rights, jurisdiction and duties of States under the Convention' and further requires that the 'Agreement shall be interpreted and applied in the context of and in a manner consistent with the Convention'. This interpretive stance had already been outlined in the mandate of the *Fish Stocks Conference*.[165]

The requirement of consistency thus functions as an overarching stricture guiding the interpretation and application of provisions concerning the conservation and management of fish stocks. As vigorously emphasized by the Chairman of the *Fish Stocks Conference* at the time of the Agreement's adoption, the text is, on the one hand, 'far-sighted, far-reaching, bold and revolutionary', while on the other, it remains firmly grounded in the principles enshrined in the Convention, with both instruments described as 'intrinsically linked and inseparable'.[166] A pertinent example is the application of MSY not as a target but as a limit reference point, subject to the precautionary approach.[167] Where used as a restorative measure, MSY may even be determined at levels exceeding those merely capable of producing sustainable yields.[168] However, owing to the progressive development of the conservation and management law brought about by the Agreement, the interpretive scope of the provisions falling within the ambit of the compatibility rule has been expanded rather than excised. This outcome was inevitable, as the Agreement was intended to give effect to provisions of the Convention that articulate only general obligations.[169] In light of the evolving imperatives of the ecosystem-based approach, the Agreement has fleshed out significant obligations under the compatibility rule to address

---

164 A/Conf.164/L.7 'Organization of work (List of issues), Republic of Korea' (10/6/1993), ¶2.

165 A/Res.47/192 (1992) *cit.* Chapter 1 (*fn*86), ¶3.

166 A/Conf.164/35 'Statement of the Chairman on 4 August 1995 upon the adoption of the Agreement for the Implementation of the Provisions of the United Nations Convention on the Law of the Sea of 10 December 1982 Relating to the Conservation and Management of Straddling Fish Stocks and Highly Migratory Fish Stocks' (20/9/1995), 1.

167 Garcia, 'The Precautionary Approach to Fisheries: Progressive review and main issues (1995–2000)', in Moore–Nordquist (Eds) *Current Fisheries Issues and the Food and Agriculture Organization of the United Nations* (2000) 508.

168 *FSA Review Conference* 2016, ¶¶85–88 and Outcomes A2(c) & 3.

169 A/Conf.164/10 'A guide to the issues before the Conference (Prepared by the Chairman)' (24/6/1993), 4(¶13).

the lacunae left by the Convention; particularly concerning the duty to cooperate in the conservation and management of fish stocks.[170] Academic literature on articles 63(2) and 64 LOSC cogently asserts that, for parties to both treaties, the Agreement constitutes not only *lex posterior* but also *lex specialis*, as it refines the Convention's provisions through more specific requirements, such as the principles of ecosystem and precautionary approach, as well as the compatibility rule.[171] Other scholars advance this view more boldly, characterizing the Agreement as a framework for the continuous interpretation of the Convention's fisheries provisions.[172] The Agreement expresses in detailed terms the necessary balance and reciprocity between the concurrent rights and obligations of DWFS and coastal States, premised on the ecosystem and precautionary approaches, principles that, although integral to effective fisheries regulation, remained implicit in the Convention. The compatibility rule, in particular, represents a pivotal legal innovation introduced by the Agreement, marking a significant evolution from articles 63(2) and 64 LOSC, which merely allude to the need for cooperation. In light of these substantive legal developments in the field of conservation and management, the modest term 'implementation' should be understood in the sense of 'change to improve'.[173]

Approaching the relationship between the Convention and the Agreement from the perspective of treaty design, it becomes readily apparent that, while the two treaties share substantive linkages, they remain formally independent instruments.[174] As a self-standing treaty, the Agreement has been largely approached in academic commentary from two core perspectives. It is regarded either as a 'subsequent agreement',[175] to be considered in conjunction

---

170 Rayfuse, 'The United Nations Agreement on Straddling and Highly Migratory Fish Stocks as an Objective Regime: A case of wishful thinking?', *Aust.* YbIL (1999) 265.

171 Harrison–Morgera, article 63, 510–511 (¶8) and article 64, 516–517 (¶9) in Proelss, *The United Nations Convention on the Law of the Sea* (2017).

172 Pulvenis de Séligny-Maurel, 'The Marine Living Resources and the Evolving Law of the Sea', *Aegean Review of the Law of the Sea and Maritime Law* (2010) 62–66.

173 *see* Hayashi, 'The 1995 Agreement on the Conservation and Management of Straddling and Highly Migratory Fish Stocks: Significance for the Law of the Sea Convention', OCM (1995) 57–58; Scovazzi, 'The Evolution of International Law of the Sea: New issues, new challenges', *Recueil des Cours* (2001) 143; and Lodge–Nandan, 'Some Suggestions Towards Better Implementation of the United Nations Agreement on Straddling Fish Stocks and Highly Migratory Fish Stocks of 1995', IJMCL (2005) 350.

174 As underscored in *Southern Bluefin Tuna Cases* (Provisional Measures) 1999, *Separate Opinion of Judge Treves*, §10.

175 VCLT 31(3)(a).

with the Convention's context when interpreting its relevant provisions, or as a 'relevant rule of international law'[176] aimed at pursuing the systemic integration of the two treaties.[177] While both stand to reason, they risk becoming moot when confronted with the interpretative constraint established in article 4 FSA, which stipulates that the Agreement must be interpreted and applied in the context of the Convention; and not *vice versa*. This is so, notwithstanding the generally stated, yet clearly applicable, observation made by ITLOS that treaties do not operate in isolation but must be 'interpreted and applied within the framework of the entire legal system prevailing at the time of the interpretation'.[178] Still, it remains unclear whether the Agreement should be regarded as furnishing rules that are 'internal' or 'external' to the Convention. The Agreement not only creates distinct rights and obligations for its parties but also has a substantial bearing on their rights and obligations under the Convention. From the perspective of non-parties to the Agreement, it may constitute a form of subsequent practice by a substantial number of parties to the Convention. All the same, this practice does not necessarily reflect a common understanding regarding the interpretation or application of the Convention's provisions.[179] In short, the Agreement was not concluded with the purpose of formally recording the parties' understanding of the relevant provisions of the Convention, and thus does not, in that technical sense, constitute *pro tanto* an 'authentic' means of interpretation.[180]

A third treaty construction approach remains compelling in treating the Agreement—along with other relevant treaties under regional regimes—within the perspective of constituting 'subsequent practice'[181] *inter partes* for

---

[176] VCLT 31(3)(c).

[177] *compare* Anderson, *Modern Law of the Sea* (2008) 368 and Ong, 'International Law of the Sea', in Bowman–Kritsiotis (Eds) *Conceptual and Contextual Perspectives on the Modern Law of Treaties* (2018) 718–720 *with* Harrison, *Making the Law of the Sea* (2011) 107–108.

[178] *Climate Change and International Law* (Advisory Opinion) 2024, §135 following the pronouncement of ICJ in *Legal Consequences for States of the Continued Presence of South Africa in Namibia (South West Africa) notwithstanding Security Council Resolution 276 (1970)* (Advisory Opinion) 1971, §35.

[179] *see Report 3 Subsequent Agreements and Subsequent Practice of States Outside of Judicial or Quasi-judicial Proceedings: Third Report for the ILC Study Group on Treaties over Time*, having the status of formal working paper but not being formally published; *qv.* Nolte (Ed.) *Treaties and Subsequent Practice* (2013) 313–315.

[180] *see* A/CN.4/Ser.A/2018/Add.1 'Draft Conclusions on Subsequent Agreements and Subsequent Practice in relation to the Interpretation of Treaties' (8/8/2018) [2018*YbILC*(II(2)) 30–32].

[181] VCLT 31(3)(b).

the purpose of interpreting the provisions of the Convention relating to the conservation and management of fish stocks. Even prior to the entry into force of the Agreement, ITLOS held that the practice of the parties to the dispute under the CCSBT *Convention* was of consequence in assessing the extent of their compliance with obligations under the Convention, explicitly referencing their duty to cooperate in the conservation and management of southern bluefin tuna stocks.[182] As an implementation treaty, the Agreement introduced significant innovations to the Convention, most notably, the incorporation of the ecosystem-based approach as a novel regulatory objective. These developments blur the distinction between treaty interpretation and the legal effects arising from the application of other relevant rules.[183] Furthermore, as will be discussed in Chapter 6, such advances in the progressive development of fisheries law have both influenced and been further incorporated into other multilateral and specialized regional treaties. The requirement of consistency holds particular significance in relation to the compatibility rule, in two key respects. First, in terms of substantive law, attention must be given to the fact that the rule directly incorporates articles 63(2) and 64(1) LOSC, thereby providing fertile ground for the recurrent debate on the adoption of compatible CMMs. The Tribunal has already forewarned more generally that, even with respect to identical or similar provisions of different treaties, the application of the rules on treaty interpretation 'may not yield the same results, having regard to, inter alia, differences in the respective contexts, objects and purposes, subsequent practice of parties and *travaux préparatoires*'.[184] There should be no doubt that, for parties to the Agreement, the obligations under these articles must now be interpreted by giving due consideration to article 7 FSA in its entirety.[185] In this respect, the guidance provided by the Tribunal—that CMMs under articles 63 and 64 LOSC, in order to be effective, 'should concern the whole stock unit over its entire area of distribution or migration routes'—is illustrative of how

182 *Southern Bluefin Tuna Cases* (Provisional Measures) 1999, §§45–52.

183 *see* Buga, 'Between Stability and Change in the Law of the Sea Convention: Subsequent practice, treaty modification, and regime interaction', in Rothwell *et al.* (Eds) *Oxford Handbook of the Law of the Sea* (2015) 53–55 and Scott, 'The LOSC: "A constitution for the oceans" in the anthropocene?', *Aust.* YbIL (2023) 277–278.

184 *see MOX Plant* (Provisional Measures) 2001, §51 and *SRFC Request* (Advisory Opinion) 2015, §57.

185 Franckx, '*Pacta Tertiis* and the Agreement for the Implementation of the Straddling and Highly Migratory Fish Stocks Provisions of the United Nations Convention on the Law of the Sea', *Tulane Journal of International and Comparative Law* (2000) 49 followed by Markowski, *The International Law of EEZ Fisheries* (2010) 52.

the requirement of consistency between the Convention and the Agreement is to be realized. This interpretation was affirmed in *SRFC Request*, where the Tribunal admitted and responded to Question 4 by 19 votes to 1,[186] with Judge Tafsir Malick Ndiaye dissenting on the ground that it should have declined jurisdiction.[187] Secondly, in terms of procedural law, the Agreement applies *mutatis mutandis* the dispute settlement procedures of Part XV LOSC and, through article 32 FSA, incorporates by reference the limitation on the compulsory settlement of fishery disputes as provided in article 297(3)(a) LOSC. Unless interpreted narrowly, this incorporation gives rise to an internal *non sequitur* with paragraph (4) of article 7 FSA. As a result, the compatibility rule stands at the crossroads of two interpretative dynamics. The first concerns the general dynamic produced by the interrelationship between substantive and procedural law, both of which are integrated within the structure of article 7 FSA. The second is the systemic dynamic driven by the requirement of consistency between the Convention and the Agreement.

2.6 CONCLUSION

The compatibility rule in article 7 FSA addresses ecological concerns stemming from the fragmented regulation of fish stocks under the Convention. However, the rule's open-ended substantive formulation regarding its spatial scope has given rise to competing interpretations. One approach, tracing its origins to the *Fishing & Conservation Convention* (1958), holds that coastal States' special interest in high seas areas justifies in-zone national measures as the basis for regulating fish stocks across their distribution and migratory range. This argument not only significantly departs from the text of articles 63(2) and 64 LOSC, but their corresponding interpretation, as provisions that are referenced in article 7(1) FSA, is productive of practical uncertainty in applying the compatibility rule following disagreements. In contrast, neutral interpretations seek to balance the legitimate interests of States by applying the compatibility criteria of article 7(2) FSA in a manner responsive to the specific circumstances and best available scientific evidence pertaining to the stocks concerned. This approach aims to preserve the biological integrity of fish stocks across their distribution and migratory range, on a case-by-case basis. Accordingly, it does not predetermine the spatial scope of the compatibility rule. It neither precludes the extension of in-zone national measures to

186 *SRFC Request* (Advisory Opinion) 2015, §219 *dispositif* (6).
187 *ibid.*, *Separate Opinion of Judge Ndiaye*, §§8–10.

high seas areas, nor does it exclude the application of international measures within EEZ. This interpretation conforms to the requirement of interpretative consistency between the Convention and the Agreement, as it qualifies, in principle, the integrity of the respective jurisdictional regimes under the Convention in light of the Agreement's object and purpose.

3. Settlement procedures for compatibility disputes in the 1995 United Nations Fish Stocks Agreement

3.1 INTRODUCTION

The regulation of straddling and highly migratory stocks is a complex process that can generate disagreements over the scope of CMMs. Although a failure to reach agreement *simpliciter* on such measures may give rise to a serious legal dispute, whether these matters fall within the remit of compulsory procedures entailing binding decisions has long been a contested issue under the Convention. A central question, bearing significant implications for the long-term conservation and sustainable use of these stocks, is how to objectively balance the legitimate interests of States when compatibility disputes emerge from unresolved disagreements. This underlying tension is further compounded by divergent interpretations of the compatibility rule and, crucially, by differing views on whether a court or tribunal seised of such disputes can exercise compulsory jurisdiction. The most contested issue in this regard concerns article 32 FSA, which incorporates by reference the fisheries limitation set out in article 297(3) LOSC. Following an analysis of the *mutatis mutandis* application of Part XV LOSC by Part VIII FSA, this chapter explores the interpretive boundaries and critically assesses the procedural consequences of the limitation, with a view to determining the extent to which compulsory settlement procedures entailing binding decisions apply to fisheries disputes involving the interpretation or application of the compatibility rule.

3.2 SCOPE OF COMPATIBILITY DISPUTES

The conservation and management of fish stocks is a complex regulatory process that entails the development of strategies and data-driven plans and actions, guided by scientific advice, to support decision-making and implement rules in pursuit of an agreed fisheries policy. This policy is typically established within an international institutional framework and developed through defined objectives aimed at collectively achieving and sustaining optimal benefits from the fishery resource. These comprise typically biological objectives (e.g. to sustain the reproductive capacity of the target stock for optimal long-term yields) and economic objectives (e.g. to maximize the income of participating stakeholders), as well as environmental objectives (e.g. to minimize the impact of fishing on the ecosystem), social objectives (e.g. to increase employment opportunities for those whose livelihood is dependent on the fishery) and even cultural objectives (e.g. to preserve the value of a fishery to the local, national or international community owing to its heritage, historical, traditional, aesthetic, educational or recreational qualities). In this context, CMMs established for high seas areas and sea areas under national jurisdiction must be compatible to ensure the sustainable regulation of fish stocks. However, the rule of compatibility does not require the adoption of identical measures. Where the regulatory areas are not subject to a unified conservation and management plan, compatibility is assessed with reference to the purpose, type, and timing of the respective measures. The aim is to ensure that such measures do not undermine one another across jurisdictional regimes and remain coherent throughout the distribution or migratory range of the stocks, thereby promoting the agreed conservation and management objectives.[1]

3.2.1 Legal Aspects of Conservation and Management Measures

Given the complexity of conserving and managing fishery resources on the basis of international cooperation, disputes concerning the interpretation or application of the compatibility rule may present interrelated legal and scientific questions. Article 7(3) FSA, in reiterating the duty to cooperate, requires States to 'make every effort to agree on compatible conservation and management measures within a reasonable period of time'. The Agreement defines elliptically the term 'conservation and management measures' as 'measures to conserve and manage one or more species of living marine resources that are adopted and applied consistent with the relevant rules of international law as reflected in the Convention and [the] Agreement'.[2]

In a technical sense, CMMs can take the form of a wide array of input/output controls in the regulatory area. From a legal perspective, ICJ examined meticulously the term in the proceedings instituted by Spain against the unilateral CMMs taken by Canada with respect to vessels fishing in the NAFO

---

1 For a detailed discussion on the multifaceted nature of fishery measures, refer to the expertly curated collection of chapters in Cochrane–Garcia (Eds) *A Fishery Manager's Guidebook* (2009).

2 FSA 1(b).

regulatory area. *In casu*, the substantive law of the Convention was of particular relevance insofar as it reflected customary international law *inter partes*, while the Agreement was treated as a persuasive primary source.[3] The Court rightly did not dwell on the question of whether a 'measure' may be legislative or executive in nature, as the term, in its ordinary sense, is sufficiently broad to encompass any act, step, or proceeding, without imposing limitations on its substantive content or intended objective. In international law and practice, a measure may be characterized as a CMM provided that its purpose is to conserve and manage fish stocks, and that it is grounded in technical requirements informed by factual and scientific criteria.[4] The Court, however, did not consider the legality of such measures to have a determinative effect on their characterization, observing that:

> The question of who may take [CMMs], and the areas to which [these] may relate, is neither in international law generally nor in [specific] agreements treated as an element of the definition [...] The authority from which such measures derive, the area affected by them, and the way in which they are to be enforced do not belong to the essential attributes intrinsic to the very concept of [CMMs]; they are, in contrast, elements to be taken into consideration for the purpose of determining the legality of such measures under international law.[5]

This line of reasoning was not unanimous on the bench. It was sternly criticized *ex cathedra* for interpreting the term based on an allegedly established normative concept of CMMs, which, by separating the technical from the legal aspects, ultimately disregarded their conformity with international law.[6] This critique is justified, given that, at the time, a substantial body of fisheries-related law already recognized legality as a prerequisite for the characterization of CMMs. On this point, Judge Raymond Ranjeva's comparative analysis of treaties defining CMMs was incontrovertible. It clearly established that legality is integral to the concept of measures. Furthermore, a historical review of official documents submitted to the *Fish Stocks Conference* demonstrated the leading role of coastal States—particularly Canada—for securing the inclusion of a reference to international law in the definition of CMMs,

---

3 *see* Davies, 'The EC/Canadian Fisheries Dispute in the Northwest Atlantic', ICLQ (1995) 933 and de la Fayette, 'The Fisheries Jurisdiction Case (Spain *v.* Canada), Judgment on Jurisdiction of 4 December 1998', ICLQ (1999) 664.

4 *Fisheries Jurisdiction* (Jurisdiction) 1998, §66.

5 *ibid.*, §70.

6 *ibid.*, *Dissenting Opinion of Judge Bedjaoui*, §§65–90 and *Separate Opinion of Judge Oda*, §§11–16.

which is now contained in article 1(b) FSA.[7] In fact, the Court itself cited the Agreement by name as an example, albeit not being applicable at the time.

Now that the Agreement is in force, the final clause of the definition, whereby measures are to be construed in relation to their conformity with international law, serves as a crucial referential context for interpreting and applying the compatibility rule. CMMs established under the Agreement must always adhere to the relevant norms, principles, and rules of international law. Essentially, a compatibility dispute may arise from disagreements over a point of law or fact, or from conflicting scientific approaches, in relation to the following: (a) the interpretation or application of the relevant rules of international law as reflected in the Agreement, in a manner consistent with the Convention and any *lex specialis* forming part of the applicable law; (b) the rights and the fulfilment of the respective obligations and duties of the parties concerning the conservation and management of fish stocks; (c) the appropriateness of CMMs to be undertaken, evaluated against the conditions set forth in the applicable legal framework and in proportion to all relevant circumstances; and (d) the necessity for all the CMM components to be applied in a timely, coherent, and comprehensive manner.

3.2.2 Subject Matter of Conservation and Management Measures

Foremost among the key conservation measures is the determination of MSY, framed by clearly defined management objectives. Under the terms of the Agreement, conservation is pronounced by and large as a long-term objective aimed at the sustainable utilization of marine living resources.[8] Sustainability, in turn, is a dynamic and flexible concept that must also be considered in light of general international law,[9] as well as contemporary political commitments, such as the SDGs. Another *par excellence* conservation measure of primary legal relevance is the subsequent attribution of fishing opportunities, which in practice is closely tied to the establishment of TAC limits and Total Allowable Fishing Effort ('TAFE') levels. Measures of this kind may give rise to manifold disagreements, both scientific and legal in nature. In practice, scientific questions must often be resolved prior to the determination of conservation measures. This is particularly relevant when establishing stock-specific target and limit reference points as part of a harvest strategy, in relation to stock biomass. It also involves reaching agreement on the actions to be taken when those

---

7 *ibid., Dissenting Opinion of Judge Ranjeva*, §24 *et seq.*

8 FSA 2.

9 *Gabčíkovo–Nagymaros Project* (Judgment) 1997, §140; and *see* further Sands, 'International Courts and the Application on the Concept of *Sustainable Development*', *Yearbook UN Law* (1999) 389*ff.*

points are exceeded, as well as on the preventive measures necessary to ensure that such thresholds are not breached in the first place. Moreover, these points must be agreed upon in determining acceptable levels of risk, as required by the application of the precautionary approach.[10] This includes, for example, accounting for potential local depletion, the spatial and temporal contraction of stocks, and scenarios involving mixed-species assemblages. Alongside scientific considerations, legal issues also arise owing to the concurrent rights of States over fish stocks and the absence of standardized formulas or agreed criteria for quota allocation across the distribution or migratory range of the stocks; a domain where political factors often play a decisive role.[11]

Complementing these are additional conservation measures aimed at: (a) adopting national TAC quotas and TAFE levels for contracting members to the regulatory scheme and making allowances as the case may be for cooperating non-contracting members; (b) designating seasonal and area closures; (c) regulating fishing appliances (e.g. fixing the size of fish aggregating devices) and gear (e.g. prohibition of large-scale pelagic driftnets and designation of technical specifications as for instance net mesh sizes employed in purse-seine operations, etc.); (d) prescribing rules for catches to be retained onboard fishing vessels, landed, exposed to or offered for sale; (e) setting standards for collecting, reporting and exchanging scientific as well as statistical data on the status of the stocks; (f) improving or increasing the fecundity of the fishery resource, which may include artificial propagation, stock transplantation, etc.; (g) conducting EFP or otherwise research programmes on the stocks (e.g. tagging and release of caught specimens); (h) keeping conservation measures under review in the light of scientific and other available data; and (i) addressing the impact of CMMs on ecologically associated fish stocks on account of the same trophic level or close spatial proximity, including measures to mitigate the mortality of non-target species (e.g. determining bycatch limits, use of approved fishing gears, etc.).

The effective implementation of conservation measures depends on corresponding management measures. However, the latter are not wholly contingent

---

10 FSA 6, in conjunction with Annex II providing detailed guidelines for a scheme of precautionary reference points in CMMs.

11 Indicative of such ramifications are the prolonged and ultimately unfruitful efforts of the North-East Atlantic Fisheries Commission to reach unanimous agreement on allocation criteria, despite years of negotiation; *qv.* Marine Stewardship Council, 'Northeast Atlantic Pelagic Fisheries—Management Challenges for Straddling Fish Stocks' (2023) 43–46; and *see* Lodge *et al.*, *Recommended Best Practices for Regional Fisheries Management Organizations* (2007) 34*ff* and Henriksen, 'Allocation of Fishing Rights: Principles and alternative procedures', in Nordquist–Moore–Long (Eds) *Challenges of the Changing Arctic* (2016) 522.

upon conservation measures; instead, the two operate as coordinated and often interdependent components, within a unified regulatory area. In many cases, CMMs must be formulated jointly as part of a coherent fishery policy. Harvest control rules—i.e. pre-agreed management actions that specify how harvesting is to be regulated based on defined stock status indicators—stand as a prime example of this type of interrelationship. The management measures most commonly used are in connection with: (a) standards, specifications, procedures and records for vessel authorization ('authorized vessels') and notification to fish, retain onboard, tranship or land catches in the regulatory area; (b) requirements on valid and uniform documentation, and the use of standard external markings and identification of authorized vessels through the implementation of Unique Vessel Identifiers based on practices such as the 1989 FAO Standard Specifications for the Marking and Identification of Fishing Vessels or by following the example of the Ship Identification Number Scheme for merchant shipping as operated by the International Maritime Organization; (c) mandatory Vessel Monitoring Systems ('VMS') to identify authorized vessels, trace and verify their operations, along procedures for access, use, as well as protecting data confidentiality; (d) compilation and dissemination of information on sightings and identifications of vessels engaging in IUU fishing; (e) marking of fixed gear or equipment or any other device, and retrieval of abandoned, lost or otherwise discarded ones; (f) mutual or reciprocal boarding and inspection procedures at sea or in port; (g) effective and impartial schemes of regional observer programmes onboard authorized vessels, with specified duties, responsibilities, rights and protection (e.g. from any delay, interference, influence, intimidation and otherwise obstruction or resistance) in the performance of their role; (h) landing or transhipment operations by, or supply services in relation to, authorized vessels fishing consistently with CMMs in the regulatory area; and (i) minimum standards of inspections at port.

Characterization by approximation of primary objectives

Management measures may exhibit compatibility aspects integral to conservation measures. By way of illustration, a widely used and cost-effective management measure to monitor conservation objectives is by means of VMS operation with coverage of the entire regulatory area. This system collects, records and reports data automatically to track and document the activities of authorized vessels. The rationale of the measure in relation to the compatibility rule is reflected in the need to maximize its functionality with the coastal State's VMS operation. So, the high seas areas are monitored not only to promote and achieve compliance with conservation measures, but also to ensure that the coastal State meets its respective obligations to maintain the integrity of these measures within sea areas under its national jurisdiction.[12] From other management measures it may be difficult to detach concomitant 'control' features, where these—especially when adopted in the package of Monitoring, Control and Surveillance ('MCS')—involve interconnecting tasks of documenting, reporting, sampling and verifying catches, as well as the exerted fishing effort.[13] The data can relate to quantification, identification and collection of caught species onboard, processed or discarded, transhipped or landed, along with corresponding information on the place and time of these operations. This composite data holds significant scientific value in the development or adjustment of harvest strategies that must be compatible with other CMMs applied within the regulatory area, including, where appropriate, on a case-by-case basis to address the specific requirements of individual stocks. Absent a scheme adopted by RFMO/As covering the entire regulatory area for monitoring and responding to compliance with CMMs, the coastal State remains solely competent to exercise control within sea areas under its national jurisdiction and to produce the relevant data concerning operations conducted at sea or in ports.

When the legal characterization of a measure as 'conservation', 'management', or 'control' is required, it will be determined by approximation to its primary objective, taking into account all relevant and surrounding factual circumstances in each case. For example, attention should be given to the evolving policies implementing international action plans to prevent, deter, and eliminate IUU fishing, developed within the framework of the *Code of Conduct for Responsible Fisheries* (1995), particularly with respect to ship-to-ship operations.[14] While some regional regimes prohibit transhipments at sea altogether, others allow them only if conducted in port or within sea areas under the national jurisdiction of a member State, and subject to authorization under that State's domestic law. In such cases, the respective RFMO/A is responsible for developing procedures to collect and verify data on transhipments within the regulatory area. In that context, decisions or actions taken by an RFMO/A that establish minimum standards for port inspection measures are typically adopted as CMMs, as they are underpinned by a two-fold

---

[12] FSA 18(4), and in practice *see* also the WCPFC, 'Statement Describing the Purpose and Principles of the WCPFC VMS' (9th Regular Session: 2–6/12/2012).

[13] *e.g.* the Agreement refers in articles 5(l), 10(h), 16(1), 18(3)(g) & 26(3)(c) FSA to MCS measures as distinct from CMMs. In some practical contexts, the former are also referred to as 'monitoring and responding to compliance' measures; *qv.* FAO, 'Technical Guidelines Code of Conduct and the Technical Guidelines No.4' (1997) and WCPFC:CMM/3–2009.

[14] *see* the Voluntary Guidelines for Transshipment, endorsed by FAO COFI (35th Session: 5-9/11/2022).

rationale. First, the effective conservation and management of fish stocks depends on accurate reporting within the regulatory area to ensure that decisions are informed by the best available scientific evidence. Second, as one of the most critical management tools,[15] transhipment regulation is particularly vulnerable to circumvention through unregulated and unreported transhipments at sea, which distort data and may abet illegal fishing activities.

Environmental and equity-related considerations

The scope of compatibility disputes will significantly expand when the relevant circumstances warrant careful consideration of the environmental consequences of CMMs or when their adoption must account for equity concerns. These considerations often become interconnected within a holistic regulatory framework aimed at balancing human and ecosystem well-being while ensuring fisheries that are biologically productive, economically viable, environmentally sustainable, and socially inclusive.[16] In this broader context, a disagreement on compatible CMMs can raise important questions on at least two key fronts.

First, measures may have an accretive effect beyond the directly targeted fish stock when they persistently impact non-target, associated, or dependent species, or other stocks inhabiting the same fishing grounds. Such outcomes may result from bycatches, discards, or the physical presence of fishing gear and equipment in the marine habitat.[17] This, in turn, may give rise to further legal and ecological questions. Specific questions include whether the conservation objective—e.g. when defined as maintaining or restoring fish populations or stock units in the EEZ to a level where their reproduction is not seriously threatened[18]should be set at a lower standard than that applied under the precautionary approach, both in EEZ and on the high seas. More general, yet equally complex, questions may arise as to whether CMMs comply with the obligation to protect and preserve the marine environment from pollution.[19] In cases where fishing activities take place in vulnerable ecosystems, such as deep-sea coral and sponge habitats, it may be necessary to integrate

15 *The "M/V Hoshinmaru" Case* (Judgment) 2007, §99.

16 Cochrane, 'Reconciling Sustainability, Economic Efficiency and Equity in Marine Fisheries: Has there been progress in the last 20 years?', *Fish and Fisheries* (2021) 298.

17 As noted in FAO, 'Input to the UN Secretary-General's Comprehensive Report for the Resumed Review Conference on the UN Fish Stocks Agreement' (2016), ¶5.

18 LOSC 61(4).

19 LOSC 192 & 194.

CMMs with habitat protection measures.[20] In light of case law addressing whether the designation of an MPA can qualify as a CMM,[21] and considering jurisprudence on pollution resulting from fishing activities,[22] the concept of 'environmentally harmful fishing' could be posed as a relevant consideration in the regulation of fish stocks.[23]

Given the ongoing promotion of area-based management tools in support of the principle of ecosystem approach, such questions are likely to assume increasing importance in the years ahead.[24] Notwithstanding that the 2023 *United Nations Agreement under the United Nations Convention on the Law of the Sea on the Conservation and Sustainable Use of Marine Biological Diversity of Areas Beyond National Jurisdiction* ('BBNJ Agreement') does not apply to fishing and fishing-related activities regulated under relevant international law,[25] it will influence both the normative content and form of CMMs in relation to compatibility requirements accruing from the holistic management of fishery resources as part of Large Marine Ecosystems.[26] Nonetheless, it should be observed that the BBNJ Agreement has appropriately departed from the so-called 'concept of adjacency',[27] which certain States had sought to invoke in order to prioritise their coastal interests in areas beyond national jurisdiction. The Agreement, on its own terms, acknowledges the 'need to avoid adverse

---

20 *see* NEAFC, 'Press Release—41st Annual Meeting' (22/11/2022).

21 *Chagos MPA* (Award) 2015, §291.

22 *South China Sea Arbitration* (Merits) 2016, §§951–961, 970 & 1203 *dispositif* B(12) finding several fishing activities in breach of LOSC 192 & 194.

23 Tanaka, 'Reflections on the Implications of Environmental Norms for Fishing: The link between the regulation of fishing and the protection of marine biological diversity', *International Community Law Review* (2020) 389.

24 *FSA Review Conference* 2016, ¶47; and *see* Scotcher, 'Tight Lines at Sea: An interdisciplinary analysis of fisheries, marine biodiversity and institutions', in Krabbe–Langlet (Eds) *Marine Bioprospecting, Biodiversity and Novel Uses of Ocean Resources* (2024) 141 and Beringen, 'International Fisheries as the "Whale in the Room" at the Negotiations for a New Instrument for Biodiversity Beyond National Jurisdiction', in Liu–Scott (Eds) *The Law of the Sea and the Planetary Crisis* (2025) 108.

25 BBNJ Agreement 10(2).

26 *see* Diz, 'The Interface Between the BBNJ Agreement and RFMOs', in Kunoy–Heidar–Yiallourides (Eds) *International Fisheries Law* (2024) 317–318.

27 For the proposal of this concept, including a relevant discussion on the distinction and affinity to the compatibility rule, *see* Oude-Elferink, 'Coastal States and MPAs in ABNJ: Ensuring consistency with the LOSC', in Freestone (Ed.) *Conserving Biodiversity in Areas Beyond National Jurisdiction* (2019) 56 and Mossop–Schofield, 'Adjacency and Due Regard: The role of coastal states in the BBNJ treaty', *Marine Policy* (2020) 103877.

impacts on the marine environment, preserve biodiversity, maintain the integrity of marine ecosystems, and minimize the risk of long-term or irreversible effects of fishing operations', and consequently, it establishes positive obligations to protect and preserve the marine ecosystem as a whole.[28]

Second, in pursuing the aims of sustainable regulation, it may be necessary for States to agree on CMMs through equitable cooperation and to implement them in an economically fair and socially just manner, thereby ensuring that both present and future generations can meet their needs. It should be noted that the Convention, whose provisions on the conservation and management of fish stocks the Agreement is tasked with implementing, recognizes the desirability of establishing, with due regard to the sovereignty of all States, a legal order for the seas and oceans that promotes the equitable and efficient utilization of their resources, the conservation of their living resources, and contributes to the realization of a just and equitable international economic order that takes into account the interests and needs of mankind as a whole, and in particular, the special interests and needs of developing countries.[29]

In that regard, particular attention must be given to the equitable criteria incorporated within the context of the compatibility rule, namely, the boundary-distribution criterion and the dependency criterion. While these criteria, as articulated in the multilateral treaty framework, reflect the progressive development of the law undertaken by the *Fish Stocks Conference*, their origins lie in broader equity norms.[30] Distinctly, the boundary-distribution criterion requires States, in determining compatible CMMs, to take into account 'the extent to which the stocks occur and are fished in areas under national jurisdiction'.[31] This element has been regarded as reflecting the precept of 'cross-boundary equity', which may constitute a key condition for reaching agreement in the negotiation of necessary measures.[32] The dependency criterion, in a related vein, requires both DWFS and coastal States to consider their 'respective dependence ...on the stocks concerned' when determining

---

[28] FSA seventh preambular recital & 5(d)–(g); and *see* Boyle–Redgwell, *International Law and the Environment* (2021) 750–751.

[29] LOSC fourth and fifth preambular recitals.

[30] *see* Goldie, 'Equity and the International Management of Transboundary Resources', *Journal of Natural Resources* (1985) 665; Oude-Elferink, 'The Determination of Compatible Conservation and Management Measures for Straddling and Highly Migratory Fish Stocks', *Yearbook UN Law* (2001) 573–577; and Hanich–Campbell–Bailey–Molenaar, 'Research into Fisheries Equity and Fairness—Addressing Conservation Burden Concerns in Transboundary Fisheries', *Marine Policy* (2015) 302.

[31] FSA 7(2)(d).

[32] FAO, *The Ecosystem Approach to Fisheries* (2003) 26 & 85–86.

compatible CMMs.[33] The equitable nature of this consideration was affirmed judicially in the *Fisheries Jurisdiction Cases*, where the Court imposed a mutual obligation on the parties to engage in good faith negotiations aimed at achieving 'an equitable solution' in the regulation of fishery resources.[34]

CMMs may also need to accommodate their application by giving full recognition to the special requirements of developing States, especially of the Least Developed Countries and Small Island Developing States and Territories ('SIDS'), which often depend heavily on fisheries for economic activity, food security, employment, and livelihoods. This is intended to prevent the direct or indirect transfer of a disproportionate burden of action onto these States.[35] To this end, RFMO/As, when adopting compatible measures, have undertaken various steps to mitigate such impacts. These include phasing in or exempting certain obligations, applying proportional or rotational approaches, and establishing compensatory or other forms of special assistance funding mechanisms.[36] Yet the concept of a 'disproportionate conservation burden' still requires further development to attain a clearer and more widely accepted understanding.[37] Sustainable regulation will remain elusive without a level playing field for all contracting parties, one that must be fostered through cooperation, including robust, well-coordinated cross-regional capacity-building initiatives and the effective, timely, and appropriate transfer of technology.[38]

ITLOS has demonstrated an evolutive approach to interpreting the Convention,[39] and specifically in relation to fisheries, has held that conservation measures must, in order to be effective, take into account the potential environmental and socio-economic impacts of climate change and ocean acidification on marine ecosystems, as well as the implications for vulnerable communities and especially affected developing States.[40] Moreover, the Tribunal has observed that articles 63 and 64 LOSC, in particular, require States to cooperate in good faith with a view to adopting the necessary measures. In

---

33 FSA 7(2)(e).

34 *Fisheries Jurisdiction Cases* (Merits) 1974, §§69–78.

35 FSA 24–26.

36 *e.g.* WCPFC:CMM/6–2013; and *see* Sinan–Bailey–Hanich–Azmi, 'Common but Differentiated Rights and Responsibilities in Tuna Fisheries Management', *Fish and Fisheries* (2022) 202.

37 *see* A/75/89 'Sustainable Fisheries, including through the 1995 FSA, and Related Instruments' (8/12/2020), 16(¶49).

38 A/Res.80/70 'Oceans and the Law of the Sea' (11/3/2025), ¶¶7 & 67.

39 *e.g. Responsibilities and Obligations of States Sponsoring Persons and Entities with Respect to Activities in the Area* (Advisory Opinion) 2011, §§57–60.

40 *Climate Change and International* Law (Advisory Opinion) 2024, §66 & §§410–414.

this regard, article 7 FSA has been recognized as providing guidance for addressing distributional changes and range shifts in stocks.[41] This builds on the approach adopted by ICJ, which has viewed the environment not as an abstraction, but as a living space intrinsically linked to food security, quality of life, and the health of human beings, including that of future generations.[42] Additionally, the Court has consistently endorsed an evolutionary approach to treaty interpretation, emphasizing the need to remain faithful to the original intent of the parties while also accounting for the development of emerging norms of general international law,[43] particularly those concerning the protection and preservation of the environment.[44]

3.2.3 Procedural Restraints

Ultimately, a compatibility dispute may be procedurally shaped by three dynamics that commonly arise in contested jurisdiction cases. The first regards the legal grounds upon which the forum is to be seised. While the scope of a dispute must be determined objectively, established jurisprudence affirms that the starting point for such determination lies with the applicant party. It is incumbent upon that party to identify the dispute it seeks to bring before the forum and to clearly set forth the claims submitted.[45] Subsequently, the forum itself may be afforded a degree of latitude in characterizing the dispute by virtue of the principle of *compétence de la compétence*, which empowers it to determine its own jurisdiction.[46] The second is by the occurrence of incidental questions on a point within the forum's jurisdiction. Although a disagreement over compatible CMMs does not constitute a dispute concerning fishing rights *stricto sensu*, issues related to 'fishing' and 'fishing-related' activities may nonetheless be interconnected,[47] giving rise to contentions between the parties

---

[41] *ibid.*, §428.

[42] *Legality of the Threat or Use of Nuclear Weapons* (Advisory Opinion) 1996, §§29–35.

[43] *Legal Consequences for States of the Continued Presence of South Africa in Namibia (South West Africa) notwithstanding Security Council Resolution 276 (1970)* (Advisory Opinion) 1971, §35.

[44] *Gabčíkovo–Nagymaros Project* (Judgment) 1997, §111.

[45] *Fisheries Jurisdiction* (Jurisdiction) 1998, §29 followed by *The "M/V Louisa" Case* (Judgment) 2013, §95; and *see* the comments *apropos* in *The "M/V Norstar" Case* (Preliminary Objections) 2016, *Joint Separate Opinion of Judges Wolfrum and Attard*, §22.

[46] LOSC 288(4).

[47] A comprehensive definition can be found in the *Convention on the Conservation and Management of High Seas Fishery Resources in the South*

that require concurrent consideration. In the latter category, such activities are defined elsewhere as 'any operation in support of, or in preparation for, fishing, including the landing, packaging, processing, transhipping, or transporting of fish that have not previously been landed at a port, as well as the provisioning of personnel, fuel, gear, and other supplies at sea'.[48] Ancillary activities that bear a sufficiently close connection to fishing have already been examined in dispute settlement proceedings, in light of State practice that has evolved since the adoption of the Convention and the Agreement. As a case in point, the conduct of transhipment operations was among the issues addressed in the *Swordfish stocks* dispute. Another example is the regulation of bunkering, a fishing-related activity considered part of the management measures available to States. The provision of bunkering services enables vessels to engage in fishing and related activities—such as onboard processing, freezing, and storage—without interruption at sea.[49] A fishing-related activity will fall within the scope of a compatibility dispute insofar as it is either directly related to the objectives and purposes of CMMs or essentially associated with the underlying reasons for an unresolved disagreement concerning such measures. The third is whether any CMM component is raised *in limine litis* as a preliminary objection,[50] whereby the jurisdiction of the forum or the admissibility of a case can be challenged without enquiring into the merits. In such cases, at least two candidate contentions are likely to arise, depending on the competing arguments. One contention may plausibly raise the question of whether the legality of a measure possesses an exclusively preliminary character, thereby justifying bifurcation of the proceedings as a matter of due process. Closely related, the other will almost certainly concern the interpretation of the fisheries limitation and its application to the specific facts of the case. Since the latter hinges on jurisdictional aspects, it will undoubtedly serve as the *pièce de résistance* in any case where a compatibility dispute is unilaterally referred to the compulsory procedures entailing binding decisions under Part VIII FSA.

Pacific Ocean (2009), 1(g), wherein 'fishing' comprises, among other things: the actual or attempted searching for, catching, taking or harvesting fishery resources; engaging in activities which can reasonably be expected to result in locating, catching, taking or harvesting fishery resources for any purpose; transhipments; and other operations at sea in support of, or in preparation for, any such activity.

48 *Agreement on Port State Measures to Prevent, Deter and Eliminate Illegal, Unreported and Unregulated Fishing* (2009), 1(d).

49 *The "M/V Virginia G" Case* (Judgment) 2014, §§214–223.

50 LOSC 294(3).

3.3 PROCEDURES UNDER PART VIII OF THE 1995 UNITED NATIONS FISH STOCKS AGREEMENT

Disagreements concerning the interpretation or application of the compatibility rule are to be resolved by the procedures set out in Part VIII FSA, which comprises articles 27–32 FSA. Paragraph (4) of article 7 FSA stipulates, in explicit terms, that if no agreement can be reached on CMMs 'within a reasonable period of time, any of the States concerned may invoke the procedures for the settlement of disputes provided for in Part VIII' of the Agreement. The general approach in Part VIII FSA underscores the essential need for cooperation to prevent disputes and is grounded in the fundamental principles of peaceful dispute resolution and the autonomy of the parties to freely choose the means of settlement.[51] Disagreements of a technical nature can be referred to an expert panel established *ad hoc* by the parties concerned.[52] This modality is all the more important in order to elucidate material aspects of a disagreement featuring mixed scientific facts and law, such as the effective implementation of the precautionary approach that remains nonetheless a justiciable matter.[53] Under Part VIII FSA, the dispute settlement provisions of Part XV LOSC apply *mutatis mutandis* to any dispute between parties to the Agreement, concerning its interpretation or application, irrespective of whether they are parties to the Convention.[54] In addition, parties to the Agreement may invoke these procedures in relation to disputes arising under the constitutive instruments of RFMO/As, or other instruments concerning straddling or highly migratory fish stocks to which they are party—including disputes over CMMs for such stocks—even if the disputing States are not parties to the Convention.[55]

3.3.1 Applying *mutatis mutandis* Part XV of the 1982 United Nations Convention on the Law of the Sea

The incorporation of Part XV LOSC at this stage warrants further examination. Given that the settlement provisions apply *mutatis mutandis* to compatibility disputes—a legal expression with its own interpretative nuances and distinctive significance—they will be examined in tandem with the relevant provisions of the Agreement, which primarily perform an adjusting function

---

51 FSA 27–28.

52 FSA 29.

53 Foster, *Science and the Precautionary Principle in International Courts and Tribunals* (2011) 137*ff*.

54 FSA 30(1).

55 FSA 30(2).

in this context. Part XV LOSC requires parties to engage in a two-tiered dispute settlement process, consisting of an initial non-binding phase followed, if necessary, by binding procedures. At the first tier, the procedures prescribed in section 1 Part XV LOSC are voluntary in nature, allowing parties the freedom to consider, at any time, alternative peaceful settlement means of their choosing, such as those envisaged in article 33(1) of the *Charter of the United Nations and Statute of the International Court of Justice* (1945) ('UNC' or the 'Charter'), or by conciliation under Annex V LOSC.[56] To avoid aggravating or prolonging the dispute, parties are obligated to proceed expeditiously in exchanging views.[57] This obligation does not require parties to enter into substantive negotiations on the merits of the dispute, but rather to exchange views on the means for its resolution.[58] Its purpose is to ensure that a dispute is not unilaterally submitted to the procedures under section 2 Part XV LOSC without the respondent party being duly informed of its existence.[59] This section represents the more 'intrusive' phase of the dispute settlement process, as it entails binding and final decisions,[60] subject to the correlative preconditions set forth in articles 281 and 282 LOSC.

The procedural preconditions in section 1 Part XV of the 1982 United Nations Convention on the Law of the Sea

Article 281 LOSC renders the compulsory procedures applicable only where the parties have failed to settle the dispute by 'a peaceful means of their own choice' and where 'the agreement between them does not exclude any further procedure'. Although this provision has been viewed as having the potential to broadly exclude the application of compulsory procedures to disputes otherwise falling within Part XV LOSC,[61] the prevailing jurisprudence has adopted a restrictive interpretative approach, thereby narrowing its applicability.

The *Southern Bluefin Tuna Cases* directly engaged this precondition, which lay at the heart of the jurisdictional dispute. Japan's objection to jurisdiction was addressed by ITLOS at the provisional measures stage, where, in assessing

56 LOSC 279 & 280 in conjunction with 284.

57 LOSC 283.

58 *Chagos MPA* (Award) 2015, §378 followed in *Arctic Sunrise Arbitration* (Merits) 2015, §151.

59 Adede, 'The Basic Structure of the Disputes Settlement Part of the Law of the Sea Convention', ODIL (1982) 129.

60 Churchill, 'The General Dispute Settlement System of the UN Convention on the Law of the Sea: Overview, context and use', ODIL (2017) 217.

61 *compare* Jia, 'The Curious Case of Article 281: A "super" provision within UNCLOS', ODIL (2015) 266 *with* Kunoy, 'Dislodging the Compulsory Dispute Settlement Mechanism: Analysis of article 281 of UNCLOS', LJIL (2024) 423.

the requirement of *prima facie* jurisdiction, the Tribunal held that recourse to arbitration under section 2 Part XV LOSC was not excluded, insofar as the dispute settlement clause in the CCSBT *Convention* did not, by itself, establish a compulsory procedure resulting in a binding decision, as required by article 282 LOSC.[62] The reasoning on this specific point gave rise to misgivings in the lesser part of the academic commentary.[63] Yet, in contrast, the arbitral tribunal declined to proceed to the merits, holding that the CCSBT *Convention*, within the meaning of article 281 LOSC, implicitly precluded recourse to any further procedure not accepted by all parties to the dispute. This conclusion was reached despite the fact that the treaty itself does not establish a dispute settlement mechanism involving compulsory jurisdiction.[64] Surprisingly, the critical role of article 282 LOSC in calibrating the interplay between multiple dispute settlement procedures available to the parties was addressed only in passing in the award. The tribunal's holding attracted significant criticism, mostly owing to the broad interpretation it afforded to article 281 LOSC by misapplying the doctrine of substantive and procedural parallelism between—and, in this case, separation of—treaties.[65] One commentator went so far as to

---

[62] *Southern Bluefin Tuna Cases* (Provisional Measures) 1999, §§52–62.

[63] *see* Evans, 'The Southern Bluefin Tuna Dispute: Provisional thinking on provisional measures?', YbIEL (2000) 7; Kwiatkowska, 'The Southern Bluefin Tuna (New Zealand *v.* Japan; Australia *v.* Japan) Cases', IJMCL (2000) 32–33; and Tanaka, 'Some Observations on the Southern Bluefin Arbitration Award', *Japanese Annual of International Law* (2001) 9.

[64] *Southern Bluefin Tuna Cases* (Jurisdiction) 2000, §§54–59 and *see* the dissenting comments on this point in *Separate Opinion of Judge Justice Sir Kenneth Keith*, §§17–30.

[65] *compare* the views held by Stephens, 'A Paper Umbrella which Dissolves in the Rain? Implications of the Southern Bluefin Tuna Case for the Compulsory Resolution of Disputes Concerning the Marine Environment under the 1982 LOS Convention', *Asia Pacific* LPJ (2001) 311; Colson–Hoyle, 'Satisfying the Procedural Prerequisites to the Compulsory Dispute Settlement Mechanisms of the 1982 Law of the Sea Convention: Did the Southern Bluefin Tuna tribunal get it right?', ODIL (2003) 68; Shany, *The Competing Jurisdictions of International Courts and Tribunals* (2003) 203–206; Rothwell–Stephens, 'Dispute Resolution and the Law of the Sea: Reconciling the interaction between the LOS Convention and other environmental instruments', 221; and Oxman, 'Courts and Tribunals: The ICJ, ITLOS, and arbitral tribunals', in Rothwell *et al.* (Eds) *Oxford Handbook of the Law of the Sea* (2015) 402 *with* the contrary view mainly expressed by Kwiatkowska, *Peaceful Settlement of Oceans and Other Environmental Disputes under International Agreements* (2002) 15–34.

describe it as 'an extraordinary reversal of the intentions underlying Part XV LOSC'.[66]

Subsequent awards have departed from this interpretation, finding it inconsistent with the intended meaning of article 281 LOSC, which is generally understood to apply only in specific circumstances where the parties have concluded an *ad hoc* and legally binding agreement to settle their dispute by alternative means, thereby excluding recourse to the procedures under Part XV LOSC.[67] In turn, article 282 LOSC permits the application of compulsory procedures leading to binding decisions under general, regional, or bilateral agreements 'in lieu' of those under section 2, provided that such procedures may be invoked at the request of any party to the dispute.

The *renvoi* mechanism embodied in articles 281 and 282 LOSC consequently renders the Convention's compulsory procedures residual in nature. However, the obligation to resort to these procedures is neither eclipsed by nor exhausted in the use of alternative general, regional, or bilateral procedures that are non-compulsory, do not lead to binding decisions, and cannot be unilaterally invoked.[68] Such mechanisms need not have been established specifically for the purposes of Part XV LOSC, as the parties may agree to resolve their dispute by means 'otherwise' agreed.[69] Both preconditions—each capable of giving rise to claims of inadmissibility—were addressed in the *South China Sea Arbitration*, where the tribunal underscored that the overarching design of Part XV LOSC establishes dispute settlement by compulsory procedures leading to a binding decision as the default rule.[70] A comprehensive analysis of what constitutes an 'agreement' within the meaning of articles 281 and 282 LOSC reveals two essential requirements: first, that the alternative procedures are capable of addressing the subject matter of the dispute; and second, that they result in a binding decision.[71]

---

[66] As aptly described by Alan Boyle in 'The *Southern Bluefin Tuna* Arbitration', ICLQ (2001) 449–450 and 'Problems of Compulsory Jurisdiction and the Settlement of Disputes Relating to Straddling Fish Stocks', in Stokke (Ed.) *Governing High Seas Fisheries* (2001) 105.

[67] *e.g. Barbados v. Trinidad and Tobago Arbitration* (Decision) 2006, §200(ii).

[68] *MOX Plant* (Provisional Measures) 2001, §§38–53 and *OSPAR Arbitration* (Decision) 2003, §118*ff.*

[69] *Maritime Delimitation in the Indian Ocean* (Preliminary Objections) 2017, §§122–134.

[70] *South China Sea Arbitration* (Jurisdiction and Admissibility) 2015, §§223–225.

[71] Roberts, 'Identifying "Exclusionary Agreements": Agreement type as a procedural limitation in UNCLOS dispute settlement', ODIL (2021) 113.

Moving on to section 2 Part XV of the 1982 United Nations Convention on the Law of the Sea

At the second tier of the settlement process, the dispute can be unilaterally referred for a binding and final decision to a court or tribunal having jurisdiction under the mechanical process of article 287 LOSC. When signing, ratifying or acceding to the Convention, or at any subsequent time, parties are free to declare in writing their choice to submit disputes either to judicial settlement by ITLOS or ICJ, or to arbitration, either before a tribunal constituted under Annex VII LOSC or a special arbitral tribunal with subject-matter expertise constituted under Annex VIII LOSC. Where no declaration has been made, or where the parties to the dispute have not previously selected the same means of dispute settlement, the matter defaults to arbitration under Annex VII LOSC. A declaration made in accordance with article 287 LOSC is deemed to extend to disputes concerning the interpretation or application of the Agreement, unless it has been expressly revoked or has lapsed. Irrespective of the means originally selected, the parties may, by special agreement, consent to submit the dispute to an alternative forum, including after the initiation of proceedings. For a party not contracted to the Convention, the corresponding declaration may be made upon signing, ratifying, or acceding to the Agreement, or at any time thereafter.[72]

Compulsory and binding procedures under the Convention are subject to the limitations and exceptions set out in section 3 Part XV LOSC.[73] Article 297 LOSC introduces specific limitations that may be invoked by parties without prior notification; these have been colloquially described as 'automatic' limitations.[74] In contrast, article 298 LOSC provides for optional exceptions, which must be declared *ex ante* by the parties. This section reflects a necessary compromise aimed at preserving the integrity of the provisions forming part of the UNCLOS/III package deal. Accordingly, no reservations[75] or additional limitations or exceptions are permitted beyond those explicitly provided for in the Convention.[76] Yet, as widely observed, a fundamental question remains as to whether the limitations and exceptions under articles 281, 282, 297, and

---

[72] FSA 30(3)–(4).

[73] LOSC 286.

[74] *see* originally in the writings of Louis Sohn in 'Peaceful Settlement of Disputes in Ocean Conflicts: Does UNCLOS/III Point the Way', LCP (1983) 198, and later in 'The Importance of the Peaceful Settlement of Disputes Provisions of the United Nations Convention on the Law of the Sea', in Nordquist–Moore (Eds) *Entry into Force of the Law of the Sea Convention* (1995) 268–269.

[75] LOSC 309.

[76] Rosenne, 'Reflections on the Final Clauses in the New Law of the Sea Treaty' (1977) VJIL 142–146.

298 LOSC should be interpreted narrowly to uphold the overarching objective of establishing a compulsory and comprehensive dispute settlement system, or broadly to reinforce the consent-based nature of jurisdiction. Jurisprudence reveals diverging approaches to the interpretation of these provisions, with significant implications for the accessibility and predictability of compulsory dispute settlement under the Convention.[77]

A forum acquiring jurisdiction over a compatibility dispute under the Agreement will exercise a broad scope of applicable law. In terms of jurisdiction *ratione materiae*, it is expected to adjudicate not only on the basis of the Agreement and the relevant provisions of the Convention, but also in light of any global, regional, or subregional fisheries-related instruments to which the parties are bound. In addition, it may apply generally accepted standards of conservation and management that are pertinent to the fish stocks in question.[78] While the Agreement does not expressly reference the competence to decide *ex aequo et bono*, article 293(2) LOSC permits such discretion where the parties so agree. Regarding jurisdiction *ratione personae*, the Agreement extends the obligations under Part XV LOSC to non-parties to the Convention, including fishing entities,[79] and integrates compulsory procedures entailing binding decisions into regional regimes that would otherwise lack such mechanisms. The practical effect of incorporating and adapting Part XV LOSC into the Agreement has been to strengthen the efficacy of dispute settlement by broadening the jurisdictional reach of the competent forum.

3.3.2 The Importance of Provisional Measures in Compatibility Disputes

Following the mechanical designation of the court or tribunal under section 2 Part XV LOSC, it became necessary to establish within the Convention a procedure for the prompt and effective prescription of binding

---

[77] *see* Allen, 'Article 297 of the United Nations Convention on the Law of the Sea and the Scope of Mandatory Jurisdiction', ODIL (2017) 313; Guilfoyle, 'The *South China Sea Award*: How should we read the UN Convention on the Law of the Sea?', *Asian* JIL (2018) 63; Oude-Elferink–Georgoula–Nguyen–Trevisanut, 'Compulsory Jurisdiction as the DNA of LOSC Dispute Settlement: An evolutionary path', IJMCL (2023) 188; and Guilfoyle–Mossop, 'The Extent and Legitimacy of the Judicial Function in UNCLOS Dispute Settlement', ICLQ (2024) 664–665.

[78] LOSC 288(2) and FSA 30(1)–(2) & (5).

[79] FSA 1(3), subject to the operational provisions and rules of the competent forum; e.g. in conjunction with ITLOS Statute 20(2), the Tribunal is open to entities other than States parties in any dispute submitted pursuant to an agreement conferring jurisdictional competence.

measures *pendente lite*. Based on article 290 LOSC, the forum to which the dispute has been duly referred may prescribe provisional measures if it determines that it has jurisdiction *prima facie* and that the circumstances present a real and imminent risk warranting action to prevent irreparable prejudice to the rights of the parties before the final decision is rendered.[80] At this preliminary stage, ITLOS has special jurisdiction to prescribe such measures, particularly where an arbitral tribunal has not yet been constituted or in the absence of agreement between the parties.[81] Within two weeks of an application being submitted, the Tribunal may issue provisional measures if it finds *prima facie* jurisdiction and considers the matter to be urgent.[82] It is further vested with the competence to prescribe measures that differ in whole or in part from those requested, including procedural safeguards designed to protect and preserve the marine environment.[83] These measures have frequently reaffirmed the duty to cooperate as a fundamental principle of both general international law and the Convention.[84]

Provisional measures are crucial in circumstances surrounding disagreements over the conservation and management of fish stocks.[85] In the *Southern Bluefin Tuna Cases*, where ITLOS exercised for the first time its special jurisdiction under article 290(5) LOSC, New Zealand and Australia sought provisional measures in alleging that Japan had failed to comply with its obligations

---

80 The respective rights sought to be protected shall be 'at least plausible' as it was found in *Delimitation of the Maritime Boundary in the Atlantic Ocean* (Provisional Measures) 2015, §58; this pronouncement is not necessarily exhausted in instances of sovereign rights to marine natural resources in disputed sea areas, but it may find application as well in clarifying and qualifying their scope within established maritime zones.

81 LOSC 290(5); and *see* Jesus, 'The Role of ITLOS in the Settlement of Law of the Sea Disputes', in Nordquist *et al.* (Eds) *The Law of the Sea Convention* (2012) 161–162.

82 On the concept of urgency as a distinctive criterion for the prescription of provisional measures by ITLOS *see* Ndiaye, 'Provisional Measures before the International Tribunal for the Law of the Sea', in Nordquist–Moore (Eds) *Current Marine Environmental Issues and the International Tribunal for the Law of the Sea* (2001) 98–99 and Wolfrum, 'Provisional Measures of the International Tribunal for the Law of the Sea', in Rao–Khan (Eds) *The International Tribunal for the Law of the Sea* (2001) 173*ff.*

83 ITLOS *Rules of the Tribunal* (as adopted on 28/10/1997 and subsequently amended), Rule 89(5).

84 *MOX Plant* (Provisional Measures) 2001, §82 and *Land Reclamation by Singapore in and around the Straits of Johor* (Provisional Measures) 2003, §§92–98.

85 *Fisheries Jurisdiction Cases* (Interim Protection) 1972, §21.

to cooperate in good faith and adopt the necessary CMMs for its nationals fishing on the high seas under articles 64 and 116–119 LOSC. The Tribunal issued six provisional measures directing the parties: (a) to prevent aggravation or extension of the dispute; (b) to prevent prejudice of the decision on the merits; (c) to keep TAC to levels last agreed; (d) to refrain from conducting EFPs; (e) to resume negotiations; and (f) to seek agreement with other States engaged in fishing the species in question. By prescribing these measures, the Tribunal advanced the precautionary approach, which—although not explicitly named—was clearly outlined. Precisely, the Tribunal, having noted that the parties did not dispute the biological depletion of the species to its historically lowest levels, and recognizing the serious concern this raised, took the view that the parties, in the circumstances, should 'act with prudence and caution to ensure that effective conservation measures are taken to prevent serious harm to the stock of southern bluefin tuna'.[86] Taking into consideration both the scientific uncertainty surrounding the appropriate CMMs and the lack of consensus among the parties on the effectiveness of those measures in improving the species' status, the Tribunal concluded that 'although it could not conclusively assess the scientific evidence presented by the parties … measures should be taken as a matter of urgency to preserve the rights of the parties and to avert further deterioration of the stock'.[87]

The Agreement has contributed to the progressive development of the law on provisional measures, particularly in the context of compatibility disputes, with the aim of preserving the respective rights of the parties in the conservation and management of fish stocks and of preventing harm to stocks at risk.[88] The recognition of the latter as a legitimate interest in its own right marks a significant advancement in furthering the Agreement's expressly stated object and purpose, namely, the sustainability of fish stocks.[89] To this end, the Agreement expressly cross-references the general availability of provisional measures in article 31 FSA, by invoking the circumstances outlined in articles 7(5) and 16(2) FSA.[90] This internal reference underscores the pivotal role of third-party dispute settlement in matters concerning CMMs and unequivocally reaffirms the applicability of the compulsory and binding procedures under Part VIII

86 *Southern Bluefin Tuna Cases* (Provisional Measures) 1999, §77.

87 *ibid.*, §80.

88 FSA 31(1).

89 As highlighted by the ITLOS President Rüdiger Wolfrum in his statements 'On Agenda Item 75(a) at the 60th Session UNGA' (28/11/2005) and 'The Potential of ITLOS in the Management and Conservation of Marine Living Resources' at the Meeting of the Friends of the Tribunal at the Permanent Mission of Germany to the UN (21/6/2007).

90 FSA 31(2).

FSA to ensure the effective regulation of fish stocks.[91] Accordingly, pending agreement on compatible CMMs, the parties, 'in a spirit of understanding and cooperation, shall make every effort to enter into provisional arrangements of a practical nature'.[92] When no agreement is reached on such arrangements, any of the parties may unilaterally seise a court or tribunal in accordance with the dispute settlement procedures set out in Part VIII FSA, for the purpose of obtaining provisional measures. However, non-parties to the Convention may withhold consent, thereby preventing ITLOS from prescribing, modifying, or revoking such provisional measures or related practical arrangements.[93] The Agreement further establishes four substantive criteria that govern both the designation of provisional arrangements of a practical nature agreed upon by the parties and the provisional measures prescribed by the competent forum. These criteria require that such arrangements or measures: (a) take into account the general principles outlined in article 5 FSA, including the application of principles of ecosystem and precautionary approach; (b) give due regard to the rights and obligations of all States concerned; (c) do not jeopardize or hamper the reaching of a final agreement on compatible CMMs; and (d) remain without prejudice to the final outcome of any dispute settlement procedure.[94]

Finally, provisional measures assume particular importance for the application of the compatibility rule in the special case of high seas enclaves, as addressed in article 16 FSA. Where DWFS and the coastal State fail to reach agreement on compatible CMMs, the compulsory dispute settlement process becomes available, including recourse to provisional arrangements or measures. This is expressly facilitated by a direct reference to the embedded clause in article 7 FSA, which provides that 'if, within a reasonable period of time', the parties are unable to agree on such measures, 'they shall ... apply article 7, paragraphs (4), (5), and (6), relating to provisional arrangements or measures'. In the interim, pending the adoption of such provisional practical arrangements or measures, each party shall, individually, adopt measures applicable

---

91 *see* Churchill, 'Trends in Dispute Settlement in the Law of the Sea: Towards the increasing availability of compulsory means', in French *et al.* (Eds) *International Law and Dispute Settlement* (2012) 158 who seems, with respect to compatibility disputes, to have departed from the earlier view expressed in 'The Jurisprudence of the International Tribunal for the Law of the Sea Relating to Fisheries: Is there much in the net?', IJMCL (2007) 389–390, namely that if article 297(3) LOSC were not applicable, then article 32 FSA would not have been necessary.

92 FSA 7(5).

93 FSA 31(3) in conjunction with article 290(5) LOSC; and *see* Rosenne, *Provisional Measures in International Law* (2005) 58–61.

94 FSA 7(6).

to its own vessels to prevent fishing activities that may undermine the stock's status in question.[95]

3.3.3 The Potential of ITLOS in Adjudicating Fisheries Disputes

The portrayal of ITLOS as the 'guardian'[96] of the Convention emerged early during UNCLOS/III. However, this characterization was never intended to establish the Tribunal as the exclusive forum for the adjudication of law of the sea disputes. Instead, as part of the broader negotiations to secure the principle of compulsory adjudication, the *Montreux formula*—conceived by Willem Riphagen (The Netherlands) and now enshrined in article 287 LOSC—grants States the autonomy to choose among the compulsory procedures entailing binding decisions under section 2 Part XV LOSC.[97] Nonetheless, the Tribunal was established under the Convention as the international judicial institution with specialised jurisdiction over disputes concerning its interpretation and application. It therefore occupies a central position not only in advancing the ideal of peaceful dispute resolution but also in contributing significantly to the progressive development of the law through its jurisprudence.[98] In this respect, the Tribunal's authority and effectiveness as an active judicial institution have been described as 'unquestionable'.[99] Several attributes underscore its unique role within the Convention's dispute settlement framework, among which the following are particularly noteworthy.

Alongside ICJ, which has historically maintained a prominent and enduring role in the peaceful resolution of law of the sea disputes, the Tribunal contributes to safeguarding the integrity and ensuring the consistent interpretation and application of the Convention as the foundational legal framework governing ocean activities. The jurisprudential respect shared between the two

---

95 FSA 16(2).

96 Anderson, *Modern Law of the Sea* (2008) 558.

97 *see* Riphagen, 'Mechanisms of Supervision in the Future Law of the Sea', in Kalshoven (Ed.) *Essays on the Development of the International Legal Order* (1980) 132 and Queneudec, 'Coup d'Oeil Retrospectif sur les Origines du Tribunal International du Droit de la Mer', in Ando–McWhinney–Wolfrum (Eds) *Liber Amicorum Judge Shigeru Oda* (2002) 624.

98 *see* 'Statement by the ITLOS President Dolliver Nelson on the Commemoration of the Twentieth Anniversary of the Opening for Signature of the 1982 LOSC at the 57th Session UNGA' (9/12/2002) and 'Statement by the ITLOS President José Luís Jesus to the Meeting of the Sixth Committee of UNGA' (4/11/2009).

99 A/75/239, 'Oceans and the Law of the Sea Report of the Secretary-General' (31/12/2020), 15(¶57).

institutions—frequently demonstrated through reciprocal citation—has facilitated their 'common goal of a mutually reinforcing corpus of international law in the settlement of international legal disputes'.[100] The universality of the Convention can only be sustained through compulsory procedures grounded in a global perspective and administered by institutions capable of developing a coherent and principled body of decisional law that promotes the long-term interests of all States.[101] The Tribunal has consistently affirmed its 'institutional responsibility' to safeguard the delicate balance struck during the package deal negotiations at UNCLOS/III.[102] Through its steady jurisprudence on the interpretation and application of the Convention's provisions, the Tribunal has fulfilled this mandate and, in doing so, has laid the groundwork for the further progressive development of the law of the sea by carefully reconciling continuity and change.

With respect to fisheries disputes in particular, the Tribunal offers considerable advantages over other courts or tribunals operating under Part XV LOSC. It is endowed with special jurisdiction to prescribe provisional measures *ex proprio motu*, enabling it—as discussed above—to act promptly in preventing harm to fish stocks. In terms of general jurisdiction, the Tribunal is competent to adjudicate disputes arising not only under the Convention and the Agreement, but also under constitutive instruments of RFMO/As and other fisheries-related treaties that confer jurisdiction upon it.[103] The growing number of instruments granting compulsory jurisdiction to the Tribunal attests to the confidence placed in both its legal expertise and administrative

---

[100] *see* the statements by the ICJ President Rosalyn Higgins 'On the occasion of the Tenth anniversary of the ITLOS' (29/9/2006) and by the ITLOS President Jin-Hyun Paik at '30th Annual Informal Meeting of the Legal Advisers' (29/10/2019).

[101] *see* the statement by the ITLOS President Rüdiger Wolfrum 'On Agenda Item 71(a) at the 61st Session UNGA' (8/12/2006).

[102] *see* the statements by the ITLOS Presidents Shunji Yanai 'On the Occasion of the Commemoration of the Thirtieth Anniversary of the Opening for Signature of the United Nations Convention on the Law of the Sea' (10/12/2012) and Albert Hoffmann 'On the Occasion of General Assembly's Commemoration of the Fortieth Anniversary of the Adoption of the United Nations Convention on the Law of the Sea' (8/12/2022), as well as the keynote address of the ITLOS President Jin-Hyun Paik at the conference *How healthy is the Ocean's Constitution? 25 Years of the United Nations Convention on the Law of the Sea* (17/10/2019).

[103] LOSC 288; and *see* Bankes, 'The Jurisdiction of the Dispute Settlement Bodies of the Law of the Sea Convention with Respect to Other Treaties', ODIL (2021) 346.

efficiency.[104] Additionally, the Tribunal offers parties considerable institutional flexibility through a variety of procedural arrangements.

While, as a general rule, cases are decided by the full bench of 21 judges, the Tribunal may address fisheries disputes in an equally authoritative but significantly more expedient manner through the constitution of one of its special-purpose chambers,[105] at the request of the parties.[106] One such chamber is the standing Chamber for Fisheries Disputes, composed of nine ITLOS judges, which is designated to hear cases concerning the conservation and management of marine living resources, should the parties agree to submit the matter to it.[107] Given the increasingly scientific and complex nature of fisheries disputes, this chamber offers the added advantage of fostering continuity and preserving subject-matter expertise within the Tribunal.[108] Another procedural option is the standing Chamber of Summary Procedure, composed of five judges—including *ex officio* the President and Vice-President—which is tasked with the prompt resolution of urgent matters.[109] Alternatively, the Tribunal may constitute an *ad hoc* chamber to address a particular dispute.[110] In the latter option, the parties benefit from greater procedural flexibility, including the ability to approve the chamber's composition and to appoint a judge of their nationality if none is already present on the Tribunal. Additionally, they may propose specific adaptations to the *Rules of the Tribunal*.[111] The potential of this *ad hoc* special chamber has been aptly described as 'arbitration within the Tribunal', since it combines the advantages of a permanent judicial body with the procedural adaptability of arbitral mechanisms, while avoiding the considerable financial costs typically associated with arbitration.[112] This option was

---

104 *see* the representative, albeit non-exhaustive, list of *Multilateral Agreements Conferring Jurisdiction on the Tribunal* {itlos.org/en/main/jurisdiction}.

105 ITLOS Statute 15(5).

106 As provided in article 15 LOSC Annex VI 'Statute of the International Tribunal for the Law of the Sea'; articles 28–31 of the 'Rules of the Tribunal' (adopted on 28/10/1997) and article 11 of the 'Resolution on the Internal Judicial Practice of the Tribunal' (adopted on 31/10/1997); and *see* Ndiaye, 'Proceedings before the International Tribunal for the Law of the Sea', *Journal of Law and Judicial System* (2018) 50*ff.*

107 ITLOS Statute 15(1).

108 ITLOS, *Interview with the President of the Chamber for Fisheries Disputes, Judge Tomas Heidar* (Newsletter 2019/1).

109 ITLOS Statute 15(3).

110 ITLOS Statute 15(2).

111 ITLOS, 'Guide to Proceedings before the International Tribunal for the Law of the Sea' (2021), 6*ff.*

112 *see* the statement by the ITLOS President Rüdiger Wolfrum 'On the Proceedings before the ITLOS' made at the Asia-Pacific Ambassadors Luncheon

first employed in the *Swordfish stocks* dispute. Chile had initially referred the matter unilaterally to Annex VII LOSC arbitration under article 287(3) LOSC, but following negotiations with the European Union, the case was jointly submitted to the Tribunal by special agreement.

3.4 THE FISHERIES LIMITATION TO COMPULSORY PROCEDURES ENTAILING BINDING DECISIONS

Compulsory procedures under the Convention that entail binding decisions may be inhibited by the limitation in article 297(3)(a) LOSC, which pertains to disputes concerning the conservation and management of marine living resources in EEZ. The provision is set out as follows:

> Disputes concerning the interpretation or application of the provisions of this Convention with regard to fisheries shall be settled in accordance with section 2, except that the coastal State shall not be obliged to accept the submission to such settlement of any dispute relating to its sovereign rights with respect to the living resources in the exclusive economic zone or their exercise, including its discretionary powers for determining the allowable catch, its harvesting capacity, the allocation of surpluses to other States and the terms and conditions established in its conservation and management laws and regulations.

This provision is introduced *en bloc* into the Agreement via article 32 FSA, which laconically stipulates that 'Article 297, paragraph 3, of the Convention applies also to this Agreement'. The immediate question, therefore, is whether a dispute concerning CMMs for straddling or highly migratory fish stocks should be dichotomized into EEZ and high seas components; in this case the fisheries limitation may thwart a settlement informed by the principle of ecosystem approach. Given that the entire dispute settlement system of the Convention emerged from sheer political compromises intended to secure consensus at UNCLOS/III,[113] the remit of these procedures was not clearly delineated.[114] In this regard, the wording of the limitation exemplifies a drafting technique that has deliberately introduced interpretative opacity, thereby

---

in Berlin (17/1/2008).

[113] As insightfully remarked by Mensah, 'The Role of Peaceful Dispute Settlement in Contemporary Ocean Policy and Law', in Vidas–Østreng (Eds) *Order for the Oceans at the Turn of the Century* (1999) 90.

[114] Well-reasoned critical perspectives on the dispute settlement system, while acknowledging its functional significance, are offered, among others, in Riphagen, 'Dispute Settlement in the 1982 United Nations Convention on the Law of the Sea', in Rozakis–Stephanou (Eds) *The New Law of the Sea* (1983) 281*ff* and Richardson, 'Dispute Settlement under the Convention on the Law of the Sea: A flexible and

necessitating a clause-by-clause analysis to elucidate its implications for compatibility disputes.

3.4.1 The Underlying Principle of Inclusive Jurisdiction

The opening clause states that 'Disputes concerning the interpretation or application of the provisions of this Convention with regard to fisheries shall be settled in accordance with section 2'. Based on a plain reading, this clause—expressed in confirmatory language—unequivocally enunciates, as a matter of principle, that fisheries disputes concerning the interpretation or application of the Convention's provisions are subject to the compulsory and binding settlement procedures under section 2 Part XV LOSC. The early literature following the adoption of the Convention contains copious commentary on this interpretation.[115] The phrase 'disputes concerning the interpretation or application of the provisions of [the] Convention' closely mirrors the language used in article 287(1) LOSC on the choice of procedure, where the grammatical preposition 'concerning' is interpreted broadly. For instance, it has been construed broadly in contexts involving the interpretation of declarations conferring jurisdiction with varying scopes.[116] Moreover, in its ordinary treaty usage, the word is expansively understood to encompass not only the directly opposed claims between the parties but also derivative questions and any related provisions.[117]

The drafting history of the provision during UNCLOS/III affirms that fisheries disputes are, in principle, subject to compulsory and binding procedures. The prospect of a general system for the compulsory adjudication of disputes was initially explored through unofficial negotiations, largely conducted within an informal working group jointly directed by Andronico Oduogo Adede (Kenya), Reynaldo Galindo Pohl (El Salvador) and Ralph Lindsay Harry (Australia), with Louis Bruno Sohn (United States) serving as the Rapporteur. The group worked intensively through informal meetings until 1976, after which the issue was opened for debate within each of the three

---

comprehensive extension of the rule of the law to ocean space', in Buergenthal (Ed.) *Contemporary Issues in International Law* (1984) 149.

115 *see* Singh, *United Nations Convention on the Law of the Sea* (1985) 137; Gamble, 'The 1982 UN Convention on the Law of the Sea: Binding dispute settlement?', *Boston University International Law Journal* (1991) 47; Casado-Raigón, *La Pesca en Alta Mar* (1994) 31–32; as well as after its entry into force by Vukas, *The Law of the Sea* (2004) 121.

116 *The "M/V Louisa" Case* (Judgment) 2013, §§80–83.

117 *South China Sea Arbitration* (Jurisdiction and Admissibility) 2015, §§174–177; and *see* Harrison, 'Defining Disputes and Characterizing Claims: Subject-matter jurisdiction in Law of the Sea Convention litigation', ODIL (2017) 270.

main Committees of the Conference, insofar as it pertained to their respective mandates. It was subsequently formalized as the Settlement of Disputes Group. By that time, significant formative discussions had already taken place during the Second Session (20/6 to 29/8/1974, Caracas), where the group produced an initial working paper containing a draft on dispute settlement procedures. The document gained prominence in the Plenary of UNCLOS/III as a co-sponsored proposal. One article within it—drafted in three alternative versions—provided for compulsory procedures leading to binding decisions in cases involving gross or persistent violations of the Convention, or alleged abuses of regulatory or enforcement jurisdiction by coastal States.[118] This core premise was retained in the document's revisions during the Third Session (17/3 to 9/5/1975, Geneva).[119]

On that basis, UNCLOS/III President Amerasinghe took the initiative to prepare an informal text comprising 18 articles and 7 annexes exclusively focused on dispute settlement, to complement the Single Negotiating Texts on substantive matters prepared concurrently by the Chairmen of the three main Committees. In spite of the considerable progress achieved during the Caracas and Geneva sessions, a plethora of substantive legal questions respecting the EEZ and CS regimes remained unresolved. In order to prevent the negotiations on those issues from stagnating, the President's text proposed a formulation that aligned with the general thrust of the negotiations, seemingly advocating for the exclusion of the entire zone from the remit of compulsory dispute settlement procedures. The proposal set out in article 18 ISNT stated that:

> Nothing contained in the present Convention shall require any Contracting Party to submit to the dispute settlement procedures (...) any dispute arising out of the exercise by a coastal State of its exclusive jurisdiction under the present Convention, except when, (...).[120]

Nonetheless, the use of negative wording was intended more to stimulate substantive discussion around a working formulation that reflected the delicate balance of interests between DWFS and coastal States, rather than to articulate

---

[118] A/Conf.62/L.7 'Working paper on the settlement of law of the sea disputes' (27/8/1974) [*UNCLOS/III Off.Records* III:85], article 11(a); for a detailed commentary *see* the insights offered by Sohn, 'Settlement of Disputes Arising out of the Law of the Sea Convention', SDLR (1975) 506*ff.*

[119] A/Conf.62/BackgroundPaper1 'Working paper on settlement of disputes' (1/5/1975) [(1975) 14ILM762].

[120] A/Conf.62/WP.9 'Informal Single Negotiating Text, Part IV (Settlement of Disputes)' (21/7/1975) [*UNCLOS/III Off.Records* V:137], article 18.

a definitive rule.[121] Viewed in its historical context, a degree of ambiguity was essential at that stage to sustain diplomatic momentum, particularly given that UNGA had initially anticipated the adoption of the Convention's final text by that time. Yet major contentious issues—especially concerning EEZ and CS disputes, including their delimitation—remained unresolved. By 1975, there was a palpable risk of UNCLOS/III being declared an 'official failure',[122] as the treaty text in ISNT, comprising the Chairmen's texts from the three main Committees, failed to reflect a consensus among delegations. This uncertain outlook was heightened by a series of remarks from the President, who oscillated between acknowledging the absence of agreement and affirming that the negotiations had not failed, as they were set to continue.[123]

Dispute settlement, as a general agenda item, was first openly debated during the plenary negotiations at the Fourth Session (15/3 to 7/5/1976, New York). In presenting the 'Informal Single Negotiating Text, Part IV', the President reaffirmed that effective dispute settlement had to serve as the assurance that 'substance and intention within the legislative language of the Convention [would] be interpreted both consistently and equitably'.[124] Yet the text had to accommodate limited exclusions in an effort to bridge the sharp divide over whether certain categories of EEZ disputes should be subject to binding dispute settlement procedures. As this was not merely a procedural or marginal issue, but one of considerable substance, any final formulation in the treaty text had to reflect the decisions reached within the relevant Committee and result from negotiation. In urging continued efforts on this point, the President added that:

> The solution may be to include third-party dispute settlement procedures for certain types of disputes whilst others are excluded, but this again is only one suggestion among the many possibilities. There is also the view that it is not an infringement of rights to ensure that the limits of those rights and the corresponding obligations in the context of the interpretation or application of the convention should be justifiable before an appropriate forum.[125]

---

121 Stevenson–Oxman, 'The Third United Nations Conference on the Law of the Sea: The 1975 Geneva session', AJIL (1975) 795*ff* and Johnson, 'A Review of Fisheries Proposals Made at the Caracas Session of LOS III', *Ocean Management* (1975) 298–299 & 302.

122 Johnston, 'The Options for LOS III: Appraisal and proposal', in Christy *et al.* (Eds) *Law of the Sea: Caracas and beyond* (1975) 357.

123 As recounted in Pickering–William, 'Recent Developments in the Law of the Sea: A synopsis', SDLR (1976) 632.

124 A/Conf.62/WP.9/Add.1 (1976), *loc. cit.* Chapter 1 (*fn*69).

125 *ibid.*, [*UNCLOS/III Off.Records* V:124(¶31)].

Finally, he exhorted coastal States not to misinterpret article 18 ISNT as casting doubt on their in the zone, clarifying that it was 'not the exclusive jurisdiction that [was] meant to be questioned', but rather 'the manner of its exercise'.[126] Several delegations voiced strong concerns over what they perceived as the propitiatory framing of the article; namely, the way its content was structured and rhetorically positioned by the President, which appeared to soften or downplay its potential challenge to coastal States' exclusive jurisdiction. Italy, in particular, categorically asserted that permitting exclusions would contravene both the principle and the very purpose of the compulsory dispute settlement system.[127]

For the remainder of the Fourth Session, the President was entrusted with revising the dispute settlement text to align with the updates made to RSNT.[128] This marked a pivotal moment in the negotiations over the legal status of EEZ, including fisheries provisions, as discussions reached their peak following broad consensus on the rights and obligations within the zone. In terms of its influence on State practice, RSNT became the most significant document on the law of the sea since the *Geneva Conventions*.[129] In the revised draft, the article outlining the limitations to the application of compulsory procedures continued to reflect the approach of balancing interests. It is noteworthy that the language shifted away from a negative formulation, which could be construed as undermining the principle of compulsory adjudication, and instead provided that:

> Disputes relating to the exercise by a coastal State of sovereign rights, exclusive rights or exclusive jurisdiction recognized by the present Convention shall be subject to the procedures specified in Section 2 only in the following cases: (...).[130]

The rephrased provision reflected the ongoing effort to strike a middle ground regarding expectations for the status of EEZ. Regarded as *sui generis* and distinct from the high seas regime, EEZ was seen by coastal States as an area wherein the principle of compulsory adjudication would represent

---

126 *ibid.*, [*UNCLOS/III Off.Records* V:124(¶32)].

127 UNCLOS/III Plenary 60th meeting (6/4/1976) [*UNCLOS/III Off.Records* V:24(¶32)].

128 A/Conf.62/WP.9/Rev.1 'Informal Single Negotiating Text, Part IV' (6/5/1976) [*UNCLOS/III Off.Records* V:185].

129 *see* Oxman, 'The Third United Nations Conference on the Law of the Sea: The 1976 New York sessions', AJIL (1977) 259 & 266–267 and Phillips, 'The Exclusive Economic Zone as a Concept in International Law', ICLQ (1977) 600–607.

130 As renumbered article 17(1) in A/Conf.62/WP.9/Rev.2 'Revised Single Negotiating Text, Part IV' (23/11/1976) [*UNCLOS/III Off.Records* VI:144].

an unwelcome constraint on their perceived sovereignty. Conversely, DWFS viewed EEZ as part of the high seas, albeit subject to certain coastal State sovereign rights, and thus advocated for robust dispute settlement language to safeguard their fishing interests. This divergence in perspectives shaped extensive debates on the scope and nature of legal rights in the zone. Nevertheless, efforts to bridge the gap often resulted in provisions so broadly worded as to be ineffectual. The level of specificity in the substantive fisheries articles became closely linked to how ambiguities might be resolved in the future. Consequently, the less confidence negotiators had in the proposed dispute settlement mechanism, the more insistent they were on precise language in the substantive provisions; and *vice versa*.[131] Against this background, the provision on limitations reverted to a negative formulation to affirm the exclusive fishing rights of coastal States in the zone. This revision appeared in the Informal Composite Negotiating Text ('ICNT'), which consolidated Parts I–IV RSNT. The corresponding provision therein stated that:

> No dispute relating to the interpretation or application of the provisions of the present Convention with respect to the living resources of the sea shall be brought before such court or tribunal unless (...).[132]

The revised text signified a clear regression in favour of coastal States compared with the broader scope envisaged in earlier drafts addressing fisheries disputes within and beyond the zone. The rewording was tentatively accepted as delineating an 'exception to the exclusion', while the principal intent of the provision remained the preservation of the general applicability of compulsory binding procedures.[133] This understanding was reflected in contemporaneous commentary by several UNCLOS/III delegates, who, in analysing the adoption of the compulsoriness principle under RSNT, underscored the need to avoid vague or ambiguous formulations.[134] Similar views were later echoed

---

131 *see* Christy, 'Transitions in the Management and Distribution of International Fisheries', *International Organization* (1977) 253–254 and Clingan, 'The Changing Global Pattern of Fisheries Management', *Lawyer of the Americas* (1978) 681.

132 As renumbered article 296(4) in A/Conf.62/WP.10 (1977), *cit.* Chapter 2 (*fn*127).

133 Brown, 'Dispute Settlement and the Law of the Sea: The UN convention regime', *Marine Policy* (1997) 22.

134 *see* Pardo, 'The Emerging Law of the Sea', in Walsh (Ed.) *The Law of the Sea, Issues in Ocean Resource Management* (1977) 59; Kimball *et al.*, 'Balancing the Interests?', *Environmental Policy and Law* (1978) 73; and Adede, 'Law of the Sea—The Integration of the System of Settlement of Disputes under the Draft Convention as a Whole', AJIL (1978) 94–95.

in the literature, which emphatically described the resulting structure of the article as comprising 'exceptions and exceptions to exceptions'.[135]

During the Seventh Session (28/3 to 19/5/1978, Geneva), several core issues remained unresolved, preventing the treaty text from being considered in its entirety under the package deal principle.[136] To address these, the Conference established seven specialized negotiating groups. Negotiating Group 5 ('NG5'), chaired by Constantinos Stavropoulos (Greece), was tasked with reviewing the provisions of ICNT relating to compulsory dispute settlement procedures, with particular attention to disputes concerning the exercise of coastal States' sovereign rights in their EEZ. In the resumed session (21/8 to 15/9/1978, New York), NG5 focused on identifying the types of disputes that could be referred to third-party procedures and the binding nature of such procedures. Although a final agreement was not reached, informal negotiations had already yielded consensus in principle on the settlement of fisheries disputes in EEZ. Ultimately, NG5 achieved a compromise, which Chairman Stavropoulos described as enjoying broad and substantial support, amounting to a conditional consensus. Under the proposed formula, disputes concerning the coastal State's sovereign rights or their exercise and discretionary powers in the zone were excluded from binding settlement. Nonetheless, such disputes would be subject to compulsory conciliation where a coastal State failed to fulfil its conservation and management obligations, including arbitrarily refusing to establish TACs, assess its own harvesting capacity, or allocate any resulting surplus. Additionally, the group formulated a general clause to prohibit abuse of rights.[137] The plenary negotiations further revised ICNT to incorporate the points of consensus embodied in the proposed compromise. Nevertheless, the broad phrasing of the initial formulation was significantly narrowed through the introduction of an explicit positive statement, which reaffirmed *ab initio* the applicability of compulsory and binding procedures to fisheries disputes by providing that:

> Unless otherwise agreed or decided by the parties concerned, disputes relating to the interpretation or application of the provisions of this Convention with

---

[135] Treves, 'The Jurisdiction of the International Tribunal for the Law of the Sea', in Rao–Khan (Eds) *The International Tribunal for the Law of the Sea* (2001) 120.

[136] Yankov, 'The Law of the Sea Conference at the Crossroads', VJIL (1977) 38–40.

[137] *see* the discussion at UNCLOS/III Plenary 105th meeting (2/4/1980) [*UNCLOS/III Off.Records* XIII:82] on A/Conf.62/RCNG/1 'Results of the work of the negotiating group on item (5) of A.CONF.62/62' (19/5/1978) [*UNCLOS III Off.Records* X:117].

> regard to fisheries shall be settled in accordance with Section 2 of Part XV of this Convention, except (...).[138]

Through this revision, UNCLOS/III reaffirmed the principle of compulsory adjudication and redirected the remaining negotiations towards refining the language of the subject-matter exceptions. The provision containing the fisheries limitation underwent its final significant redrafting in the context of the Informal Draft Convention on the Law of the Sea.[139] This draft was issued by the collegium of the Conference, which comprised the President of the Conference, the Chairmen of the three main Committees, the Rapporteur-General, and the Chairman of the Drafting Committee. At this stage, the provision was further narrowed by refining the introductory statement on the applicability of compulsory procedures entailing binding decisions, producing the current wording of the first clause of article 297(3)(a) LOSC, which clarifies that:

> Disputes concerning the interpretation or application of the provisions of this Convention with regard to fisheries shall be settled in accordance with section 2.

As observed by Bernard Oxman (United States), who served as Chairman of the English Language Group of the Drafting Committee, the provision is structured to affirm compulsory jurisdiction over all fisheries disputes, while carving out an exclusion only in relation to the sovereign rights of coastal States concerning the living resources in EEZ.[140] The opening clause thus reflects the Convention's conclusive position that fisheries disputes, in principle, fall within the remit of the compulsory procedures entailing binding decisions set out in section 2 Part XV LOSC.

3.4.2 Qualifying the Scope of Fisheries Disputes

The second and third clauses comprise the remainder of the provision, wherein the operative body of the subject-matter limitation is laid. The second clause, introduced with the word 'except', imposes a limitative effect on the remit of compulsory procedures entailing binding decisions by providing that 'coastal States shall not be obliged to accept the submission to such settlement of any

---

138 A/Conf.62/WP.10/Rev.1 'Revised Informal Composite Negotiating Text' (28/4/1979) [(1979) 18ILM686].

139 A/Conf.62/WP.10/Rev.3 'Draft Convention on the Law of the Sea (Informal Text)' (22/9/1980) [(1980) 19ILM1129].

140 Oxman, 'The Third United Nations Conference on the Law of the Sea: The tenth Session (1981)', AJIL (1982) 19.

dispute relating to its sovereign rights with respect to the living resources in the exclusive economic zone or their exercise'. The stipulation 'shall not be obliged to accept', framed in passive language, implies that the coastal State must expressly invoke this clause to challenge the jurisdiction of the forum, or the admissibility of the case, and suspend proceedings on the merits. In any event, the provision does not operate as an *ipso facto* objection.[141] Turning to its substantive content, the view favoured by coastal States holds that the exclusions are so broadly drafted as to counterpoise the inclusive jurisdiction conferred in the first clause over fisheries disputes.[142] However, such a reading proves unpersuasive, as both the wording and the underlying intent of the provision necessitate a proper characterization of the dispute.[143] Whether a particular dispute falls outside the scope of the limitation or is excluded from the compulsory and binding procedures of section 2 Part XV LOSC is a determination to be made by the forum itself.[144]

While there is no doubt that the second clause denotes the sovereign nature of the coastal State's economic rights in its exclusive zone, as provided in articles 56, 61 and 62 LOSC, the spatial scope of these rights in the context of fisheries disputes requires careful qualification. The syntactical construction of these clauses repays study, particularly in how the third clause—introduced by the conjunctive term 'including'—functions gerundively as a coordinating continuation of the second clause. In doing so, it delineates two categories of disputes concerning the exercise of sovereign rights by the coastal State in its EEZ: those involving access to fisheries, and those concerning the applicable terms and conditions set out in the coastal State's conservation and management laws and regulations. If the second clause were interpreted broadly so as to exclude all fisheries disputes, it would amount to a sweeping generalization,

---

[141] *The "M/V Saiga" Case (No.2)* (Judgment) 1999, §§44–45; *qv.* Eiriksson, *The International Tribunal for the Law of the Sea* (2001) 139 as it pertains to the fisheries-related elements of the dispute. Followed, among other cases, in *South China Sea Arbitration* (Jurisdiction and Admissibility) 2015, §358.

[142] *see* Orrego-Vicuña, *The Exclusive Economic Zone* (1989) 129; this was a proposition originally made by Shabtai Rosenne as a comment on the compromise formula of NG5 in 'Settlement of Fisheries Disputes in the Exclusive Economic Zone', AJIL (1979) 97–99 and rehearsed as an assumption by Dahmani, *The Fisheries Regime of the Exclusive Economic Zone* (1987) 121–122 and Treves, 'Fisheries Disputes: Judicial and arbitral practice since the entry into force of UNCLOS', in Wolfrum–Seršić–Šošić (Eds) *Contemporary Developments in International Law* (2016) 331.

[143] Rao–Gautier, *The International Tribunal for the Law of the Sea* (2018) 95–96.

[144] Tanaka, *The Peaceful Settlement of International Disputes* (2018) 239.

thereby rendering the third clause redundant. This is because, as defined in article 55 LOSC, the zone extends the national jurisdiction of the coastal State to sea areas beyond and adjacent to its territorial sea, within which article 56(1)(a) LOSC explicitly affirms its sovereign rights for the purposes of exploring, exploiting, conserving, and managing marine living resources. Furthermore, the third clause reiterates obligations already detailed elsewhere in the Convention, most notably in articles 61 and 62 LOSC, which address, respectively, the conservation and optimum utilization of the marine living resources in the zone. Therefore, the second clause should be interpreted narrowly, limiting its exclusionary scope to those sovereign rights whose exercise by the coastal State occurs strictly 'in' the zone, with the grammatical preposition itself warranting a restrictive construction tied to the spatial scope of the provision. A plain reading of these two clauses together would readily lead one to describe the excluded disputes as those concerning fishing in the economic zone,[145] a sea area where fishing is, unequivocally, a sovereign right reserved for the coastal State. The proposed construction gains further support from article 297(3)(b) LOSC, which specifies the grounds upon which the two categories of disputes mentioned *eo nomine* in the third clause of article 297(3)(a) LOSC are to be submitted to compulsory conciliation. These cover allegations that the coastal State has either manifestly failed to fulfil its obligation to ensure that the resources within its EEZ are not seriously endangered,[146] or has arbitrarily declined to determine TAC, assess its harvesting capacity, and allocate any declared surplus.[147] These fisheries disputes are deemed nonjusticiable and—absent the coastal State's consent—constitute the 'automatic' exclusions from the compulsory and binding procedures under section 2 Part XV LOSC, in accordance with the exhaustive terms of article 297 LOSC.

Given the interpretative opacity that continues to surround the fisheries limitation, compatibility disputes may give rise to a continuum of additional critical questions. One such question is whether the limitation applies solely to so-called 'domestic' stocks, i.e. those situated entirely within EEZ. Under the general consensus reached during UNCLOS/III, the compulsory and binding procedures under section 2 Part XV LOSC were understood to apply to all fisheries disputes, with exclusions limited to those relating to the coastal State's sovereign rights, or their exercise, with respect to the living resources

145 *e.g.* Ball, 'Law of the Sea: Expression of solidarity', SDLR (1982) 471.

146 LOSC 297(3)(b) *lit.*(i); *qv.* Lavalle, 'Conciliation under the United Nations Convention on the Law of the Sea: A critical overview', *Austrian Review of International and European Law* (1997) 36–38.

147 LOSC 297(3)(b) *litt.*(ii)–(iii); *qv.* Sohn, 'Settlement of Law of the Sea Disputes', IJMCL (1995) 210.

'in' the zone.[148] On the basis of this spatial construction, disputes concerning straddling and highly migratory fish stocks cannot, in principle, be excluded, as the coastal State's sovereign rights do not extend into high seas areas adjacent to or beyond the zone.[149]

By way of example, it cannot be logically defended that the limitation applies in blanket fashion to disputes concerning stocks of highly migratory species, since article 64 LOSC places the States concerned under a *de contrahendo* obligation to cooperate 'with a view to ensuring the conservation and promoting the objective of optimum utilization of such species throughout the region, both within and beyond the [EEZ]'. The essence of this argument was postulated in the *Southern Bluefin Tuna Cases*, where the tribunal, in interpreting article 281 LOSC, acknowledged that there is 'a certain balance in the rights and obligations of DWFS and coastal States in respect of settlement of disputes arising from events occurring within the respective EEZ and on the high seas', which it has been deliberately established in the Convention.[150] Even commentary favouring coastal State interests concedes that, if the scope of the limitation is construed so broadly as to exclude the EEZ components of such fisheries disputes, it will, by extension, also remove—at least in part—the corresponding high seas components from the remit of the compulsory and binding procedures under section 2 Part XV LOSC. A dispute of this kind 'is not simply about fishing for resources that are subject to the sovereign rights of the coastal State', and therefore, when the coastal State invokes these procedures, it will probably be required to submit its own in-zone national measures for review as well.[151]

At this point, it is instructive to revisit the *Swordfish stocks* dispute by examining the introductory legal arguments put forward by the parties, as these typify the nature of the substantive disagreements that may give rise to a compatibility dispute. In its role as coastal State, Chile requested the Chamber to determine whether its sovereign rights and obligations to prescribe CMMs within its national jurisdiction, and to ensure their implementation through port State measures, could be challenged on the basis of the Convention. Chile

---

[148] 1995 *Virginia Commentary* III ¶VII.8(34–36).

[149] *see* Brown, *The International Law of the Sea* (1994) 228 and de Yturriaga, *The International Regime of Fisheries* (1997) 170–171.

[150] *Southern Bluefin Tuna Cases* (Jurisdiction) 2000, §62; *qv.* Romano, 'The Southern Bluefin Tuna Dispute: Hints of a world to come … like it or not', ODIL (2001) 332.

[151] *see* the analysis by William Burke in 'Highly Migratory Species in the New Law of the Sea', ODIL (1984) 289 and 'The Law of the Sea Convention Provisions on Conditions of Access to Fisheries Subject to National Jurisdiction', OLR (1984) 118–119.

further questioned whether the European Union had complied with its duty under the Convention to ensure conservation of the species in the high seas areas adjacent to the Chilean EEZ and, if not, whether such failure amounted to an abuse of flag State rights. Conversely, the European Union requested the Chamber to find that Chile's unilateral CMMs on the high seas constituted a breach of, inter alia, articles 87, 89, and 116–119 LOSC. These questions critically engaged the duty of cooperation under article 64 LOSC. In particular, Chile further questioned whether the European Union had fulfilled the obligations to cooperate directly in the conservation of swordfish stocks in the high seas areas adjacent to the Chilean EEZ, specifically by reporting its catches and other relevant information. The European Union, seemingly aiming to argue that Chile's unilateral adoption of CMMs had already breached the *pactum de contrahendo* by being undertaken *mala fide*, further questioned whether such an exercise of jurisdiction conformed with article 300 LOSC and, if not, whether the duty to negotiate and agree on the implicated measures under article 64 LOSC remained extant.[152]

Notwithstanding that the proceedings were discontinued at the joint request of the parties, two key legal conclusions can be extrapolated from the case. First, although both parties invited the Chamber to adjudicate the issues under the Convention—with reference to the extent that they were subject to the compulsory procedures entailing binding decisions under Part XV LOSC—it is notable that the fisheries limitation was not invoked. On the contrary, Chile cited the positive obligation to submit a dispute concerning the exercise of its sovereign rights or jurisdiction as provided in article 297(1)(b) LOSC.[153] Second, and relatedly, in the interest of the long-term conservation and management of swordfish stocks, the parties ultimately resolved their disagreement in a manner guided by the principles of ecosystem and precautionary approach. To this end, the agreed settlement terms included, among other measures, the conduct of fishing at catch levels that maintain stocks at or near MSY, consistent with the objective of ensuring the sustainability of these resources and the protection of the marine ecosystem, as well as a firm commitment to the effective promotion of cooperation for the conservation and management of these stocks throughout their migratory range.[154]

Returning to the question of spatial application, if the limitation were not confined solely to domestic stocks, could any EEZ component of a dispute involving straddling stocks, or associated and dependent species, then fall

152 *Conservation and Sustainable Exploitation of Swordfish Stocks* (Constitution of Chamber) 2000, 2§3(a)–(h).

153 *ibid.*, 2§3(d).

154 *Understanding Concerning the Conservation of Swordfish Stocks in the South-Eastern Pacific Ocean* (2009), ¶¶1, 5(a), 8–9 & 13.

within the remit of the compulsory and binding procedures of section 2 Part XV LOSC, consistent with the principle of inclusive jurisdiction as set out in the first clause of article 297(3)(a) LOSC? This is a substantive issue which commentary generally favourable to coastal State interpretations nonetheless acknowledges, noting that a ruling may be required on the scope of sovereign rights as a preliminary matter in disputes concerning their exercise,[155] not least because it supports an equality of arms argument, reinforcing the procedural balance between the parties in a case.[156] That question could arise in a dispute considering the fulfilment by the coastal State of its obligation to cooperate with RFMO/As in establishing in-zone CMMs, as required under article 61(2) LOSC, consistent with the evolving ecosystem approach and the emerging demand in general international law for holistic ocean management.[157] The Convention has expressly imposed this obligation on coastal States as a conceptual constraint on traditionally perceived sovereign rights over marine living resources. These rights must be reconciled with the individual non-economic interests of other States in the zone and the broader collective interest in sustainable regulation.

It must be further noted that the coastal State is subject to the obligation of due regard when exercising its sovereign rights and performing its duties within the zone, requiring it to act in a manner compatible with the Convention.[158] The obligation serves as an indispensable condition for exercising functional jurisdiction in a zone that is exclusive solely in economic terms, as it qualifies these rights by ensuring they are not exercised without restraint and that the coastal State remains accountable.[159] To assert that this obligation excludes fishing rights is inconsistent not only with the Convention's original intent, but also with the subsequent implementation of its provisions through State practice.[160]

---

155 *see* Infante, 'The Settlement of Disputes regarding the Law of the Sea and its Bearing on the Legal Nature of the Exclusive Economic Zone', in Orrego-Vicuña (Ed.) *The Exclusive Economic Zone, A Latin American Perspective* (1984) 165–166.

156 Carstensen, 'A Re-Internationalisation of Dispute Settlement in Law of the Sea', ZaöRV (2002) 74.

157 Kirk, 'The Ecosystem Approach and the Search for an Objective and Content for the Concept of Holistic Ocean Governance', ODIL (2015) 33.

158 LOSC 56(2).

159 *see* Wolfrum, 'Law of the Sea: An example of the progressive development of international law', in Tomuschat (Ed.) *The United Nations at Age Fifty* (1995) 320–321; Gavouneli, *Functional Jurisdiction in the Law of the Sea* (2007) 102–103; and Holst, *Change in the Law of the Sea* (2022) 55*ff.*

160 *compare* Pierce, 'Dispute Settlement Mechanisms in the Draft Convention on the Law of the Sea', *Denver Journal of International Law and Policy* (1981) 337*ff with* Noyes, 'Compulsory Third-Party Adjudication and the 1982 United

The Convention clearly states that the objective of optimum utilization in the zone is to be pursued 'without prejudice to the aim of conservation'.[161] An interpretation that narrowly qualifies the scope of sovereign rights in EEZ was foreshadowed in the arbitration of *Affaire concernant le Filetage à l'Intérieur du Golfe du Saint-Laurent* between Canada and France, which was the first case to address this issue following the conclusion of the Convention. The tribunal observed that while articles 56, 61, and 62 LOSC bestow sovereign rights upon the coastal State to exploit and manage fisheries in the zone, these rights need to be consistently exercised with the aim of conservation. In this regard, the management powers are presented primarily as an administrative function which—although best exercised by the coastal State—nonetheless remains a matter of general interest to all States and therefore calls for a balanced approach. Notably, as remarked by the tribunal regarding the coastal State's motivations in the instant case:

> [...] ce souci majeur de protéger les ressources biologiques par une gestion équilibrée de celles-ci a également été celui qui a animé l'Etat Canadien dans chacune des étapes qu'il a parcourues dans l'extension progressive de ses limites de pêche.[162]

This fundamental understanding of the need to conserve and manage fish stocks had been broadly shared among States during UNCLOS/III. Based on the fishery provisions in RSNT, many coastal States began unilaterally enacting legislation that reflected the principles embodied in the negotiating drafts. In that context, the extension of coastal fisheries jurisdiction to 200 nm was undertaken with the recognition that effective CMMs were imperative to prevent overexploitation of stocks in EEZ.[163] Although the award was met with mixed reactions at the time—particularly respecting its treatment of the role of management measures under the Convention[164]—it remains especially pertinent in contemporary terms, given the systemic integration of environmental norms into the aims and objectives to ensure sustainable fisheries.

Nations Convention on the Law of the Sea', *Connecticut Journal of International Law* (1989) 693–695 and Scovazzi, '*Due Regard* Obligations, with Particular Emphasis on Fisheries in the Exclusive Economic Zone', IJMCL (2019) 68–70.

161 LOSC 62(1).

162 *Affaire concernant le Filetage à l'Intérieur du Golfe du Saint-Laurent* (Decision) 1986, §50.

163 Moore, 'National Legislation for the Management of Fisheries under Extended Coastal State Jurisdiction', JMLC (1979) 159–160.

164 *e.g.* Burke, 'Coastal State Fishery Regulation under International Law: A comment on the *La Bretagne* Award of July 17, 1986 (The Arbitration between Canada and France)', SDLR (1988) 518.

The Convention, through articles 192 and 193 LOSC, positively imposes on all States the obligation 'to protect and preserve the marine environment' when exercising their sovereign rights to exploit natural resources within national jurisdiction. It is worth noting that the terms 'sovereign rights' and 'jurisdiction', as referred to respectively in subparagraphs (a) and (b) of article 56(1) LOSC, do not share the same legal scope, and their interrelationship may vary within the context of the EEZ regime.[165] This obligation has been repeatedly emphasized by ITLOS as applicable to all maritime areas, including EEZ, pointedly recalling that the conservation of fishery resources forms an integral part of marine environmental protection.[166] Similarly, scholarly views are increasingly recurring on the need for holistic ocean governance to support the sustainable regulation of fisheries both within and beyond areas of national jurisdiction.[167] Obligations derived from Part XII LOSC can serve to supplement and contextualize the obligations of coastal States under Part V LOSC with respect to fisheries,[168] and may ultimately substantiate interpretations ordained by the normative evolution of sustainability and the progressive development of conservation and management law.[169] Some reasonably argue that these should be recognized as *erga omnes* obligations—or at the very least, *inter partes*[170]—and, in the same vein, others advocate that the question of whether a coastal State complies with its obligation to ensure the sustainability of fisheries could fall under the parameters for invoking the compulsory

---

165 Vrancken, *State Ocean Jurisdiction* (2023) 12–17.

166 *see* in *Southern Bluefin Tuna Cases* (Provisional Measures) 1999, §70; *SRFC Request* (Advisory Opinion) 2015, §120; and *Climate Change and International Law* (Advisory Opinion) 2024, §§169 & 409.

167 *see* Paik, 'ITLOS at Twenty: Reflections on its contribution to dispute settlement and the rule of law at sea', 205–208 in Nordquist–Moore–Long (Eds) *Legal Order in the World's Oceans* (2017) and Lanovoy, 'Dispute Settlement and Ocean Governance', in Borg *et al.* (Eds) *Research Handbook on Ocean Governance Law* (2023) 81–84.

168 Bankes, 'Legislative and Enforcement Jurisdiction of the Coastal State with respect to Fisheries in the Exclusive Economic Zone', in Øystein (Ed.) *The Development of the Law of the Sea Convention* (2021) 91.

169 Tanaka, 'Toward Sustainable Management of Marine Natural Resources', in Kotzur *et al.* (Eds) *Sustainable Ocean Resource Governance* (2018) 110.

170 *see* the views held by Rüdiger Wolfrum in 'Realization of Sustainable Management/Development under the Law of the Sea Convention?', in Kotzur *et al.* (Eds) *Sustainable Ocean Resource Governance* (2018) 104–105 and 'Implementation and Enforcement of Community Interest-Related Treaties by Judicial Means: Procedural limitations, chances and prospects', IJMCL (2023) 282.

dispute settlement procedures of the Convention.[171] Disputes over cooperation in this respect may arise when straddling stocks, or associated and dependent species, are put at risk by the coastal State owing to factors such as unregulated aquaculture or fish farming, persistent IUU fishing, habitat degradation, or other circumstances in which the coastal State fails to prevent serious threats to the reproductive viability of associated or dependent species.

From the foregoing, it follows that fisheries disputes concerning CMMs for straddling and highly migratory fish stocks should be distinguished from those automatically excluded under the second and third clauses. Whereas the latter genus of disputes are to be settled by means of compulsory conciliation, disputes in the former category must be referred for impartial characterization to a court or tribunal having jurisdiction under the compulsory procedures set forth in section 2 Part XV LOSC.[172] This proposed interclausal construction aligns with the underlying principle of Part XV LOSC to preserve the balanced approach established by the package deal and, in this specific context, to prevent EEZ from being treated as either territorial sea or high seas. It is also consistent with the general principle prohibiting the abuse of rights, jurisdiction, and freedoms as recognized in the Convention.[173] Accordingly, in such situations, it is envisaged that the forum having jurisdiction under article 287 LOSC—and to which an application is submitted concerning a dispute as referred to in article 297 LOSC—shall determine, at the request of a party or *ex proprio motu*, 'whether the claim constitutes an abuse of legal process or whether *prima facie* it is well founded'.[174] In this regard the award in the *Southern Bluefin Tuna Cases* stressed that the proceedings were constructive and did not amount to an abuse of process.[175] This view has been strongly supported in subsequent academic commentary.[176]

---

171 Mossop, 'Reimagining the Procedural Aspects of Part XV of the United Nations Convention on the Law of the Sea', IJMCL (2023) 388–391.

172 *see* also the comments on article 296 ICNT in A/Conf.62/WP.10 (1977), *cit.* Chapter 2 (*fn*127), containing at that time the provisions on the fisheries limitation, in Bernhardt, 'Compulsory Dispute Settlement in the Law of the Sea Negotiations: A reassessment', VJIL (1978) 88–90.

173 LOSC 300.

174 LOSC 294(1).

175 *Southern Bluefin Tuna Cases* (Jurisdiction) 2000, §65.

176 *e.g.* Treves, 'Preliminary Proceedings in the Settlement of Disputes under the United Nations Law of the Sea Convention: Some observations', in Ando–McWhinney–Wolfrum (Eds) *Liber Amicorum Judge Shigeru Oda* (2002) 749.

3.4.3 The Pronouncements in the Chagos Marine Protected Area Arbitration

Important questions concerning the interpretation and application of the fisheries provisions in the Convention were posed in the dispute over the proclamation by the United Kingdom of a MPA in the whole of the 200 nm zone around the Chagos Islands. Mauritius sought to challenge its legality, claiming it to be in breach of the United Kingdom's substantive and procedural obligations under the Convention, including articles 2, 55, 56, 63, 64, 194 and 300 LOSC, as well as article 7 FSA. In response, the United Kingdom raised jurisdictional objections on the grounds that the designation of MPA was a CMM, against which the Convention excluded the application of compulsory procedures entailing binding decisions under section 2 Part XV LOSC. The tribunal, in determining its jurisdiction *ratione materiae*, rightly observed that the central issue dividing the parties was the interpretation and application of article 297 LOSC, which in turn, necessitated the proper characterization of the dispute within that context.[177]

By a majority of three to two, the tribunal agreed with the United Kingdom that the proclaimed MPA served, among other purposes, the objective of fishery regulation by consolidating under one framework both environmental and resource conservation goals, which had previously been addressed separately through the overlapping Environmental Protection and Preservation Zone and the Fisheries Conservation and Management Zone. Neither of the zones extended beyond the 200 nm limit from the baselines of the Chagos Archipelago, having been established under the sovereign rights conferred by article 56(1)(a) LOSC over natural resources in EEZ, as further defined by articles 61 and 62 LOSC concerning conservation and utilization of living marine resources.[178] With respect to the regulation of fisheries within the area, the United Kingdom had enacted a ban on commercial fishing by vessels of other States, including those of Mauritius, which had previously been licensed for in-zone fishing.[179] As part of the respective arguments, the tribunal was required to examine the fisheries limitation under article 297(3)(a) LOSC, upon which its line of reasoning culminated in the passage contained in operative paragraph 301 of the award, holding that:

---

177 *Chagos MPA* (Award) 2015, §283.

178 PCA, 'Hearing on jurisdiction and the merits, in the matter of arbitration between the Republic of Mauritius and the United Kingdom of Great Britain and Northern Ireland' (22/4–9/5/2014) ('*Chagos MPA Arbitration Final Transcript*'), 792:9–12.

179 *ibid.*, 1274:22–3 and 1278:7–11.

> The Tribunal is aware of the view, advanced in certain academic settings, that article 297(3) [LOSC] should be construed narrowly in its application to article 63 and article 64 [LOSC] and to the 1995 [FSA] on the grounds that the entire purpose of the special regime for these species is to enable populations to be managed as a unified whole, and that this object and purpose is potentially frustrated by providing distinct dispute resolution regimes for such species in the exclusive economic zone and in the high seas. However desirable this purpose may be as a matter of policy, the Tribunal can see no textual basis for such a construction in either the Convention or the 1995 [FSA]. The latter agreement afforded ample opportunity to remedy any ambiguity of drafting in the earlier Convention, but nevertheless expressly provides that 'Article 297, paragraph 3, of the Convention applies also to this Agreement'.

The tribunal interpreted article 297(3)(a) LOSC by broadening the scope of exclusions in the second clause, which led to its declaration of being without jurisdiction to address articles 63 and 64 LOSC, as well as article 7 FSA. Conversely, on this precise point, Judges James Kateka and Rüdiger Wolfrum reflecting *quod arguendo* on the application of the fisheries limitation adopted a differentiated stance by observing that the second and third clauses of article 297(3)(a) LOSC must be construed more narrowly than the first clause.[180] This position has been supported by academic scholars, who likewise argue that it is difficult to accept that the general rules of interpretation permit a conclusion that the fisheries limitation excludes all EEZ fisheries-related matters under the Convention,[181] and even more so in the context of the Agreement, which treats a straddling or highly migratory stock as a single biological unit.[182]

While the *Chagos MPA* arbitration has prompted scholarly discussion about its likely bearing on EEZ-related fisheries disputes under the Convention, little attention has been paid to assessing the tribunal's statement through the lens of the Agreement. Despite this noticeable gap, academic commentary seems to generally acknowledge that the tribunal has, to varying degrees, clarified the exclusionary scope of article 297(3) LOSC.[183] Be that as it may, it is this

180 *Chagos MPA* (Award) 2015, *Dissenting and Concurring Opinion by Judge Kateka and Judge Wolfrum*, §58.

181 Kunoy, 'The Scope of Compulsory Jurisdiction and Exceptions thereto under the United Nations Convention on the Law of the Sea', *Canad. YbIL* (2021) 110–111.

182 Churchill–Vaughan–Sander, *The Law of the Sea* (2022) 570.

183 *see* Klein, 'Expansions and Restrictions in the UNCLOS Dispute Settlement Regime: Lessons from recent decisions', CJIL (2016) 410–411; Nguyen, 'The Chagos Marine Protected Area Arbitration: Has the scope of LOSC compulsory jurisdiction been clarified?', IJMCL (2016) 135–140; Talmon, 'The Chagos Marine Protected Area Arbitration: Expansion of the jurisdiction of UNCLOS Part XV courts and tribunals', ICLQ (2016) 945–946; Cho, 'Is Compulsory Jurisdiction

author's opinion, as already argued elsewhere in more detail, that the award on this particular matter has bequeathed to case law a rather doubtful authority in terms of its subsequent application to disputes relating to the conservation and management of fish stocks under the Agreement.[184] The dispute in the instant case did not arise from a disagreement over the interpretation or application of article 7 FSA. Mauritius sought to correlate articles 56 and 297 LOSC with a view to drawing a distinction between the effect of the provisions on the coastal sovereign rights and the rights of third States in EEZ.[185] As this was put in its written submissions, and accentuated further in the oral pleadings, Mauritius, in averring United Kingdom's 'failure to respect Mauritian fishing rights in the EEZ', contended that 'the dispute [was] not based on the purported sovereign rights of the [United Kingdom] as a coastal State in relation to the living resources in the EEZ' but, to the contrary, it concerned the fishing 'rights of Mauritius in the EEZ of the Chagos Archipelago'.[186] Hence, Mauritius' fishery claim was not only irrelevant to article 7 FSA but, in fact, not even germane to articles 63(2) and 64(1) LOSC. None of these articles confer sovereignty or establish a legal basis for accessing fisheries within another State's EEZ. Article 7 FSA was entangled in the arguments by Mauritius having reserved its position as a 'State fishing for stocks' in the high seas areas adjacent to the Chagos Archipelago's EEZ, and from that position seeking to assert procedural obligations of consultation and cooperation owed to it.[187] Citing selectively article 7(3) FSA on the obligation of States to make 'every effort to agree on compatible conservation and management measures within a reasonable period of time',[188] Mauritius overlooked that this obligation is specifically framed in the context of disputes arising from disagreements over compatible CMMs, as contemplated in article 7(2) FSA. A genuine disagreement would involve the United Kingdom, acting as the coastal State in this case, bringing an action against Mauritius on the grounds that fishing activities on the high seas adjacent to and beyond the MPA undermined the moratorium imposed in the area, which served as a conservation measure aiming at the recovery of commercially exploited pelagic stocks. The fact that these stocks

---

Under UNCLOS Part XV Expanding? A Critical Review of the Claims', ODIL (2024) 185; and Sun, *Finding a Balance in the Exclusive Economic Zone* (2025) 61–62.

184 Ntovas, 'Interpreting the Dispute Settlement Limitation on Fisheries after the Chagos Marine Protected Area Arbitration', in Minas–Diamond (Eds) *Stress Testing the Law of the Sea* (2018) 225.

185 *Chagos MPA Arbitration Final Transcript*, 1119–1122.

186 *ibid.*, 322:3 & 477:19–25.

187 *ibid.*, 335:6–20, 321:16–20, 335:17–20 & 336.

188 *ibid.*, 335:7–9.

were under heavy pressure from poorly regulated fishing in the Indian Ocean had been acknowledged by both parties in the proceedings. In that scenario, Mauritius would have contested the compatibility of the measure—credibly on grounds of its reasonableness or proportionality within the MPA—alongside procedural obligations concerning consultation, as governed by the applicable regimes in articles 63(2) and 64 LOSC. Yet, this was not the manner in which the case was presented to the tribunal. At the time the proceedings were initiated, no material dispute had arisen between the parties concerning compatible CMMs.

Turning to the award itself, the dictum contained in the passage quoted earlier is derived from a reductive reading of the fisheries provisions in question. The tribunal's reasoning on this point was developed in operative paragraph 300 of the award by invoking the *Barbados v. Trinidad and Tobago Arbitration*, which was also debated between the parties regarding its bearing on the procedural obligations under consideration. That case marked the first jurisprudential application of article 297(3) LOSC. The claim relevant to fishing had been made by Barbados in the form of an *infra petita* remedy request during the oral pleadings for the right to access *Exocoetidae* (Flyingfish) stocks within the EEZ of Trinidad and Tobago.[189] In that case, the tribunal found itself lacking jurisdiction to pronounce on the fisheries access regime in the zone, as the claim pertained to coastal State rights and duties, which were excluded under article 297(3)(a) LOSC. Trinidad and Tobago had made it clear that it neither consented to nor considered itself obliged to submit these matters to the jurisdiction of the tribunal.[190] It is difficult to understand why the tribunal in the *Chagos MPA* arbitration allowed itself to be carried along by the parties in invoking this authority within an operative paragraph dealing indistinguishably and in a broad-brush fashion with 'articles 63 and 64 LOSC (as well as the 1995 Fish Stocks Agreement)', and in doing so to conclude that 'in their application in the EEZ [these articles] are subject to the exclusion article 297(3)(a) LOSC'.[191] With respect to the substantive provisions at issue, this authority has been partially misapplied, as it was decided in relation to a type of fish stock under article 63(1) LOSC—addressing so-called domestically 'shared' stocks[192]—and therefore does not pertain to straddling or highly migratory fish stocks, which are the exclusive focus of the Agreement.

[189] *Barbados v. Trinidad and Tobago Arbitration* (Decision) 2006, §§272–279; and *see* Kwiatkowska, 'The 2006 Barbados/Trinidad and Tobago Maritime Delimitation (Jurisdiction and Merits) Award', in Ndiaye–Wolfrum (Eds) *Law of the Sea, Environmental Law, and Settlement of Disputes* (2007) 944.

[190] *ibid.*, §276.

[191] *Chagos MPA* (Award) 2015, §300.

[192] *Barbados v. Trinidad and Tobago Arbitration* (Decision) 2006, §283.

Last but not least, from a procedural law perspective, the tribunal's jurisdiction *ratione materiae* in the *Chagos MPA* arbitration was not engaged with any material dispute of fact falling under the Agreement. On this exact point, the remarks by the United Kingdom at pleading the inapplicability of articles 63(2) and 64 LOSC and article 7 FSA were precise.[193] It is only in the arguments of Mauritius, at the stage of filing its Memorial, when a brief reference—in a footnote—was made alluding to jurisdiction under article 30(1) FSA in relation to article 7 FSA.[194] This was followed later by a more assertive statement mentioning that 'article 30 of the Agreement provides that the dispute settlement provisions of the [Convention] apply to disputes regarding the interpretation or application of the Agreement'.[195] This reference was not ignored by the United Kingdom, which in turn noted in filing its preliminary objections—also in a footnote—that '[t]he Tribunal is not a court or tribunal to which a dispute has been submitted under Part VIII FSA'.[196] This position was reiterated in the Counter-Memorial.[197] No further reference to the dispute settlement procedures under the Agreement was made by either party thereafter. The subject-matter jurisdiction was firmly fixed and unanimously founded by the tribunal upon article 288(1) LOSC.[198]

While the tribunal commendably had recourse to the interpretation of the fisheries limitation based on a textual construction, confirmed by the drafting history of the provisions, this methodology was applied exclusively within the scope of article 297 LOSC.[199] If article 7 FSA had any bearing on the dispute, the tribunal should have undertaken an analogous interpretative exercise between the treaty texts of the Convention and the Agreement, in order to properly construe the limitation set out in article 297(3)(a) LOSC in light of its incorporation by reference in article 32 FSA. Since no transtextual exercise of that kind was undertaken in the tribunal's reasoning, it leaves no doubt that the tribunal interpreted the limitation solely—and entirely in terms of jurisdiction—by reference to the Convention. Although the effect of the award has

193 *Chagos MPA Arbitration Final Transcript*, 893–895.

194 *Chagos MPA*; *Memorial of the Republic of Mauritius* (1/8/2012), 87(¶5.6; footnote n378).

195 *ibid.*, 98(¶5.35).

196 *Chagos MPA*; *Preliminary Objections to Jurisdiction submitted by the United Kingdom* (31/10/2012), 39(¶4.2; footnote n113).

197 *Chagos MPA*; *Counter-Memorial submitted by the United Kingdom* (15/7/2013), 132(¶5.16; footnote n382).

198 *Chagos MPA* (Award) 2015, §§204 & 547 *dispositif* A(3).

199 *ibid.*, §§301, 308, 317 & 378.

been largely negated by UN political developments,[200] and its legal reasoning on the nature and scope of the sovereign rights addressed in the case has been overtaken by subsequent determinations on the territorial and maritime rights of Mauritius in the Chagos Archipelago,[201] its pronouncements regarding the Agreement should be regarded as mere *obiter dicta*, if only to avoid the perception that the tribunal overstepped its jurisdiction.

3.5 INTERPRETING THE FISHERIES LIMITATION

As has already been noted in the scholarship, although the exclusions set out in the fisheries limitation may lend themselves to both broad and narrow interpretations, nothing in the Convention supports—or even implies—a blanket exclusion of fisheries disputes in EEZ from the remit of compulsory procedures entailing binding decisions.[202] To the contrary, this category of disputes has been recognized as presenting some of the most contentious issues in the implementation of the Convention.[203] Admittedly, UNCLOS/III left this question in abeyance as part of the political compromises shrouded in the opaque formulation of article 297 LOSC, set against the backdrop of articles 63(2) and 64 LOSC, in which coastal States exhibited a latent tendency to broadly interpret the exclusionary scope of the limitation. Still, the analysis does not support any argument suggesting that the limitation, in principle, excludes disputes concerning compatible CMMs from section 2 Part XV LOSC. The uncertainty inhabiting the text of the limitation in article 297(3)(a) LOSC has nonetheless been carried over into its interpretation under article 32 FSA, creating fertile ground for the recurrent emergence of opposing debates between DWFS and coastal States under the Convention, thereby generating two principal approaches regarding the applicability of the compulsory procedures entailing binding decisions under Part VIII FSA.

---

200 A/Res.73/295 'Advisory Opinion of the International Court of Justice on the Legal Consequences of the Separation of the Chagos Archipelago from Mauritius in 1965' (22/5/2019).

201 *see* in *Legal Consequences of the Separation of the Chagos Archipelago from Mauritius in 1965* (Advisory Opinion) 2019 and *Delimitation of the Maritime Boundary in the Indian Ocean* (Merits) 2023.

202 *see* de Mestral, 'Compulsory Dispute Settlement in the Third United Nations Convention on the Law of the Sea: A Canadian perspective', in Buergenthal (Ed.) *Contemporary Issues in International Law* (1984) 183 and Oxman, 'The Rule of Law and the United Nations Convention on the Law of the Sea', EJIL (1996) 368.

203 *see* Sharma, 'Framework of Likely Disputes Under the Law of the Sea Convention—Some Thoughts', ZaöRV (1985) 490 and Erasmus, 'Dispute Settlement in the Law of the Sea', *Acta Juridica* (1986) 22–23.

3.5.1 The Broad Approach Expanding the Scope of Exclusions

One approach interprets the scope of exclusions broadly, practically barring the application of compulsory adjudication to compatibility disputes arising within EEZ. Such disputes are viewed as inherently linked to the coastal State's exercise of sovereign rights and discretionary powers over marine living resources in the zone. Consequently, the terms and conditions governing in-zone national CMMs for fish stocks are considered exempt from the compulsory procedures entailing binding decisions under Part VIII FSA. The broadness of this approach derives its impulse mainly from academic commentary analysing the text of the limitation under the Convention.[204] This premise is typically expounded by positions that regard the dispute settlement procedures under the Agreement as a form of leverage afforded to coastal States in the context of a compatibility dispute. According to this interpretation, a coastal State may invoke article 7(4) FSA to challenge the international CMMs adopted by DWFS or RFMO/A for the high seas, when the latter are not compatible with those established for the same stock in its EEZ. Conversely, this procedural right is not equally extended in the opposite direction, owing—putatively—to the operation of the limitation as set out in article 297(3)(a) LOSC.[205] Reflecting on the cross-reference to the Convention's fisheries limitation in article 32 FSA, adherents to this view maintain that the sovereign fishing rights of the coastal State within EEZ remain unaffected, as to exclude their exercise from the compulsory and binding procedures under the Agreement.[206] This view is far from uniformly expressed, even among

---

[204] The most frequent references are made to Andronico Adede's views contained in 'Prolegomena to the Disputes Settlement Part of the Law of the Sea Convention', NYUJ *Int'l LP* (1977) 386; 'Streamlining the System for Settlement of Disputes under the Law of the Sea Convention', *Pace Law Review* (1980) 23–24; and *The System for Settlement of Disputes under the United Nations Convention on the Law of the Sea* (1987) 254.

[205] Meltzer, *The Quest for Sustainable International Fisheries, Regional Efforts to Implement the 1995 United Nations Fish Stocks Agreement* (2009) 206–207, although it is noted that the brilliance of article 7 FSA lies in its ability to downplay the contentious issue of prescriptive dominance between DWFS and coastal States.

[206] *see* Boyle, 'Dispute Settlement and the Law of the Sea Convention: Problems of fragmentation and jurisdiction', ICLQ (1997) 43–44; McDorman, 'The Dispute Settlement Regime of the Straddling and Highly Migratory Fish Stocks Convention', *Canad. YbIL* (1997) 65–67; Ørebech–Sigurjonsson–McDorman, 'The 1995 United Nations Straddling and Highly Migratory Fish Stocks Agreement: Management, enforcement and dispute settlement', IIJMCL (1998) 135–136; and Orrego-Vicuña, *The Changing International Law of High Seas Fisheries* (1999) 282–287.

like-minded commentators. While some firmly maintain that disputes concerning fish stocks are definitely excluded from the procedures provided in section 2 Part XV LOSC, they remain silent on the implications of the *mutatis mutandis* application in disputes between parties to the Agreement.[207] Others generally contend that only the EEZ components of a fisheries dispute are excluded; however, where a disagreement concerns the same stock on the high seas, the forum seised should declare itself competent. Yet, such analyses do not engage directly with compatibility disputes under article 7 FSA.[208] Those who address the dispute settlement procedures under both the Convention and the Agreement tend to eschew the fisheries limitation altogether,[209] while only a few openly make this argument in connection with straddling fish stocks but not in relation to stocks of highly migratory species.[210]

Hence this approach most frequently complements interpretations of the compatibility rule favouring the extension of coastal State interests seawards, by obligating DWFS to adopt CMMs consistent with those established for the same fish stock in EEZ. This position maintains, in plain terms, that the forum competent to hear a compatibility dispute—although bound by the fisheries limitation—will defer to the in-zone national measures as the benchmark, not least because their effectiveness must not be undermined.[211] This view appears to be based on reserving for coastal States a special interest in high seas areas adjacent and beyond their national jurisdiction, as previously that concept had been provided in the *Fishing & Conservation Convention* (1958). It is viewed therefore that tipping the balance on the side of coastal State interests is beneficial, since they are more invested in the sustainable regulation of fish stocks

[207] Orellana, 'The Law on Highly Migratory Fish Stocks: ITLOS jurisprudence in context', *Golden Gate University Law Review* (2004) 463–465.

[208] *see* Casado-Raigón's analysis of 'Le Règlement des Différends' in the collective monograph by Vignes *et al.*, *Le Droit International de la Pêche Maritime* (2000) 352*ff* followed by García-Revillo, *The Contentious and Advisory Jurisdiction of the International Tribunal for the Law of the Sea* (2015) 176–185, apart from occasional references separately addressing requests for provisional measures on pages 247–249.

[209] *e.g.* Rengifo-Lozano, *International Regime Theory and the Law of the Sea* (2011) 207–218.

[210] *see* Zumwalt, 'Straddling Stocks Spawn Fish War on the High Seas', *University of California Davis International Law and Policy* (1997) 56 and Tanaka, 'Jurisdictional Issues in the Settlement of Disputes Concerning Straddling Fish Stocks Under UNCLOS', in Kunoy–Heidar–Yiallourides (Eds) *International Fisheries Law* (2024) 194–195.

[211] Henriksen–Hønneland–Sydnes, *Law and Politics in Ocean Governance* (2006) 55.

than DWFS, and presumably more motivated to promote the aim of long-term conservation.[212]

Yet, the majority of unsustainable fisheries occur in EEZs due to IUU fishing. Although it may seem improbable for a coastal State to enforce stringent CMMs on the high seas while neglecting similar measures within its own EEZ, this remains a frequent grievance raised by DWFS.[213] In contrast to the conformist stance typically adopted by coastal States, the critical remarks of a prominent academic authority—otherwise known for advocating their interests—are particularly striking, as he observes that:

> The perception prior to 1975 that global fisheries were in trouble because of lack of a jurisdictional means for adequate management has been shown to be only a partial and faulty analysis. The widespread adoption of the EEZ has not by any means resulted in adequate management. Overfishing, overcapitalization, dissipation of benefits, and ineffective regulation continue to be characteristic of management all over the world. The fault lies with the individual coastal States for inept measures relating to the stocks wholly within their jurisdiction, exacerbated to some degree by the jurisdictional quagmire of straddling stocks.[214]

Indeed, there have been numerous instances where coastal States have failed to implement effective CMMs for fish stocks in their EEZ.

The acute crisis in the north Atlantic fisheries that impelled Canada to unilateral measures in the NAFO regulatory area was ascribed also to its own national fishery policy following the declaration of the 200 nm zone in 1977 amidst the negotiations on RSNT at UNCLOS/III. Having overestimated the biomass dynamics of stocks, Canada in the 1980s set excessive national TACs. The harvesting and processing capacity was correspondingly increased, which was a decision preferred also at that time on account of maximizing national employment in these sectors. Commentators have drawn attention to the fact that just in the period from 1986 to 1992, the Canadian fleet was responsible for depleting several straddling stocks within the zone while other

---

212 Martin, 'Fisheries Conservation and Management of Straddling Stocks and Highly Migratory Stocks under the United Nations Convention on the Law of the Sea', GIELR (1995) 766.

213 *see* Davies–Redgwell, 'The International Legal Regulation of Straddling Fish Stocks', BYBIL (1997) 204 & 247; Scheiber, 'Ocean Governance and the Marine Fisheries Crisis: Two decades of innovation—and frustration', *Virginia Environmental Law Journal* (2001) 129–132; and Christie, 'It Don't Come EEZ: The failure and future of coastal state fisheries management', *Journal of Transnational Law and Policy* (2004) 1.

214 Burke, 'Importance of the 1982 UN Convention on the Law of the Sea and its Future Development', ODIL (1996) 2–3.

NAFO members had reduced their catches.[215] On the other side of the Atlantic Ocean, the North-East Atlantic Fisheries Commission ('NEAFC') has experienced persistent mismanagement practices for pelagic and demersal stocks following frequent failures among its coastal member States to agree on TAC limits and TAFE levels, which have increased pressure to undertake radical institutional reforms for addressing the need to adopt CMMs in line with the principle of ecosystem approach.[216]

Regarding highly migratory species, the findings of the Forum Fisheries Agency of the Pacific Islands ('FFA') are particularly revealing. Its review of CMMs adopted by the SPFF coastal member States found that many of these measures failed to meet the standards of the precautionary approach and were inadequate to prevent the overexploitation of tuna species within sea areas under national jurisdiction.[217] Moreover, SIDS of the Pacific in the past developed a discourse of 'resource nationalism' that sought to increase political intervention in the tuna fishing industry. Acting out on such political discourse entailed a domestication approach by encouraging constant expansion in the fishing effort that only led to high economic costs and unsustainable fishing practices.[218] Similarly, CCSBT had urged Indonesia—an important coastal State due to the critical role of its national jurisdictional waters in the spawning and recruitment necessary for rebuilding southern bluefin tuna stocks—to address several pressing issues. In 2001, it expressed grave concerns about IUU fishing conducted in the Indonesian EEZ, which were diminishing the effectiveness of CMMs across the biological range of the stock and despite calls for cooperation, no action had been taken.[219] In 2014, it was noted that additional efforts should be made to implement, in sea areas under the national jurisdiction of Indonesia, spatiotemporal restrictions, compatible with the rest

---

215 *see* Schrank, 'Extended Fisheries Jurisdiction: Origins of the current crisis in Atlantic Canada's fisheries', *Marine Policy* (1995) 285; Mitchell, 'Fisheries Management in the Grand Banks, 1980–1992 and the Straddling Stock Issue', *Marine Policy* (1997) 97; and Joyner, 'On the Borderline? Canadian Activism in the Grand Banks', in Stokke (Ed.) *Governing High Seas Fisheries* (2001) 212.

216 *e.g. NEAFC plans 'performance review' in 2024 amid pelagics mismanagement* (23/11/2023) {undercurrentnews.com} and *NEAFC advances ecosystem-based management measures, but overfishing concerns persist for blue whiting, mackerel, and herring* {seafoodsource.com}.

217 FFA, 'A Strategic Framework for the Effective Management and Conservation of the Tuna Resources of FFA Member Countries' (1994).

218 Stephens, 'Fisheries-led Development in the South Pacific: Charting a "Pacific Way" to a sustainable future', ODIL (2008) 257.

219 CCSBT, 'Decision Regarding Indonesia Pursuant to the 2000 Action Plan' in the *Report of the Eighth Annual Meeting* (19/10/2001).

of the management strategy applied by CCSBT.[220] With the same concerns remaining unabated in 2021, the organization is now monitoring scientific data on this matter closely as part of its long term action plan at least until 2028.[221]

3.5.2 The Narrow Approach Advancing the Principle of Compulsory Adjudication

Another interpretation firmly advocates a restrictive construction of article 32 FSA, reading the fisheries limitation narrowly and asserting that both the EEZ and high seas components of CMMs remain subject to the compulsory procedures entailing binding decisions under Part VIII FSA. This approach, in principle, corresponds to neutral interpretations of the compatibility rule. The jurisprudential nascence of this approach can be traced in the *Southern Bluefin Tuna Cases*, where ITLOS, in prescribing provisional measures, deliberately refrained from specifying their spatial scope so as to circumvent the controversy regarding the effect of the limitation on the procedures set forth in section 2 Part XV LOSC.[222] While the arbitral tribunal later resigned its jurisdiction to the fact that Part XV LOSC 'falls significantly short of establishing a truly comprehensive regime of compulsory jurisdiction', it nevertheless concluded by remarking that when the Agreement is in force, it should, for States parties to it, 'not ... go far towards resolving procedural problems'.[223] A coherent argument, based on the integrative reading of the Convention and the Agreement, would further posit that if the obligation to conserve fish stocks were to be imposed solely on the high seas, then the fundamental objective of the Agreement would either be potentially frustrated or its application would inevitably transform the procedural limitation into a substantive allocational preference predisposed in favour of coastal States.[224] Even scholars who broadly interpret article 297(3)(a) LOSC often hesitate to extend such

---

220 CCSBT, 'Performance Review Panel Report: 2014', ¶40.

221 CCSBT, 'Performance Review Panel Report: 2021', ¶¶28–29 & 57 and *see* CCSBT, 'Strategic Plan for the Commission for the Conservation of Southern Bluefin Tuna 2023–2028' (2023).

222 *see* the comments by Robin Churchill in 'International Tribunal for the Law of the Sea the Southern Bluefin Tuna Cases (*New Zealand v. Japan; Australia v. Japan*): Order for Provisional Measures of 27 August 1999', ICLQ (2000) 987–988 and 'Some Reflections on the Operation of the Dispute Settlement System of the UN Convention on the Law of the Sea During its First Decade', in Freestone–Barnes–Ong, *The Law of the Sea, Progress and Prospects* (2006) 407–409.

223 *Southern Bluefin Tuna Cases* (Jurisdiction) 2000, §§62 & 71.

224 Oxman, 'Complementary Agreements and Compulsory Jurisdiction', AJIL (2001) 294–295.

reasoning to the Agreement, especially in the context of fisheries disputes of this kind. For instance, drawing from commentary on the case readily at hand, the *obiter dictum* statement made by the tribunal has been read to advance the principle of compulsory adjudication, in implying—inevitably, as a corollary connotation—that the limitation is not to affect compatibility disputes.[225]

Further support for the normative basis of this approach has been drawn from the emphasis on construing the fisheries limitation in the context of disputes concerning compatible CMMs, particularly as informed by the precautionary approach. It should be recalled that, with respect to both articles 6 and 7 FSA, the Agreement unequivocally affirms its applicable scope also 'within areas under national jurisdiction'.[226] The forum seised with a compatibility dispute should rely on this express *ratione loci* provision to assert, consistently with the principle of inclusive jurisdiction, its compulsory jurisdiction over disputes concerning the interpretation or application of these articles. In the specific case of the compatibility rule, the textual inclusion of the procedural stipulation embedded in article 7(4) FSA not only necessitates the qualification of article 32 FSA but, *in extenso*, compels a restrictive construction of the latter, thereby overriding any contrary interpretation of the limitation's text as contained in article 297(3)(a) LOSC.[227] A construction of this kind is buttressed further by a purposive argument that construes the procedural incorporation by reference of the limitation in a consonant manner with the substantive *lex specialis* of the Agreement. The effectiveness of the provisions governing the precautionary approach and the compatibility rule relies on the available compulsory settlement procedures to elaborate on the content of the obligations of States and maintain cooperation when there is disagreement over CMMs. This construction fulfils to the fullest extent the requirement of interpretative consistency in article 4 FSA in accommodating the balance of interests between DWFS and coastal States by providing the forum with compulsory jurisdiction to safeguard their respective rights in the context of such fisheries disputes.[228]

The narrow approach maintains that the ambiguity inhabiting the fisheries limitation does not confer any doctrinal primacy upon coastal States, which would in turn endorse the biased political narrative propagating views that

---

225 Kwiatkowska, 'The Australia and New Zealand *v.* Japan—Southern Bluefin Tuna (Jurisdiction and Admissibility) Award of the First Law of the Sea Convention Annex VII Arbitral Tribunal', IJMCL (2001) 278.

226 FSA 3(1).

227 Freestone–Oude-Elferink, 'Flexibility and Innovation in the Law of the Sea—Will the LOS Convention Amendment Procedures Ever be Used?', in Oude-Elferink (Ed.) *Stability and Change in the Law of the Sea* (2005) 195–196.

228 Klein, *Dispute Settlement in the UN Convention on the Law of the Sea* (2005) 204.

only these States hold a genuine and sincere interest in the conservation and management of fish stocks throughout their distribution and migratory range. On the contrary, from a legal standpoint, it would constitute a *contra legem* interpretation to read expansively a limitation that was specifically adopted to ringfence the sovereign rights of coastal States within their EEZs and to shield them from frivolous or vexatious claims alleging breaches of their obligations. It is important to recall that the institution of EEZ—as a zone *sui generis* per se—did not merely represent a significant extension of the coastal State's sovereign rights beyond the territorial sea. By the same token, it codified exhaustively any prior concept of special interest or preferential rights by proportionately assimilating the coastal State's corollary obligations to undertake effective CMMs, and by circumscribing the limits of the exercise of those sovereign rights in both spatial and functional terms.[229] Thus, the restrictive construction of the fisheries limitation is congruent with one of the principal motivations behind the elaboration of a universal fisheries regime—first under the Convention and subsequently through its implementation in the Agreement—namely, to eliminate any scope for creeping jurisdiction within and beyond the zone.[230] In contrast, the broad interpretation of the limitation does not merely exclude DWFS from fishing activities within the zone; rather, it inversely results in the expansion of the coastal State's sovereign fishing rights into high seas areas.

Preserving the balance of interests between DWFS and coastal States stands as a necessary precondition for qualifying the exclusionary scope of the limitation, as previously discussed in relation to the proper characterization of the second clause's reference to the 'sovereign rights with respect to the living resources in the exclusive economic zone'. To this end, it is advisable to delve here further into some important aspects in the historical development of the EEZ regime through the arguments set forth in the *Declaration of Santo Domingo* (1972). The undersigning Latin American States sought to pre-empt UNCLOS/III by proclaiming the concept of the 'patrimonial sea' prior to the commencement of the negotiations. This concept reserved for the coastal State 'sovereign rights over the renewable ... natural resources, which are found in the waters ... of an area adjacent to the territorial sea ... [which] ... should

[229] *see* Dupuy, *The Law of the Sea* (1974) 22–23; Ahwoi, 'The Exclusive Economic Zone: An analysis of the concept as contained in the *Revised Single Negotiations Text*', *University of Ghana Law Journal* (1977) 143–145 & 160–161; and Attard, *The Exclusive Economic Zone in International Law* (1987) 32–69.

[230] Higgenson, 'The Law of the Sea Convention and the Protection of Fisheries', GIELR (1995) 771–772.

not exceed a maximum of 200 nm'.[231] Let it be recalled first that, while the Declaration borrowed the words 'sovereign rights' from article 2(1) CSC, it did not envisage a 200 nm zone equating to the legal status of territorial waters through, for example, the rigid concept of *ad dominium* sovereignty.[232] But very different from that notion, it was intended to attribute a specialized and functional character to the competence of the coastal State to exercise *imperium* sovereign rights on fisheries within the confines of the zone.[233] These rights were to be exclusive in that no other State would exploit the resources of the zone, except with the express consent of the coastal State.[234] This is readily discernible in various proposals put forward by coastal States, such as the 'Draft articles on fisheries in national and international zones in ocean space' submitted by the delegations of Ecuador, Panama and Peru.[235] The specific proposal asserting the special interest of the coastal State in high seas areas adjacent to and beyond the zone on a functional basis also provided that any relevant dispute be made subject to the settlement procedures of the Convention. The same rationale was reflected in other proposals submitted by Canada, India, Kenya and Sri Lanka,[236] USSR,[237] and Japan,[238] with perhaps

---

231 *cit.* Chapter 2 (*fn*29); and *see* Infante, 'South America', in Kraska–Ryou-Ellison (Eds) *Cultural Influences on the Law of the Sea* (2025) 209*ff.*

232 *e.g.* as it was employed by the Organization of African Unity in A/AC.138/89 'Declaration on the issues of the law of the sea' (2/7/1973) [(1973) 12ILM1200], ¶7 and by Ecuador, Panama and Peru in A/AC.138/SC.II/L.27/Corr.2 'Draft articles for inclusion in a convention on the law of the sea' (13/7/1973) [(1973) 12ILM1224], article 1.

233 *see* Nelson, 'The Patrimonial Sea', ICLQ (1973) 677; Aguilar, 'The Patrimonial Sea or Economic Zone Concept', SDLR (1974) 579 and Ganz, 'The United Nations and the Law of the Sea', ICLQ (1977) 1; and on the negotiating attitude of the 'patrimonialist' States *see* specifically Kildow, 'The Law of the Sea: Alliances and divisive issues in international ocean negotiations', SDLR (1974) 558.

234 Pollard, 'The Exclusive Economic Zone: The elusive consensus', SDLR (1974) 609*ff.*

235 A/AC.138/SC.II/L.54 (10/8/1973) [(1973) 12ILM1267], ¶(I) & ¶(M).

236 A/AC.138/SC.II/L.38 'Draft articles on fisheries' (16/7/1973) [(1973) 12ILM1239], articles 8, 10 & 13.

237 A/AC.138/SC.II/L.6 'Draft articles on fishing' (18/7/1972) [(1973) 12ILM36], providing in article 7 for arbitration at the request of any party, unless the parties to the dispute agree to have recourse to any other means of their choice.

238 A/AC.138/SC.II/L.12 (1972), *cit.* Chapter 2 (*fn*72), wherein ¶¶4(1)–6(2) explicitly excluded highly migratory stocks from any special or preferential regime and put them under the scope of compulsory dispute settlement procedures very similar to those provided in the *Fishing & Conservation Convention* (1958).

the most pronounced advancement of compulsory dispute settlement offered by the United States, which proposed that any dispute be submitted to procedures similar to those provided in the *Fishing & Conservation Convention* (1958).[239]

Linked to this functional conceptualization of the zone was the blueprint for comprehensive dispute settlement, initially proposed during the Second Session (1974).[240] This envisaged both general and special—or functional, as they were described at the time—procedures. The latter category addressed specific types of disputes, such as fisheries, which, although remaining within the remit of the compulsory procedures, were, in certain instances, subject to limitations.[241] Two years later, the plenary debates at the Fourth Session (1976) in New York during the negotiations on RSNT revealed that the scope of the exclusions had become the most intricate and contentious issue. It was evident that if exclusions were broadly defined, the value of the entire dispute settlement system would be seriously weakened, if not entirely nullified.[242] As discussed earlier, the wording of the fisheries limitation—irrespective of the various formulations it underwent during the negotiations—consistently maintained a distinctive phraseology aimed at forestalling abusive interpretations by carefully enumerating specific disputes that could be excluded from the compulsory procedures. Such exclusions were condoned only owing to the UNCLOS/III *modus operandi* of reaching agreement by consensus, and because these were already significant, they were intended to be express and limited.[243] Likewise, the specific references regarding the applicability of section 2 Part XV LOSC—albeit made in relation to the scope of optional exceptions under article 298 LOSC—underline the drafting intention at

---

239 A/AC.138/SC.II/L.9 'Revised draft article on fisheries' (4/8/1972) [(1973) 12ILM42], ¶IX; and *see* Sohn, 'US Policy Toward the Settlement of the Law of the Sea Disputes', VJIL (1976) 9.

240 A/Conf.62/L.7 (1974), *supra cit.* (*fn*118); and *see* Stevenson–Oxman, 'The Third United Nations Conference on the Law of the Sea: The 1974 Caracas session', AJIL (1975) 1.

241 *see* Adede, 'Settlement of Disputes Arising under the Law of the Sea Convention', AJIL (1975) 798.

242 Gamble, 'The Law of the Sea Conference: Dispute settlement in perspective', *Vanderbilt Journal of Transnational Law* (1976) 329–331 and Saxena, 'Limits of Compulsory Jurisdiction in Respect of the Law of the Sea Disputes', in Anand (Ed.) *Law of the Sea: Caracas and beyond* (1980) 335.

243 *see* Lehoux, 'La Troisième Conférence sur le Droit de la Mer et le Règlement Obligatoire des Différends', *Canad. YbIL* (1981) 49 and Mensah, 'The Role of Peaceful Dispute Settlement in Contemporary Ocean Policy and Law', in Vidas–Østreng (Eds) *Order for the Oceans at the Turn of the Century* (1999) 83.

UNCLOS/III to keep all exclusions to a minimum,[244] an approach consistently upheld by case law, which interprets exclusions narrowly, within the limits expressly permitted.[245]

Support for a narrow approach to the fisheries limitation in the context of compatibility disputes is also evidenced by the submission to amend article 63 ICNT by States traditionally regarded as 'coastal' in terms of their interests in straddling stocks. The most salient feature protruding from the text of the amending proposal is what this book identifies as a pattern of clausal embeddedness (see Box 2.1).[246] It is worth noting that this pattern was reminiscent of proposals previously featured in UNCLOS/II whereby the extension of fishery jurisdiction was amenable to dispute settlement procedures equivalent to those under the *Fishing & Conservation Convention* (1958).[247] The legislative background reveals that coastal States intensified their efforts to amend paragraph (2) of article 63 ICNT by introducing a procedural stipulation for the compulsory settlement of disputes in cases where DWFS and coastal States failed to agree on CMMs.[248] This proposal was tabled at the eleventh hour of the negotiations, as a last resort following the failure of an earlier Canadian–Argentine proposal that sought to significantly enhance coastal State rights over fish stocks occurring both in EEZ and in adjacent high seas areas. The fact that UNCLOS/III ultimately rejected this amendment demonstrates the outright dismissal of any special coastal State interest in this regard and strongly suggests that any such assertion would constitute an interpretation contrary to the drafting intentions underlying articles 63(2) and 116 LOSC.[249]

Furthermore, the restrictive construction of article 32 FSA remains consistent with the principle of inclusive jurisdiction underpinning article 297(3)(a) LOSC and does not amount to a reinterpretation of the Convention, as disputes

---

244 1989 *Virginia Commentary* V ¶298.13(115).

245 *see*, for example, the narrow construction adopted in *Arctic Sunrise*, both by ITLOS in satisfying the *prima facie* jurisdiction requirement in (Provisional Measures) 2013, §45 and by the arbitral tribunal in subsequently upholding jurisdiction in (Jurisdiction) 2014, §72.

246 A/Conf.62/L.114 (1982), *cit.* Chapter 2 (*fn*129).

247 *see* the joint proposals by Canada and the United States A/Conf.19/C.1/L.10 (1960), *cit.* Chapter 1 (*fn*52), article 4 and A/Conf.19/L.11 (22/4/1960), article 4 [*UNCLOS/II Off.Records* I:173]; and the accompanying comments by the delegation representatives at UNCLOS/II Committee of the Whole 21st meeting (8/4/1960) [*UNCLOS/II Off.Records* II:408–417].

248 1993 *Virginia Commentary* II ¶63.8(644) & ¶63.11(645–646).

249 *see* Oxman, 'The Third United Nations Conference on the Law of the Sea: The ninth session (1980)', AJIL (1981) 234–235 and Momtaz, 'L'Accord Relatif à la Conservation et à la Gestion des Stocks de Poissons Chevauchants et de Grands Migrateurs', *Annuaire Français* (1995) 685–687.

over fish stocks were never excluded *expressis verbis* from the remit of the compulsory procedures entailing binding decisions under section 2 Part XV LOSC. A historical anecdote, highly evocative of the rationale underlying the narrow approach, recounts the submission made by the Rapporteur to the Second Committee of UNCLOS/III, pondering on the intended role of limitations as these were contained at that point in article 18 ISNT, and for which it was stated that:

> The procedures for the settlement of disputes arising from the interpretation and application of the convention [are] essential and should be an integral part of the convention. The new convention [will] be a delicately balanced compromise and there [will] inevitably be widely divergent interpretations of its provisions. The procedures for the settlement of disputes must therefore be prompt, final and of universal application [...] Uniform interpretation of the convention [is] also essential in order to give effect and meaning to its provisions [...] [My] delegation also [has] reservations regarding the exception provisions contained in article 18(2), because they [are] too broad and ambiguous. Such a broad range of exceptions could result in wide disagreement on the extent of the exclusions. It would also exclude from the dispute settlement procedures many disputes which by their very nature should be the subject of prompt compulsory settlement. Exceptions, if any, should be restricted to the absolute minimum and spelt out with great clarity.[250]

This statement belongs to Ambassador Nandan, who played a seminal role in drafting ISNT,[251] while some 18 years later in his capacity as the Chairman of the *Fish Stocks Conference* was personally responsible for the core negotiating drafts that paved the way to the final text of the Agreement.

In addition to the drafting history, the narrow approach comports with the jurisprudence on the qualification of exclusions from treaty obligations and with the technical rule that exceptions to general rules must be construed narrowly, as encapsulated in the maxim *exceptio est strictissimae applicationis*. This consideration must be borne in mind principally when interpreting the fisheries limitation, as CMMs involve the exercise of sovereign rights by coastal States in their EEZ and by all States in the context of the freedom of fishing and the exercise of exclusive flag jurisdiction on the high seas. Jurisprudence points to an emerging doctrinal approach that limitations affecting sovereign rights cannot be presumed or implied, nor interpreted *a priori* in favour of a less restrictive application against the primary rule on which the exception

250 UNCLOS/III Plenary 64th meeting (9/4/1976) [*UNCLOS/III Off.Records* V:48–49].

251 Eiriksson, 'Satya N. Nandan's Role in Drafting the Informal Single Negotiating Text: Aspects of the preparatory work for UNCLOS', in Lodge–Nordquist (Eds) *Peaceful Order in the World's Oceans* (2014) 37.

pertains.[252] Where, despite all pertinent considerations, the vagueness of an exclusion nonetheless gives rise to arguments for invoking the counterbalancing technical rule of *in dubio mitius*, and such doubt about the scope of the rule is objectively established, its interpretation should be based on a normative core that can be deduced with sufficient certainty. This applies insofar as the construction does not yield a meaning or result contrary to the plain terms and intentions of the parties at the time the treaty was concluded, as reflected by its text and any other *lex specialis* requirements.[253]

Finally, it is also important to take into account subsequent treaty practice concerning the adoption of compatible CMMs by RFMO/As. For instance, a regulatory body that promotes a neutral approach to the implementation of article 7 FSA is the Western and Central Pacific Fisheries Commission ('WCPFC').[254] In reviewing the performance of the Commission against the requirements of the Convention and the Agreement, it has been observed that in areas where the coastal State exercises 'sovereign rights', as distinguished from areas of 'sovereignty', the compatibility rule has not given rise to controversy.[255] Even more instructive in support of a restrictive construction of the limitation in article 297(3)(a) LOSC are the findings arising from the review of the NAFO *Convention*, which are likely to carry authoritative weight in any future dispute within that organization. It was recalled that fisheries

---

[252] As evidenced in pronouncements across judicial [*e.g.* PCIJ in *S.S. "Wimbledon"* (Judgment) 1923, pp.24–25; *Certain German Interests in Polish Upper Silesia* (Merits) 1926, pp.21 & 76 *et seq* and in *Territorial Jurisdiction of the International Commission of the River Oder* (Judgment) 1929, p.26; ICJ in *Dispute regarding Navigational and Related Rights* (Judgment) 2009, §48 and ITLOS in *Arctic Sunrise* (Provisional Measures) 2013, §45], arbitration/conciliation [*e.g.* by the commission in *Différend concernant l'Interprétation de l'Article 79(6)(c) du Traité de Paix* (Décisions) 1952/1954/1955, pp.395–399], as well as quasi-judicial [*e.g.* WTO DSB in *Restrictions on Imports of Cotton and Man-Made Fibre Underwear—Costa Rica* (Panel) 1996, §§7.18–7.21 and *Measures Affecting Trading Rights and Distribution Services for Certain Publications and Audiovisual Entertainment Products–China* (Appellate) 2009, §§410–411] forums.

[253] *see* notably in Jennings–Watts, *Oppenheim's International Law* (1992) 1278; yet the preponderance of the scholarly arguments in more recent analysis [*e.g.* by Merkouris, 'In Dubio Mitius', 259 and by Solomou, 'Exceptions to a Rule Must be Narrowly Construed', 359 in Klingler–Parkhomenko–Salonidis (Eds) *Between the Lines of the Vienna Convention?* (2018), and by Fahner, 'In Dubio Mitius: Advancing clarity and modesty in treaty interpretation', EJIL (2021) 835] remain far more sceptical as to the doctrinal clarity of the rule.

[254] *see* further in Chapter 6 (pp. 250–254).

[255] WCPFC, 'Performance Review Panel Report: 2012', pp. 17 & 76–80.

disputes—with the exception of disputes over domestic fish stocks—are to be settled through the compulsory and binding procedures of section 2 Part XV LOSC.[256]

Following the analysis above, the second and third clauses of article 297(3)(a) LOSC should be given a narrow construction, limiting their scope to EEZ domestic fish stocks or to aspects of CMMs that are reserved entirely to the prescriptive authority of the coastal State. Accordingly, disputes concerning disagreements over compatible CMMs addressing the biological integrity of fish stocks throughout their distribution and migratory range are not excluded by the fisheries limitation. The limitation nevertheless remains applicable to matters falling outside the scope of disagreements governed by the compatibility rule as provided in article 7 FSA, subject to any further qualification arising from applicable *lex specialis* between the parties and following its proper characterization by the forum seised. Such matters may concern: (a) fishing activities *stricto sensu* undertaken by the coastal State within its own EEZ, unless these activities are regulated as part of the agreed CMMs; (b) management measures concerning the optimum utilization of fish stocks in EEZ, insofar as such measures do not prejudice the aim of conservation across the biological range of the stocks; (c) the exercise of the coastal State's discretion on matters that are purely of EEZ concern; and (d) law enforcement activities by the coastal State in areas under its sovereignty or national jurisdiction to ensure compliance with agreed CMMs, or actions sanctioned by article 73 LOSC or excepted through a valid declaration pursuant to article 298(1)(b) LOSC.

Components of compatibility disputes that are qualified to fall within these categories—whether automatically excluded or optionally excepted—may nonetheless be submitted to the compulsory procedures under section 2 Part XV LOSC by the express agreement of the parties.[257] As has been remarked in the literature, a critical question arises as to whether the forum seised would require an agreement between the parties, on the premise that it must have jurisdiction over the dispute *in concreto* so as to resolve that effectively.[258] Such a request would most likely arise following the adjudication of preliminary objections, with bifurcation being granted as a matter of course in the proceedings. In the absence of an agreement between the parties, components that are automatically excluded under the terms of the fisheries limitation are to be submitted, *faute de mieux*, to compulsory conciliation.

256 NAFO, 'Performance Review Panel Report: 2011', pp. 9, 39 & 176–180.
257 LOSC 299(1); and *see* the *Delimitation of the Maritime Boundary in the Bay of Bengal* (Judgment) 2012, *Declaration of Judge Treves*, §12.
258 Serdy, Article 299, 1935 (¶7) in Proelss, *The United Nations Convention on the Law of the Sea* (2017).

3.6 CONCILIATION PROCEDURE

Conciliation has long been a traditional means of peacefully resolving disputes in diplomatic practice and had proven effective in the law of the sea even before the Convention's entry into force. It showed particular promise in addressing complex issues, such as fisheries and maritime boundary delimitation, as seen in the case of Iceland and Norway, where a commission was tasked with dividing an undelimited CS area in the Arctic Ocean.[259] This flexible approach avoided setting binding legal precedent or applying case law not reflective of the parties' preferences, while allowing broader economic and social factors to be considered.[260]

The Agreement makes conciliation readily available through the Convention in the case of fisheries disputes.[261] While the Convention does not define conciliation, it is understood as a method for addressing disputes of any nature, whereby a commission established by the parties undertakes an impartial examination of the issues and seeks either to define terms for a settlement likely to be accepted by them or to assist in reaching a resolution.[262] The Agreement does not elaborate on the procedure to be followed, as this is already provided under the Convention, with Part XV LOSC applying *mutatis mutandis* for this matter as well. Either party to a fisheries dispute may initiate the process by inviting the other party, pursuant to article 284 LOSC, to refer the dispute to conciliation under either the procedure set out in section 1 Annex V LOSC, or another procedure available *inter se*.[263] If the invitation is not accepted, or if the parties cannot agree on the procedure to be followed, the conciliation process is deemed to have ended. However, once the dispute has been duly

---

259 *Conciliation Commission on the Continental Shelf Area between Iceland and Jan Mayen: Report and recommendations to the governments of Iceland and Norway* (Decision) 1981.

260 Linderfalk, 'The Jan Mayen Case (Iceland/Norway): An example of successful conciliation', in Tomuschat *et al.* (Eds) *Conciliation in International Law* (2017) 193.

261 FSA 27.

262 Wolfrum, 'Conciliation under the UN Convention on the Law of the Sea', in Tomuschat *et al.* (Eds) *Conciliation in International Law* (2017) 182, reflecting a generally accepted definition in international law as employed by the Annuaire de l'Institut de Droit International, 'Resolution 49(II) On International Conciliation' (11/9/1961).

263 *e.g.* UN, 'Model Rules for the Conciliation of Disputes Between States' (1995), or the specialized procedures of PCA in the 'Optional Rules for Conciliation' (1996) and 'Optional Rules for Conciliation of Disputes Relating to Natural Resources and/or the Environment' (2002).

submitted, the proceedings may be terminated only under the agreed conciliation procedure, unless otherwise agreed.[264]

While conciliation is initially offered as a voluntary option, it compulsorily applies to disputes excluded by article 297(3)(a) LOSC or those excepted under article 298 LOSC by means of an optional declaration. Compulsory conciliation was invoked for the first time under the latter article in the maritime boundary dispute between Australia and Timor–Leste.[265] Where compulsory conciliation is applicable, either party may unilaterally initiate proceedings by written notification addressed to the other party, following the procedure set out in section 2 Annex V LOSC. Failure to reply to the notice or to engage in the conciliation proceedings does not prevent the process from continuing. To avoid frustration of the process, the Convention provides for matters including, inter alia, the constitution of the conciliation commission, its procedure, competence, and functions, thereby enabling the proceedings to conclude with the issuance of a report.[266]

Conciliation is also mandated for EEZ fisheries disputes excluded by article 297(3)(a) LOSC from the binding procedures set out in section 2 Part XV LOSC. Regarding the use of compulsory conciliation in the law of the sea, scholarly literature often traces its origins to the *Vienna Convention on the Law of Treaties* (1969).[267] Largely forgotten, the very notion—and the term itself—actually dates back to the ILC's deliberations on the regimes of the territorial sea and high seas. Special Rapporteur François had referred to 'compulsory conciliation' in contemplating a process that, while not akin to arbitral proceedings, could nevertheless lead to a decision independent of the parties' attitudes.[268] This 'lapidary phrase', as praised by Commissioner Sergei Borisovich Krylov (USSR), aptly described additional means for settling fisheries disputes.[269] In the present context, as already discussed, disputes involving allegations of *abus ou détournement de pouvoir* by a coastal State remain subject to compulsory submission to the procedures of section 2 Annex V LOSC, at the request of any party, as a residual and final mechanism for

---

264 LOSC 284.

265 *Timor Sea Conciliation* (Report and Recommendations) 2018.

266 *Timor Sea Conciliation* (Decision on Competence) 2016.

267 VCLT 66(b) and in the Annex; *see* among others, Treves, 'New Trends in the Settlement of Disputes and the Law of the Sea Convention', in Scheiber (Ed.) *Law of the Sea* (2000) 70–71 and Yee, 'Conciliation and the 1982 UN Convention on the Law of the Sea', ODIL (2013) 315–316.

268 A/CN.4/97 'Report on the Questions Relating to the Regime of the High Seas and the Territorial Sea' (27/1/1956) [1956*YbILC*(II) 1]; and *see* further in Chapter 4 (p. 155).

269 ILC 354th meeting (28/5/1956) [1956*YbILC*(I) 111].

settling non-justiciable fisheries disputes. Although the conciliation report is non-binding and the commission cannot override the coastal State's discretion, the process offers a basis for renewed consideration of the dispute through further cooperation.[270]

3.7 CONCLUSION

Compulsory and binding procedures for the settlement of compatibility disputes under Part VIII FSA serve a vital function in ensuring effective cooperation for the conservation and management of fish stocks. Yet, significant interpretative difficulties arise concerning article 32 FSA. Interpretations that favour extending coastal interests into high seas areas adopt a broad approach, expansively reading the exclusionary scope of the fisheries limitation to remove EEZ components from the compulsory jurisdiction of the competent court or tribunal to which a compatibility dispute is to be referred. Upon examining the *mutatis mutandis* application of Part XV LOSC, it is maintained here that, so long as the cumulative procedural preconditions under section 1 Part XV LOSC are satisfied, the proper characterization of a compatibility dispute precludes any *a priori* exclusion of questions relating to EEZ components from the procedures established in section 2 Part XV LOSC. By deconstructing the text of article 297(3)(a) LOSC within its original context and subsequently placing its literal reading within the Agreement, it is further proposed that a restrictive construction of the limitation more faithfully reflects the progressive development of conservation and management law under the Agreement and more coherently integrates the texts of both treaties. The narrow approach to the scope of exclusions thus corresponds with, and further refines, neutral interpretations of the compatibility rule, reinforcing the principle of compulsory adjudication.

[270] LOSC 297(3)(c); *qv.* Miron, 'Compulsory Conciliation Proceedings Regarding Disputes on Conservation and Management of Fish Stocks', in Kunoy–Heidar–Yiallourides (Eds) *International Fisheries Law* (2024) 190–191.

4. Drafting antecedents of embedded clauses in the Law of the Sea

4.1 INTRODUCTION

The *Fishing & Conservation Convention* (1958) progressively developed fisheries law by allowing coastal States with a special interest in maintaining the productivity of the living resources to prescribe CMMs in high seas areas adjacent to and beyond national jurisdiction. Since the legal rights associated with the concept of special interest were functional rather than absolute, and given the concept's substantive ambiguity, procedural safeguards were established to assure its objective interpretation and application in the event of disagreements between DWFS and coastal States. For that reason, the convention put in place a system of compulsory procedures for the settlement of what can be seen as an early notion of compatibility fisheries disputes. As the treaty text features extensively the pattern of clausal embeddedness, the relevant provisions constitute the necessary point of departure to critically examine the procedural stipulation in article 7(4) FSA. The first part of this chapter reflects on several years of deliberations, beginning with the development of the *ILC Draft Articles*, and examines in depth the negotiations leading to their final endorsement by UNCLOS/I. In its second part, the chapter's structural analysis turns to the Convention and the progressive development of the EEZ/CS delimitation rule, as reflected in articles that emulate the pattern of clausal embeddedness. The formula employed in the delimitation rule defers the interpretation and application of the substantive law to the littoral States concerned and, in case of disagreement, to compulsory adjudication, subject to optional exceptions. Building on the findings from the legislative work of ILC and UNCLOS/I, this study examines the extensive negotiations at UNCLOS/III on the drafting of embedded clauses and considers the influence of subsequent jurisprudence on their interpretation.

4.2 THE 1958 UNITED NATIONS FISHING & CONSERVATION CONVENTION

The value of the compulsory dispute settlement scheme under the *Fishing & Conservation Convention* (1958) has been substantially underappreciated in much of the literature published following the conclusion of UNCLOS/III. While the Convention is rightly lauded for having integrally incorporated compulsory procedures to the substantive body of its text, such commendation is often accompanied by commentary that either neglects the *Fishing & Conservation Convention* (1958) or disregards altogether its seminal role in the progressive development of compulsory procedures entailing binding decisions in law of the sea disputes. Admittedly, one reason for this marginalization—especially in comparison with the other *Geneva Conventions*—is that scholarly and institutional focus has often shifted towards the separately adopted and largely unratified *Optional Protocol* (1958), which conferred compulsory jurisdiction on ICJ.[1] However, it should be stressed that the undertakings according to article II *Optional Protocol* (1958) cover all the provisions of the *Geneva Conventions* except articles 4, 5, 6, 7 and 8 FCC, to which articles 9, 10, 11 and 12 FCC apply. The fact that the compulsory dispute settlement scheme of the convention has not been invoked to date in legal proceedings is another reason why it has understandably fallen into historical oblivion. Moreover, it has been superseded *inter partes* by the Convention since 1994, upon the latter's entry into force.[2]

Yet, for one of the *Geneva Conventions* to be considered no longer applicable as a treaty in force, the most conclusive indications of the parties' intention to that effect are required.[3] Accordingly, its potential residual effect remains unclear in circumstances where a coastal State has not declared EEZ or an equivalent functional zone. Renewed State engagement with formalities

---

1 *e.g.* Saario, 'Legal Aspects on the Peaceful Settlement of International Disputes', NJIL (1965) 77–78; Jacovides, 'The Law of the Sea – Where Now? A Comment', LCP (1983) 201; Jaenicke, 'Dispute Settlement under the Convention on the Law of the Sea', ZaöRV (1983) 815; Oda, 'Dispute Settlement Prospects in the Law of the Sea', ICLQ (1995) 863; and Caminos, 'The Jurisdiction and Procedure of the International Tribunal for the Law of the Sea: An overview', in Van Dyke *et al.* (Eds) *Governing Ocean Resources* (2013) 259.

2 LOSC 311(1), although the specific provision does not resolve all treaty-law questions that may conceivably arise between the Convention and the *Geneva Conventions*; *qv.* Sinclair, *The International Law Commission* (1987) 48 and Watts, *International Law Commission 1949–1998* (1999) 28.

3 *Anglo-French Continental Shelf Arbitration* (Decision) 1977, §47 and *see* O'Connell, *The International Law of the Sea* (1982) 22–24.

affecting substantive matters,[4] or other procedural aspects,[5] suggests that the successive codification and progressive development in the Convention and the Agreement may not have rendered it entirely obsolete. Whatever its practical implications may ultimately prove to be, the *Fishing & Conservation Convention* (1958) retains doctrinal significance, as its text represents the outcome of a robust law-making process by ILC and UNCLOS/I, furnishing the foundational blueprint for key fisheries provisions in both the Convention and the Agreement.

4.2.1 The Dispute Settlement Process

Of the limited references made to the *Fishing & Conservation Convention* (1958) in recent writings, those commenting on the recognition of a special interest for coastal States in maintaining the productivity of fisheries in adjacent high seas areas are the least surprising.[6] Fewer still go further to note that this special interest did not confer upon coastal States the right to undertake enforcement actions against other States, but rather established 'a convoluted procedure', whereby unilateral CMMs in high seas areas could be qualified in cases where negotiations with other interested States had failed.[7] In that context, a court or tribunal vested *ex-ante* with compulsory jurisdiction would serve—as has been astutely described elsewhere—as the 'conservator of last resort' where States could not reach agreement on the necessary measures.[8] Although this procedure was never invoked, it is instructive to examine its core aspects in greater depth for two principal reasons. First, as discussed earlier, the notion of compatibility disputes began to develop in its provisions concerning a special interest in high seas areas adjacent to and beyond national jurisdiction. The second reason is because the construction of clausal embeddedness

---

[4] Bosnia and Herzegovina on 12/1/1994 [CN.35.1994:XXI-3], Serbia on 12/3/2001 [CN.152.2001:XXI-3] and Montenegro on 23/12/2006 [CN.888.2006:XXI-3] proceeded to acts of treaty succession following the dissolution of Yugoslavia, while Congo acceded on 5/12/2012 [CN.693.2012:XXI-3].

[5] Colombia withdrew from the *Optional Protocol* (1958) on 20/9/2017 [CN.521.2017:XXI.5].

[6] *e.g.* Henriksen, 'Revisiting the Freedom of Fishing and Legal Obligations on States Not Party to Regional Fisheries Management Organizations', ODIL (2009) 90.

[7] *e.g.* Rothwell–Stephens, *The International Law of the Sea* (2023) 349.

[8] Oxman, 'Conservation and Management of the Living Resources of the Sea: The role of dispute settlement procedures', in Brower *et al.* (Eds) *By Peaceful Means* (2024) 420–423.

originally appeared in this treaty. For ease of reference, the key provisions to be discussed in the analysis that follows are contained in Table 4.3.

The law governing the dispute settlement process is set out in article 9(1) FCC, which constitutes a compound arbitration clause based on the concept of judicial arbitration as the designated default procedure.[9] This method of adjudication was extensively discussed by ILC and at UNCLOS/I, especially as it formed an indispensable condition for the acceptance of the substantive fisheries provisions. In elaborating the compulsory dispute settlement procedures, the Commission drew upon this novel concept,[10] which had been developed in hindsight of important dicta on the procedural frustration of *ad hoc* arbitrations.[11] The clause operates on two successive levels, triggered under the provisos of 'at the request of any of the parties' and 'unless the parties agree to'. Only when the disputing parties fail to reach agreement on the means of settlement does the clause mechanically shift from a general dispute resolution provision into a specific compulsory arbitration clause, triggered at the request of one party. Thus, whether compulsion arises depends entirely on the spirit with which the parties approach the fisheries dispute.[12] In the event that no agreement is reached within the fixed time-limits for selecting a mutually acceptable method from those provided by article 33 UNC, article 9(1) FCC mandates compulsory arbitration through a Special Commission procedure. The subsequent paragraphs of the article comprehensively address issues regarding the constitution of the Special Commission as a five-member body, capable of being formed and operating even without the cooperation of one of the parties to the dispute. The settlement process is further supported by article 10 FCC, which provides the applicable criteria for the determination of disputes and the prescription of provisional measures. Matters concerning the binding nature of decisions, as well as the grounds for their revision, are governed respectively by articles 11 and 12 FCC, offering supplementary rules to assist the forum in resolving the dispute.

---

9 A/A/CN.4/58Corr.1 'ILC Report on the Work of its Fourth Session' (4/6 to 8/8/1952) [1952*YbILC*(II) 60].

10 A/CN.4/76 'ILC Report on the Work of its Fifth Session' (1/6 to 14/8/1953) [1953*YbILC*(II) 202].

11 *see* notably the pronouncements in *Interpretation of Peace Treaties (Second phase)* (Advisory Opinion) 1950, concerning the constitution of treaty-based forums, and *Ambatielos Case* (Merits) 1953, regarding the obligation to adjudicate.

12 Opsahl, 'Towards the Rule of International Law in High Seas Fisheries', NJIL (1957) 282–306.

4.2.2 The Construction of Embedded Clauses

The textual pattern of clausal embeddedness reflects a normative imperative to prevent interpretative or applicative abuses of progressively developed legal rules. This imperative was highlighted from the outset of the ILC deliberations, when Special Rapporteur François, in his report on the *Regime of the High Seas*, emphasized that should coastal States be granted special or exclusive rights for the protection of fisheries up to 200 nm, these rights had to be exercised in accordance with procedural safeguards to preclude potential abuses.[13] While acknowledging that his statement served only as a 'principle of orientation' for the Commission's work, he proposed—without designating a specific court or tribunal—that such safeguards could only be ensured by subjecting the exercise of those rights to the compulsory jurisdiction of an adjudicative forum agreed upon by the parties.[14] This proposal was grounded in his acceptance of the principle of 'arbitration in advance', which resonated with the concept of judicial arbitration developed contemporaneously by Commissioner Georges Scelle (France) in his capacity as Special Rapporteur on Arbitral Procedure.[15] The formulation advanced by Special Rapporteur François proved to have an enduring formative effect upon the development of the fisheries provisions. In setting a morphology of articles being patterned upon the intertwinement of substantive and procedural provisions, it exerted a decisive influence on the construction of embedded clauses. The resulting structure of clausal embeddedness was thus designed not merely as a stylistic device, but as a functional expression of the principle of compulsory adjudication, which was intended to secure the fulfilment of assumed obligations and to guard against the abuse of rights.

The foundational premise underlying the construction of embedded clauses within various provisions was preserved throughout the remainder of the Commission's work on the fisheries regime. However, opposition to the principle of compulsory adjudication gradually intensified, culminating in sustained debates over the inclusion of such procedural stipulations. This resistance reached a peak during the Eighth Session, when the Commission, prior to adopting the final *ILC Draft Articles*,[16] deemed it necessary to conduct an article-by-article review. This was prompted primarily by objections to the procedural clauses embedded in the provisions addressing the concept of special interest. At that stage, the relevant content was codified in articles

13 A/CN.4/42 'Deuxième Rapport sur La Haute Mer' (10/4/1951) [1951*YbILC*(II) 75].

14 *ibid.*, 88.

15 ILC 117th meeting (4/7/1951) [1951*YbILC*(I) 301].

16 *cit.* Chapter 1 (*fn*39).

28 and 29 of the *Provisional Articles Concerning the Regime of the High Seas* (‘*Provisional Articles*’), as per the ILC numbering.[17] Specifically, at its 338th meeting, the Chairman, Commissioner Francisco Víctor García-Amador (Cuba), upon concluding the general discussion on the conservation of living resources of the high seas, remarked that, ‘[i]n view of the number and scope of the amendments [proposed] to articles 24–33, it would be advisable to defer discussion until … the Commission had had time to digest their significance’. At the 350th meeting, when deliberations on this topic resumed, the Chairman invited the Commission to undertake a detailed, article-by-article examination of the *Provisional Articles*, in conjunction with a substantial submission by Commissioner Douglas Edmonds (United States), who had proposed significant changes to the fisheries provisions outside the context of special interest (see Table 4.1).[18] Owing to the scope and nature of the submission, the Commission treated it as a proposal effectively seeking an article-by-article amendment to the *Provisional Articles*.[19]

Following extensive deliberations, only two paragraphs were ultimately submitted for a vote as finalized draft texts. These were, in procedural sequence, paragraph (3) of article 29—identical in both the *Provisional Articles* and the revised proposals—and paragraph (2) of article 26, which had been redrafted by Commissioner Edmonds to include criteria for determining the necessity and appropriateness of CMMs. With respect to the former, the first two paragraphs of article 29 were accepted in principle as formulated in the *Provisional Articles*, although it was agreed that they should be referred to the Drafting Committee for clarification of their wording.[20] In contrast, the procedural stipulations contained in paragraph (3) of article 29 provoked significant objections from two members of the Commission. These objections are noteworthy, as they targeted the structural formation of the provision itself. In response, two counterproposals were submitted: one by Commissioner Sergei Borisovich Krylov (USSR), who had consistently expressed a preference for non-binding means of dispute settlement,[21] and another by Commissioner Luis Padilla-Nervo (Mexico).

Commissioner Krylov urged the Commission to discard the embedded dispute settlement provisions, arguing that they imposed an ‘unnecessarily stringent and formal machinery’ and should instead be replaced with a reference to the general procedures laid down in article 33 UNC. In his view, these

---

17 A/CN.4/94 ‘ILC Report Covering the Work of its Seventh Session’ (2/5 to 8/7/1955) [1955*YbILC*(II) 29–30].

18 ILC 338th meeting (2/5/1956) [1956*YbILC*(I) 22–26].

19 ILC 350th meeting (19/5/1956) [1956*YbILC*(I) 82¶7].

20 ILC 352nd meeting (24/5/1956) [1956*YbILC*(I) 96¶35].

21 *see* Chapter 3 (*fn*269).

Table 4.1 *Amendments proposed by Commissioner Edmonds to Articles 26 and 27 of the ILC Provisional Articles Concerning the Regime of the High Seas*

Provisional Articles Concerning the Regime of the High Seas (1953)	Redrafted Articles on Fisheries, as submitted by Commissioner Edmonds (1956)
Article 26	Article 26
1. If the nationals of two or more States are engaged in fishing in any area of the high seas, these States shall, at the request of any of them, enter into negotiations in order to prescribe by agreement the measures necessary for the conservation of the living resources of the high seas. 2. (P) *If the States concerned do not reach agreement within a reasonable period of time, any of the parties may initiate the procedure envisaged in article 31.* (P)	1. If the nationals of two or more States are engaged in substantial fishing of the same stock or stocks of fish in any area or areas of the high seas, these States shall, at the request of any of them, enter into negotiations in order to prescribe by agreement the measures necessary for the conservation of such stock or stocks of fish. 2. (P) *If these States do not, within a reasonable period of time, reach agreement upon the need for conservation or as to the appropriateness of conservation measures proposed by any of them, any of the parties may initiate the procedure contemplated in article 31,* (P) in which case the arbitral commission shall make one or more of the following determinations depending upon the nature of the disagreement: (a) whether conservation measures are necessary to make possible the maximum sustainable productivity of the concerned stock or stocks of fish; (b) whether the specific measure or measures proposed are appropriate for this purpose, and if so which are the more appropriate, taking into account particularly: (i) the expected benefits in terms of maintained or increased productivity of the stock or stocks of fish; (ii) the cost of their application and enforcement; and (iii) their relative effectiveness and practicability; (c) whether the specific measure or measures discriminate against the fishermen of any participating State as such. 3. Measures considered by the arbitral commission under paragraph 2(b) of this article shall not be sanctioned by the arbitral commission if they discriminate against the fishermen of any participating State as such.

Provisional Articles Concerning the Regime of the High Seas (1953)	Redrafted Articles on Fisheries, as submitted by Commissioner Edmonds (1956)
Article 27	Article 27
1. If, subsequent to the adoption of the measures referred to in articles 25 and 26, nationals of other States engage in fishing in the same area, the measures adopted shall be applicable to them. 2. (P) *If the States whose nationals take part in the fisheries do not accept the measures so adopted and if no agreement can be reached within a reasonable period of time, any of the interested parties may initiate the procedure envisaged in article 31.* (P) Subject to paragraph 2 of article 32, the measures adopted shall remain obligatory pending the arbitral decision.	1. If, subsequent to the adoption of the measures referred to in articles 25 and 26, nationals of other States engage in fishing the same stock or stocks of fish in any area or areas of the high seas, the measures adopted shall be applicable to them. 2. (P) *If the States whose nationals are so engaged in fishing do not accept the measures so adopted and if no agreement can be reached within a reasonable period of time, any of the interested States may initiate the procedure contemplated in article 31, in which case the arbitral commission shall make one or more of the determinations set forth in paragraph 2 of article 26 of these articles, depending upon the nature of the disagreement.* (P) Subject to the provisions of paragraph 2 of article 33 of these articles, the measures adopted shall be obligatory pending the arbitral decision. 3. Where the maximum sustainable yield is, within reasonable limits, already being obtained from any stock of fish, and the maintenance and further development of such yield is dependent on the conservation programme, involving research, development and conservation being carried on by all the States whose nationals are fishing such stock substantially, States not fishing such stock substantially or which have not done so within a reasonable period of time, excepting the coastal State adjacent to the waters in which this stock occurs, shall abstain from fishing such stock. In the event of disagreement as to whether a particular stock meets the above qualifications for abstention, the matter shall be referred for decision to an arbitral commission to be set up as provided in article 31. 4. The arbitral commission shall reach its decision and make its recommendations under paragraph 3 of this article on the basis of the following criteria: (a) whether the stock is subject to reasonably adequate scientific investigation with the object of establishing and taking the measures required to make possible the maximum sustainable yield; (b) whether the stock is under reasonable regulation and control for the purpose of making possible the maximum sustainable yield, and whether such yield is dependent upon the programme of regulation and control; and (c) whether the stock is, within reasonable limits, under such exploitation that an increase in the amount of fishing will not reasonably be expected to result in any substantial increase in the sustainable yield.
Article 28	Article 28
1. A coastal State having a special interest in the maintenance of the productivity of the living resources in any area of the high seas contiguous to its coasts is entitled to take part on an equal footing in any system of research and regulation in that area, even though its nationals do not carry on fishing there. 2. (P) *If the States concerned do not reach agreement within a reasonable period of time, any of the parties may initiate the procedure envisaged in article 31.* (P)	1. A coastal State having a special interest in the maintenance of the productivity of the living resources in any area of the high seas contiguous to its coasts is entitled to take part on an equal footing in any system of research and regulation in that area, even though its nationals do not carry on fishing there. 2. (P) *If the States concerned do not reach agreement within a reasonable period of time, any of the parties may initiate the procedure envisaged in article 31.* (P)

Provisional Articles Concerning the Regime of the High Seas (1953)	Redrafted Articles on Fisheries, as submitted by Commissioner Edmonds (1956)
Article 29	Article 29
1. A coastal State having a special interest in the maintenance of the productivity of the living resources in any area of the high seas contiguous to its coasts may adopt unilaterally whatever measures of conservation are appropriate in the area where this interest exists, provided that negotiations with the other States concerned have not led to an agreement within a reasonable period of time. 2. The measures which the coastal State adopts under the first paragraph of this article shall be valid as to other States only if the following requirements are fulfilled: (a) that scientific evidence shows that there is an imperative and urgent need for measures of conservation; (b) that the measures adopted are based on appropriate scientific findings; (c) that such measures do not discriminate against foreign fishermen. 3. (P) *If these measures are not accepted by the other States concerned, any of the parties may initiate the procedure envisaged in article 31.* (P) Subject to paragraph 2 of article 32, the measures adopted shall remain obligatory pending the arbitral decision.	1. A coastal State having a special interest in the maintenance of the productivity of the living resources in any area of the high seas contiguous to its coasts may adopt unilaterally whatever measures of conservation are appropriate in the area where this interest exists, provided that negotiations with the other States concerned have not led to an agreement within a reasonable period of time. 2. The measures which the coastal State adopts under the first paragraph of this article shall be valid as to other States only if the following requirements are fulfilled: (a) that scientific evidence shows that there is an imperative and urgent need for measures of conservation; (b) that the measures adopted are based on appropriate scientific findings; (c) that such measures do not discriminate against foreign fishermen. 3. (P) *If these measures are not accepted by the other States concerned, any of the parties may initiate the procedure envisaged in article 31.* (P) Subject to paragraph 2 of article 32, the measures adopted shall remain obligatory pending the arbitral decision.

Notes: The symbol '(P)' has been used to highlight the textual disposition of clausal embeddedness.

procedural stipulations across multiple articles introduced unnecessary complexity. He noted that '[a]s a matter of drafting, it would be preferable to deal with the settlement of disputes in a single article, so as to remove the somewhat clumsy repetition which occurred in, for example, articles 26, 27, 28 and 29'.[22] Likewise, Commissioner Padilla-Nervo expressed scepticism regarding the necessity of detailed procedural clauses. He argued that the likelihood of disputes was overstated, given the degree of technical specificity built into article 29, which in his view would obviate controversy among States acting in good faith. As a result, he proposed that paragraph (3) of article 29, along with articles 31–33, be replaced by a simplified formulation reading as follows:

> If these measures are not accepted by the other States concerned, the parties to the dispute shall seek a settlement by negotiation, inquiry, mediation, conciliation, arbitration, judicial settlement, reference to regional bodies or by other peaceful means of their choice.[23]

The Chairman, expressing regret that difficulties had arisen over a matter that appeared to have been satisfactorily resolved in previous sessions, acknowledged that Commissioner Krylov's general proposal to eliminate the procedural provisions, along with Commissioner Padilla-Nervo's specific proposal to replace compulsory arbitration with alternative peaceful settlement mechanisms, necessitated a vote on both amendments.[24] Following their rejection—Krylov's proposal by 10 votes to 4 with 1 abstention, and Padilla-Nervo's by 9 votes to 4 with 2 abstentions—*Provisional Article* 29(3) was retained intact. On voting against these paragraphs, Commissioner Krylov expressed doubts regarding the nature of the proposed arbitral commission even though much of what they contained was otherwise sound.[25] A similar conclusion regarding the Commission's support for embedded clauses may be inferred from the outcome of the vote on *Provisional Article* 26(2). Although the version tabled by Commissioner Edmonds was not adopted, the embedded clause within it remained unaffected, despite the article having been submitted to a vote as a concrete text.[26] The procedural stipulations embedded in *Provisional Articles* 27(2) and 28(2) were subsequently endorsed without further vote.[27]

In sum, ILC did not endorse any inherent right of coastal States to unilaterally extend their regulatory authority into high seas areas. Rather, it required

22 ILC 352nd meeting (28/5/1956) [1956*YbILC*(I) 97¶¶44–45].
23 *ibid.*, 99.
24 ILC 353rd meeting (25/5/1956) [1956*YbILC*(I) 101].
25 *ibid.*, 104–105.
26 ILC 356th meeting (30/5/1956) [1956*YbILC*(I)122].
27 ILC 357th meeting (31/5/1956) [1956*YbILC*(I)127*ff*].

cooperation between DWFS and coastal States, and where such cooperation could not be achieved, the unilateral exercise of regulatory powers by coastal States was to be subjected to scrutiny through compulsory arbitration in the event of a disagreement.[28]

4.2.3 The Functional Morphology of Clausal Embeddedness

During the ILC deliberations on the *Provisional Articles*, two notable characterizations of clausal embeddedness emerged. Commissioner Krylov had criticized the procedural provisions as mere 'clumsy repetition'. In contrast, Commissioner Edmonds, while introducing his own amendments, acknowledged that 'the draft articles did not constitute an ideal text', yet contended that 'on the whole they formed a consistent pattern of provisions safeguarding the interests of all States concerned'.[29] At first glance, the structure of clausal embeddedness may appear as a grotesque textual architecture. However, its substantive value is undeniable. Dismissing this drafting technique, as Commissioner Krylov suggested, overlooks the significance of a legal text meticulously crafted by distinguished legal experts and debated extensively by the Commission over nearly four years. Furthermore, this text was subsequently adopted by the universal conference of UNCLOS/I after rigorous scrutiny by the Third Committee. It would be imprudent to regard this seemingly abstruse textual presence as a drafting redundancy or otherwise superfluity without examining the functionality of its pattern within a signifying morphological context. The analysis that follows demonstrates that clausal embeddedness serves two mutually reinforcing functions on the construction of a principled rule. First, it integrates the principle of compulsory adjudication as a constituent procedural element directly into the substantive articles. Second, as a corollary of this integration, it acts decisively as a safeguard against any potential abuse in the interpretation or application of these articles. Accordingly, the structure of clausal embeddedness was crafted to guarantee the objective interpretation and consistent application of substantive rules through compulsory third-party adjudication whenever a disagreement occurs.

The principle of compulsory adjudication

Further valuable evidence supporting the correlation between the textual pattern of clausal embeddedness and the principle of compulsory adjudication emerges from an analysis of the objections raised against their inclusion in

---

[28] Bishop, 'International Law Commission Draft Articles on Fisheries', AJIL (1956) 631.

[29] ILC 351st meeting (23/5/1956) [1956*YbILC*(I) 89¶23].

the treaty text during UNCLOS/I. Notably, embedded clauses were opposed by the USSR, whose stance was backed by other States within the communist bloc, as well as by a number of Latin American States, predominantly those with coastal interests in the Pacific Ocean. The latter group included States that, at the time, had made excessive claims of sovereignty over marine living resources in sea areas extending well beyond the prevailing customary 3 nm limit. This coalition was unusual in the context of negotiating an international legal instrument on fisheries, given that the USSR policy was mostly aligned with the interests of DWFS. Albeit for different reasons, the objections advanced by both blocs converged on a shared opposition to the principle of compulsory adjudication. Their arguments militating strongly against embedded clauses, especially in the articles conferring to coastal States a special interest, are identified and discussed in turn below.

An unambiguous exposition of the rationale behind the objections to embedded clauses was presented in a Mexican proposal seeking to replace the scheme of compulsory arbitration altogether by inserting in its place article 33 UNC (see Box 4.1).[30] Although the purported aim of the proposal was 'merely to simplify the complicated procedure' set out in *ILC Draft Article* 57 (see Table 4.3), Mexico openly acknowledged the inevitable dilution of the compulsoriness principle, justifying it on the basis that the overwhelming majority of States would not accept these procedural stipulations.[31] A similar stance had been adopted by the Mexican Commissioner Padilla-Nervo during the ILC deliberations, where he had emphasized that 'no one could entertain any illusions about the possibility of securing acceptance of compulsory arbitration', insisting that the only kind of durable settlement would be one voluntarily accepted by the parties, or by recourse to one of the processes enumerated in article 33 UNC.[32] Furthermore, Mexico sought to downplay the significance of compulsory procedures, arguing that 'disputes of an infinitely graver nature than those connected with fisheries were solved by means of peaceful settlement freely chosen according to the circumstances', and therefore there 'was absolutely no need to impose a particular means of settlement which, by its very inflexibility, was likely to give rise to more serious problems than the minor ones it sought to solve'.[33]

---

30 A/Conf.13/C.3/L.1 (7/3/1958) [*Geneva Off.Records* V:134].

31 Third Committee 29th meeting (12/4/1958) [*Geneva Off.Records* V:80–81].

32 ILC 338th meeting (2/5/1956) [1956*YbILC*(I) 25–26].

33 Third Committee 28th meeting (11/4/1958) [*Geneva Off.Records* V:74–75].

BOX 4.1 SUBMISSION BY MEXICO AT UNCLOS/I TO AMEND ARTICLES 52 TO 59 OF THE ILC ARTICLES CONCERNING THE LAW OF THE SEA

A/Conf.13/C.3/L.1 (7 March 1958)

Articles 57 to 59

Replace the text of the draft articles 57, 58 and 59 prepared by the International Law Commission by the following text:

> Disputes concerning the matters to which the present provisions relate shall be settled by the States concerned by the modes of pacific settlement provided for in Article 33 of the Charter of the United Nations.

If the above proposal is approved, it will be necessary to make the corresponding changes in articles 52 to 56 inclusive.

In seeking to replace compulsory procedures, Mexico further specified that 'if the above proposal [was] approved, it [would] be necessary to make the corresponding changes in [*ILC Draft Articles*] 52 to 56 inclusive'. Clearly, this particular aspect of the amendment was directed at dismantling the structure of clausal embeddedness, as it can be reasonably inferred that Mexico deemed it essential to eliminate any procedural provision that could be construed as substantively reinforcing the principle of compulsory adjudication within the respective articles. This understanding was unequivocally corroborated by Bulgaria, which expressed its inability to support paragraphs containing procedural stipulations, noting that the corresponding procedures proposed for the settlement of disputes in *ILC Draft Articles* 53–57 were predicated on the principle of compulsory arbitration. Unable to accept this outcome, Bulgaria accordingly supported the Mexican proposal.[34] Ecuador, another State backing the Mexican initiative, similarly opposed the principle of compulsory adjudication. Shortly after the Mexican proposal was tabled, Ecuador appended a statement endorsing the dispute settlement scheme proposed therein, alleging that 'the system for the settlement of disputes described in [*ILC Draft Article*]

---

[34] Third Committee 10th meeting (18/3/1958) [*Geneva Off.Records* V:23–24].

57 had been widely criticized, and it seemed that States were not prepared to accept compulsory arbitration'.[35]

In particular, regarding the Mexican proposal—being the only explicit reference made by Ecuador to the settlement procedures—it aimed not merely to replace the compulsory arbitration clause with a general dispute settlement clause, the efficacy of which remained uncertain in securing compulsory and binding procedures, but also to demand the withdrawal of all procedural stipulations from the substantive articles. The ulterior motive behind the Mexican proposal was thus to erode the compulsoriness principle, which underpinned the acceptance of the entire fisheries regime. To this end, it sought not only to amend the central dispute settlement provision by replacing it with a vague reference to article 33 UNC, but also to reformat comprehensively the substantive *ILC Draft Articles* containing embedded clauses. The Mexican proposal was subsequently put to the vote and—contrary to the Ecuadorean assertion regarding the prevailing will of States at UNCLOS/I against the compulsory adjudication of fisheries disputes—it was rejected by 32 votes to 19, with 12 abstentions.[36]

In the same vein, another proposal was submitted by Ecuador jointly with Chile, Costa Rica, and Peru.[37] A comparison of this proposal with the corresponding *ILC Draft Articles* reveals immediately that the embedded clauses were purposefully excised from the text of the respective articles (see Table 4.2). The intention to dismantle the textual pattern of clausal embeddedness becomes even more conspicuous when considering other proposals concerning the articles on special interest, which, although consolidating the provisions into a single article, nevertheless preserved the procedural stipulations intact.[38] Particularly noteworthy in Ecuador's position is its consistent tendency to defer the prospect of dispute settlement to voluntary arrangements and methods external to the treaty text. For instance, all the methods suggested by Ecuador would, at best, fall within the ambit of article 33 UNC, as similarly proposed in the Mexican amendment. Further reflecting Ecuador's disinclination towards compulsory procedures is the background to a statement in which it maintained that:

> It was the object of safeguarding the resources of the maritime areas off their coasts that Chile, Ecuador and Peru had, by the Declaration of 1952, laid down

---

35 Third Committee 9th meeting (17/3/1958) [*Geneva Off.Records* V:18¶8].

36 Third Committee 30th meeting (12/4/1958) [*Geneva Off.Records* V:83].

37 A/Conf.13/C.3/L.41 (27/3/1958) [*Geneva Off.Records* V:147].

38 *e.g.* the proposals submitted by Yugoslavia A/Conf.13/C.3/L.13 (21/3/1958) [*Geneva Off.Records* V:138] and South Korea A/Conf.13/C.3/L.45 (28/3/1958) [*Geneva Off.Records* V:148].

Table 4.2 *Joint submission by Chile, Costa Rica, Ecuador, and Peru at UNCLOS/I to Amend Articles 54 and 55 of the ILC Articles Concerning the Law of the Sea*

Articles Concerning the Law of the Sea adopted by the ILC (1956)	A/Conf.13/C.3/L.41 (27 March 1958)
Article 54 1. A coastal State has a special interest in the maintenance of the productivity of the living resources in any area of the high seas adjacent to its territorial sea. 2. A coastal State is entitled to take part on an equal footing in any system of research and regulation in that area, even though its nationals do not carry on fishing there. 3. (P) *If the States concerned do not reach agreement within a reasonable period of time, any of the parties may initiate the procedure contemplated by article 57.* (P) Article 55 1. Having regard to the provisions of paragraph 1 of article 54, any coastal State may, with a view to the maintenance of the productivity of the living resources of the sea, adopt unilateral measures of conservation appropriate to any stock of fish or other marine resources in any area of the high seas adjacent to its territorial sea, provided that negotiations to that effect with the other States concerned have not led to an agreement within a reasonable period of time. 2. The measures which the coastal State adopts under the previous paragraph shall be valid as to other States only if the following requirements are fulfilled: (a) that scientific evidence shows that there is an urgent need for measures of conservation; (b) that the measures adopted are based on appropriate scientific findings; (c) that such measures do not discriminate against foreign fishermen. 3. (P) *If these measures are not accepted by the other States concerned, any of the parties may initiate the procedure contemplated by article 57. Subject to paragraph 2 of article 58, the measures adopted shall remain obligatory pending the arbitral decision.* (P)	Replace articles 54 and 55 by a single text as follows: 1. In addition to the right established by article 49 to make use of the resources of the sea, a coastal State has a special right, inherent in its geographical situation, to adopt measures for their conservation and to regulate and control their exploitation in an area of sea adjacent to its territorial sea or its contiguous zone, as the case may be. 2. Other States may not object to the provisions which a coastal State may adopt for that area, based on the conservation of the living resources of the sea and on the subsistence and economic development needs of its population, provided such provisions do not exclude foreign fishermen who comply therewith from the exploitation of the resources. 3. The provisions referred to in the preceding paragraph shall be based on scientific investigations and findings showing the need for them. 4. When it is necessary to restrict the scale of fishing, a coastal State applying measures of conservation may require the nationals of other States who have been authorized to fish in that area to refrain entirely or partially from doing so for the period of time which may be necessary for a sufficient yield to be restored. 5. When, in order to conserve the living resources of an area of sea, co-operation with other States is deemed desirable, because they are coastal States bordering the same area, the coastal State shall institute negotiations with those States with a view to the adoption of joint conservation measures. Should no agreement be reached, however, the measures adopted by the coastal State by virtue of the preceding paragraphs shall continue in force. 6. A coastal State shall have the right to join any research organization and to participate in any system of investigation set up by other States or international organizations in respect of the said area of the sea.

Notes: The symbol '(P)' has been used to highlight the textual disposition of clausal embeddedness.

> a common policy for the conservation, development and rational exploitation of those resources and set up joint machinery for the regulation of fishing in the areas in question. That declaration was not an isolated case; other countries had enacted regulations concerning the use or conservation of the living resources of the high seas.[39]

The declaration cited in the statement pertains to the *Santiago Declaration* (1952), adopted in conjunction with the *Convenio sobre la Organización de la Comisión Permanente de la Conferencia sobre Explotación y Conservación de las Riquezas Marítimas del Pacífico Sur* (1952), which established the *Comisión Permanente del Pacífico Sur* ('CPPS'). This regional regime stood in tension with the principles envisioned in the fisheries provisions of the *ILC Draft Articles* under negotiation at UNCLOS/I. Contemporaneous commentary from that period confirms that the undersigning States expressed strong disapproval of compulsory dispute settlement procedures concerning fisheries. Notably, at that time, Chile, Ecuador, and Peru were firmly entrenched in their extended unilateral claims and refused to resolve related disputes through negotiation or other available international mechanisms, including recourse to ICJ.[40]

Turning to the USSR, its approach to law-making treaties during this period consistently resisted the settlement of disputes through arbitration or judicial means. Instead, it ardently supported direct diplomatic negotiations and non-binding procedures, even if these ultimately led to a stalemate. This stance was evident during UNCLOS/I, where the USSR negotiated with marked aversion to any matters that implicated by necessity the prospect of compulsory adjudication of disputes.[41] Lending generous support to the Mexican proposal, the USSR emphasized that, since the Charter itself contained no clause dictating compulsory recourse to arbitration or ICJ, it was difficult to comprehend 'why, in the matter of fishing, it was necessary to confer on such recourse a compulsory character which the authors of the Charter had not seen fit to accord'.[42] It considered the dispute settlement provisions to be 'out of place' and asserted that their deletion 'would improve the chances of reaching agreement on the articles embodying the substance of contemporary international law of the

---

39 Third Committee 9th meeting (17/3/1958) [*Geneva Off.Records* V:18¶[3].

40 *see* Johnston, *The International Law of Fisheries* (1965) 334*ff* and Goldie, 'The Oceans' Resources and International Law – Possible Developments in Regional Fisheries Management', CJTL (1969) 31*ff.*

41 Regarding the Soviet doctrine during that period *see* Triska–Slusser, *The Theory, Law, and Policy of Soviet Treaties* (1962) 381–388 and Lissitzyn, *International Law Today and Tomorrow* (1965) 47–62.

42 Third Committee 28th meeting (11/4/1958) [*Geneva Off.Records* V:77].

sea'.[43] Accordingly, the USSR proposed to amend *ILC Draft Articles* 54 and 55, which addressed the special interest of coastal States, by merging them into a single article and omitting the embedded clauses.[44] The intention to disintegrate the compulsoriness principle was evident, as transpired from the introductory statement by Sergei Krylov, who, on that occasion, served as the USSR delegate to UNCLOS/I. In his argument, it was claimed that eliminating the dispute settlement provisions would be consistent with the recommendations of the *1955 Rome Conference*.[45] However, this understanding did not accord with the actual recommendations made. These had been carefully considered during ILC's deliberations, notably by Commissioner Edmonds, who argued that:

> The Rome Conference had linked the granting of special rights to coastal States with the obligation to resort to arbitration in the case of any dispute arising out of the exercise of those rights, and the Commission ... had accepted that principle, which was the cornerstone of the whole system. The Commission had therefore agreed that international law should recognize the grant of certain additional rights to the coastal State, but that in exchange for such rights the coastal State should accept arbitration if any measure it imposed was objected to by another interested State.[46]

The position held by the USSR at UNCLOS/I remained unchanged from that previously taken by Commissioner Krylov during the ILC deliberations. It is worth recalling one of his statements, which, at this point, merits being quoted in full:

> [He] was surprised that lawyers of such distinction should expect governments to commit themselves to compulsory arbitration when machinery for the peaceful settlement of disputes was provided by article 33 [UNC]. Without in any way wishing to be intransigent, [he] urged the Commission to drop the provisions concerning compulsory arbitration...which it would be difficult for States to accept, and to substitute for that unnecessarily stringent and formal machinery a provision for the settlement of disputes in accordance with the [Charter's] procedures ... In conclusion, as a matter of drafting, it would be preferable to deal with the settlement of disputes in a single article, so as to remove the somewhat clumsy repetition which now occurred in, for example, [*ILC Draft Articles*] 26, 27, 28 and 29.[47]

---

43 Third Committee 6th meeting (12/3/1958) [*Geneva Off.Records* V:8].

44 A/Conf.13/C.3/L.42 'USSR: Proposal (articles 54 and 55)' (28/3/1958) [*Geneva Off.Records* V:147].

45 Third Committee 24th meeting (9/4/1958) [*Geneva Off.Records* V:58].

46 ILC 353rd meeting (25/5/1956) [1956*YbILC*(I) 102¶10].

47 ILC 352nd meeting (24/5/1956) [1956*YbILC*(I) 97¶44].

This statement portrays very well the requisite context in which the functional morphology of embedded clauses can be fully revealed. He himself had personified numerous times the Soviet aversion to the principle of compulsory adjudication in various international fora.[48]

Evidently, the series of proposals submitted by the USSR at UNCLOS/I were directed at severing the procedural stipulations from the substantive rules in the relevant articles, thereby seeking to relegate the compulsoriness principle within the treaty text.[49] This approach reflected the USSR's prevailing resistance at the time to binding dispute settlement mechanisms. Notably, following the adoption of the *Fishing & Conservation Convention* (1958) by UNCLOS/I, it concluded or acceded to a number of fisheries-related treaties, which none of them contained dispute settlement procedures. This policy had been in effect since 1956, in response to the *ILC Draft Articles*, beginning with the conclusion of the bilateral *Convention Concerning the High Seas Fisheries of the Northwest Pacific Ocean* (1956) with Japan. Subsequently, the USSR entered into the *Convention Concerning Fishing in the Black Sea* (1959) with Bulgaria and Romania, and, in the interim, acceded to multilateral regional arrangements under the *Convention for the Regulation of the Meshes of Fishing Nets and the Size Limits of Fish* (1946) and to the *International Convention for the Northwest Atlantic Fisheries* (1949). It goes without saying that the USSR refrained from signing the *Fishing & Conservation Convention* (1958) principally because of the objections to the compulsory settlement of fisheries disputes, nor, indeed, did it become a party to the *Optional Protocol* (1958).[50]

Overall, the arguments considered above demonstrate the rationale underlying the pattern of clausal embeddedness, which establishes an inextricable link between the substantive rules and the dispute settlement procedures of the *Fishing & Conservation Convention* (1958), effectively associating the principle of compulsory adjudication with disputes concerning the interpretation and application of the respective articles. Therefore, those States opposed to the compulsoriness principle resisted from the outset the construction of

---

48 *see* Kulski, 'The Soviet Interpretation of International Law', AJIL (1955) 518 and Zile, 'A Soviet Contribution to International Adjudication: Professor Krylov's jurisprudential legacy', AJIL (1964) 360.

49 *e.g.* A/Conf.13/C.3/L.30 'Poland and USSR: Proposal (article 56)' (26/5/1958) [*Geneva Off.Records* V:143] and A/Conf.13/C.3/L.61 'USSR: Proposal (articles 57 to 59)' (7/4/1958) [*Geneva Off.Records* V:152].

50 *see* Butler, *The Soviet Union and the Law of the Sea* (1971) 189–194; Hayashi, 'Soviet Policy on International Regulation of High Seas Fisheries', *Cornell International Law Journal* (1972) 131; and Henkin *et al.*, *International Law* (1980) 870.

embedded clauses in the treaty text. To put this more prosaically, it was a common understanding among the participating States in the UNCLOS/I negotiations that these clauses were nothing short of a positive textual representation of the compulsoriness principle within the articles in which they were embedded.

The impressed perception of inseparableness between substance and procedure

Despite facing numerous deconstructive proposals during both the ILC deliberations and the UNCLOS/I negotiations, the pattern of clausal embeddedness remained intact up to the final adoption of the *Fishing & Conservation Convention* (1958). A cursory examination of the articles adopted by UNCLOS/I reveals that, aside from minor amendments, they closely mirror those previously adopted by the Commission (see Table 4.3). As aptly noted by a former ILC member and ICJ judge, while the honour of adopting these rules in the form of an international convention belongs to UNCLOS/I, the merit of their elaboration rests entirely with the Commission.[51] The functional rationale underlying the construction of clausal embeddedness was propounded early on in the second report of the Special Rapporteur François. Throughout their textual evolution in the drafts of the Commission and the Third Committee, consistent statements—whether in approval or critique—highlighted the instrumental role of this clausal construction. Its primary purpose was to consolidate the principle of compulsory adjudication concerning disputes over the interpretation or application of the substantive rules contained within the fisheries provisions.

The stipulations contained in the embedded clauses were inseparable from the substance of the fisheries provisions, thereby possessing not only procedural significance but also substantive importance by upholding the principle against the abuse of rights. The recognition of the imperative need for the progressive development of the law—without unsettling traditional freedoms yet providing States with a legal framework for cooperation—was a defining feature of the *Fishing & Conservation Convention* (1958). The convention advanced beyond prior customary law by granting coastal States new regulatory rights in high seas areas adjacent to and beyond their territorial sea, specifically for conservation purposes. Concurrently, it acknowledged that the exercise of these rights could be susceptible to abuse, necessitating that DWFS be afforded adequate guarantees for the efficient and equitable

---

51 Spiropoulos, 'The Contribution of the International Law Commission to the Codification of the Law on Fishing and Conservation of the Living Resources of the High Seas', in Alfaro *et al.* (Eds) *Varia Juris Gentium* (1959) 332*ff.*

Table 4.3 *The consistent pattern of clausal embeddedness in the drafting history of the 1958 UN Fishing & Conservation Convention*

Articles Concerning the Law of the Sea adopted by the ILC (1956)	Text adopted by the Third Committee of UNCLOS/I (1958)	Final Text of the *Fishing & Conservation Convention* (1958)
Article 54	Article 54	Article 6
1. A coastal State has a special interest in the maintenance of the productivity of the living resources in any area of the high seas adjacent to its territorial sea. 2. A coastal State is entitled to take part on an equal footing in any system of research and regulation in that area, even though its nationals do not carry on fishing there. 3. (P) *If the States concerned do not reach agreement within a reasonable period of time, any of the parties may initiate the procedure contemplated by article 57.* (P)	1. A coastal State has a special interest in the maintenance of the productivity of the living resources in any area of the high seas adjacent to its territorial sea. 2. A coastal State is entitled to take part on an equal footing in any system of research and regulation for conservation purposes in that area, even though its nationals do not carry on fishing there. 3. A State whose nationals are engaged in fishing in any area of the high seas adjacent to the territorial sea of a coastal State shall, at the request of that coastal State, enter into negotiations with a view to prescribing by agreement the measures necessary for the conservation of the living resources of the high seas in that area. 4. A State whose nationals are engaged in fishing in any area of the high seas adjacent to the territorial sea of a coastal State shall not enforce conservation measures in that area which are opposed to those which have been adopted by the coastal State but may enter into negotiations with the coastal State with a view to prescribing by agreement the measures necessary for the conservation of the living resources of the high seas in that area. 5. (P) *If the States concerned do not reach agreement, with respect to conservation measures, within twelve months, any of the parties may initiate the procedure contemplated by article 57.* (P)	1. A coastal State has a special interest in the maintenance of the productivity of the living resources in any area of the high seas adjacent to its territorial sea. 2. A coastal State is entitled to take part on an equal footing in any system of research and regulation for purposes of conservation of the living resources of the high seas in that area, even though its nationals do not carry on fishing there. 3. A State whose nationals are engaged in fishing in any area of the high seas adjacent to the territorial sea of a State shall, at the request of that coastal State, enter into negotiations with a view to prescribing by agreement the measures necessary for the conservation of the living resources of the high seas in that area. 4. A State whose nationals are engaged in fishing in any area of the high seas adjacent to the territorial sea of a coastal State shall not enforce conservation measures in that area which are opposed to those which have been adopted by the coastal State, but may enter into negotiations with the coastal State with a view to prescribing by agreement the measures necessary for the conservation of the living resources of the high seas in that area. 5. (P) *If the States concerned do not reach agreement with respect to conservation measures within twelve months, any of the parties may initiate the procedure contemplated by article 9.* (P)

Articles Concerning the Law of the Sea adopted by the ILC (1956)	Text adopted by the Third Committee of UNCLOS/I (1958)	Final Text of the *Fishing & Conservation Convention* (1958)
Article 55	Article 55	Article 7
1. Having regard to the provisions of paragraph 1 of article 54, any coastal State may, with a view to the maintenance of the productivity of the living resources of the sea, adopt unilateral measures of conservation appropriate to any stock of fish or other marine resources in any area of the high seas adjacent to its territorial sea, provided that negotiations to that effect with the other States concerned have not led to an agreement within a reasonable period of time. 2. The measures which the coastal State adopts under the previous paragraph shall be valid as to other States only if the following requirements are fulfilled: (a) that scientific evidence shows that there is an urgent need for measures of conservation; (b) that the measures adopted are based on appropriate scientific findings; (c) that such measures do not discriminate against foreign fishermen. 3. (P) *If these measures are not accepted by the other States concerned, any of the parties may initiate the procedure contemplated by article 57. Subject to paragraph 2 of article 58, the measures adopted shall remain obligatory pending the arbitral decision.* (P)	1. Having regard to the provisions of paragraph 1 of article 54, any coastal State may, with a view to the maintenance of the productivity of the living resources of the sea, adopt unilateral measures of conservation appropriate to any stock of fish or other marine resources in any area of the high seas adjacent to its territorial sea, provided that negotiations to that effect with the other States concerned have not led to an agreement within six months. 2. The measures which the coastal State adopts under the previous paragraph shall be valid as to other States only if the following requirements are fulfilled: (a) that there is a need for urgent application of conservation measures in the light of the existing knowledge of the fishery; (b) that the measures adopted are based on appropriate scientific findings; (c) that such measures do not discriminate against foreign fishermen. 3. These measures shall remain in force pending the settlement, in accordance with the pertinent provisions of this convention, of any disagreement as to their validity. 4. (P) *If the measures are not accepted by the other States concerned, any of the parties may initiate the procedure contemplated by article 57. Subject to paragraph 2 of article 58, the measures adopted shall remain obligatory pending the decision of the special commission.* (P) 5. The principles of geographical demarcation as defined in articles 12 and 14 shall be adopted when coasts of different States are involved.	1. Having regard to the provisions of paragraph 1 of article 6, any coastal State may, with a view to the maintenance of the productivity of the living resources of the sea, adopt unilateral measures of conservation appropriate to any stock of fish or other marine resources in any area of the high seas adjacent to its territorial sea, provided that negotiations to that effect with the other States concerned have not led to an agreement within six months. 2. The measures which the coastal State adopts under the previous paragraph shall be valid as to other States only if the following requirements are fulfilled: (a) that there is a need for urgent application of conservation measures in the light of the existing knowledge of the fishery; (b) that the measures adopted are based on appropriate scientific findings; (c) that such measures do not discriminate in form or in fact against foreign fishermen. 3. These measures shall remain in force pending the settlement, in accordance with the relevant provisions of this Convention, of any disagreement as to their validity. 4. (P) *If the measures are not accepted by the other States concerned, any of the parties may initiate the procedure contemplated by article 9. Subject to paragraph 2 of article 10, the measures adopted shall remain obligatory pending the decision of the special commission.* (P) 5. The principles of geographical demarcation as defined in article 12 of the Convention on the Territorial Sea and the Contiguous Zone shall be adopted when coasts of different States are involved.

Articles Concerning the Law of the Sea adopted by the ILC (1956)	Text adopted by the Third Committee of UNCLOS/I (1958)	Final Text of the *Fishing & Conservation Convention* (1958)
Article 56	Article 56	Article 8
1. Any State which, even if its nationals are not engaged in fishing in an area of the high seas not adjacent to its coast, has a special interest in the conservation of the living resources in that area, may request the State whose nationals are engaged in fishing there to take the necessary measures of conservation. 2. (P) *If no agreement is reached within a reasonable period, such State may initiate the procedure contemplated by article 57.* (P)	1. Any State which, even if its nationals are not engaged in fishing in an area of the high seas not adjacent to its coast, has a special interest in the conservation of the living resources in that area, may request the State or States whose nationals are engaged in fishing there to take the necessary measures of conservation, under articles 51 and 52 respectively, at the same time mentioning the scientific reasons which in its opinion make such measures necessary, and indicating its special interest. 2. (P) *If no agreement is reached within twelve months, such State may initiate the procedure contemplated by article 57.* (P)	1. Any State which, even if its nationals are not engaged in fishing in an area of the high seas not adjacent to its coast, has a special interest in the conservation of the living resources of the high seas in that area, may request the State or States whose nationals are engaged in fishing there to take the necessary measures of conservation under articles 3 and 4 respectively, at the same time mentioning the scientific reasons which in its opinion make such measures necessary, and indicating its special interest. 2. (P) *If no agreement is reached within twelve months, such State may initiate the procedure contemplated by article 9.* (P)
Article 57	Article 57	Article 9
1. Any disagreement arising between States under articles 52, 53, 54, 55 and 56 shall, at the request of any of the parties, be submitted for settlement to an arbitral commission of seven members, unless the parties agree to seek a solution by another method of peaceful settlement. 2. [...] 3. [...] 4. [...] 5. [...]	1. Any disagreement arising between States under articles 52, 53, 54, 55 and 56 shall, at the request of any of the parties, be submitted for settlement to a special commission of five members, unless the parties agree to seek a solution by another method of peaceful settlement, as provided for in Article 33 of the Charter of the United Nations. 2. [...] 3. [...] 4. [...] 5. [...] 6. [...] 7. [...]	1. Any dispute which may arise between States under articles 4, 5, 6, 7 and 8 shall, at the request of any of the parties, be submitted for settlement to a special commission of five members, unless the parties agree to seek a solution by another method of peaceful settlement, as provided for in Article 33 of the Charter of the United Nations. 2. [...] 3. [...] 4. [...] 5. [...] 6. [...] 7. [...]

Notes: The symbol '(P)' has been used to highlight the textual disposition of clausal embeddedness.

settlement of disputes. Thus, the principle of compulsory adjudication became an integral component of the substantive rules, forming an essential part of a 'package deal' that predated UNCLOS/III. Without this inclusion, the convention could not have been adopted.[52] This particular aspect was elucidated by Commissioner Gerald Fitzmaurice (United Kingdom) during the ILC deliberations, in stressing that:

> The condition by which the whole scheme was made acceptable to other States was the provision of arbitration machinery as an essential element in the whole project, so that other countries which found the measures instituted by the coastal State unacceptable could have some means of appeal. There had been general agreement that the arbitration provisions were indispensable ... Any suggestion of dropping [these] would largely destroy the value of the whole draft with which, in the main, all could agree.[53]

Embedded clauses acquire, hence, an immanent and interrelated substantive quality derived from their deliberate placement within the structure of the articles. Acknowledging this constitutive rationale, Commissioner Krylov objected to the morphology of the articles, asserting that 'the settlement of disputes was quite a different issue from the establishment of substantive rules'.[54] Accordingly, he advocated—albeit unsuccessfully, as already discussed—for the removal of this clausal construction. Similarly, Commissioner Padilla-Nervo proposed replacing the pattern of clausal embeddedness with a single dispute settlement clause akin to article 33(1) UNC. Nevertheless, the majority of commissioners held a contrary view. In particular, Commissioner Fitzmaurice, in his continued defence of the structure of clausal embeddedness, underlined that:

> [Proposals] to delete the provisions for compulsory arbitration, reaffirmed [the] conviction that they formed an essential part of the draft and that without them many States would find it impossible to subscribe to articles conferring extensive unilateral rights on the coastal State.[55]

Clearly, the doctrinal intent dictating the textual pattern of clausal embeddedness was to maintain the inseparability of substantive and procedural elements of the rules within each article. This functional design imparted a distinctive character to their structure, evident in the unique morphology of the articles.

---

[52] McDougal–Burke, *The Public Order of the Oceans* (1962) 996 followed by Schaefer, 'Some Recent Developments Concerning Fishing and the Conservation of the Living Resources of the High Seas', SDLR (1970) 381.

[53] ILC 337th meeting (1/5/1956) [1956*YbILC*(I) 31].

[54] ILC 352nd meeting (24/5/1956) [1956*YbILC*(I) 62–63].

[55] ILC 338th meeting (2/5/1956) [1956*YbILC*(I) 25¶21].

Thus, embedded clauses should not be viewed as meaningless or redundant provisions within the treaty text. Rather, their rationale resides in the deliberate intent of the drafters to forge an inseparable legal nexus between the substantive and procedural elements in the formulation of the rule, an interrelation structurally manifested in the very formation of the article in which the rule is articulated.

The voting outcomes further substantiate the perception of functional inseparability embedded within the morphology of the articles. Prior to reflecting on this dynamic aspect of the law-making process,[56] it is essential to highlight the procedural approach adopted by the Third Committee. To facilitate the adoption of the final articles before their submission to the UNCLOS/I Plenary for approval, the Committee implemented a specific *modus operandi*. Given the disagreements concerning the textual status of embedded clauses, the Chairman proposed deferring decisions on the procedural stipulations within the articles until determinations had been made on *ILC Draft Articles* 57, 58, and 59, which addressed the dispute settlement process. These were designated as 'outstanding paragraphs'. Pursuantly, the Committee conducted provisional votes to identify any necessary amendments that might mitigate objections to the substantive content of the articles, after which negotiations continued with the consideration of proposed changes, culminating in a second and definitive vote on the articles.[57] While delegations did not formally oppose the deferral of decisions regarding embedded clauses until consensus was achieved on the substantive provisions, they reserved the right to debate or vote at various stages of the proceedings. This reservation was predicated on the understanding that the issue of dispute settlement was 'inseparable' from any article in which it was referenced.[58]

The results of the voting taken on a paragraph-by-paragraph, as summarized in Table 4.4, show that the number of votes cast against the embedded clauses increased in between the first and second reading. This comparison indicates that the attitude of the States opposing the adoption of these clauses became

---

[56] On voting-related aspects to inform legal analysis *see* Todd, 'The "Law-making" Behavior of States in the United Nations as a Function of their Location within Formal World Regions', *International Studies Quarterly* (1971) 297 and Sohn, 'Voting Procedures in United Nations Conferences for the Codification of International Law', AJIL (1975) 310.

[57] *see* the comments upon the procedure at the 13th (21/3/1958) [*Geneva Off. Records* V:34], 20th (1/4/1958) [*Geneva Off.Records* V:48], and 33rd (15/4/1958) [*Geneva Off.Records* V:93] Third Committee meetings.

[58] *see* the statements recorded by Norway and the United States at the 14th (24/3/1958) [*Geneva Off.Records* V:35–36] and by Norway during the 20th (1/4/1958) [*Geneva Off.Records* V:48] Third Committee meetings.

*Table 4.4 Voting outcomes in the UNCLOS/I Third Committee**

ILC Draft Article	Votes on embedded clauses: first reading in favour	Votes on embedded clauses: first reading against	Votes on embedded clauses: first reading abstain	Votes on embedded clauses: second reading in favour	Votes on embedded clauses: second reading against	Votes on embedded clauses: second reading abstain	Article as a whole in favour	Article as a whole against	Article as a whole abstain
52 paragraph 2 *(4 paragraph 2)***	41	0	15	47	11	5	–	–	–
52 as a whole	–	–	–	–	–	–	48	8	7
53 paragraph 2 *(5 paragraph 2)***	42	7	6	40	12	6	–	–	–
53 as a whole	–	–	–	–	–	–	45	9	7
54 paragraph 3 *(6 paragraph 5)***	44	1	11	40	2	11	–	–	–
54 as a whole	–	–	–	–	–	–	41	8	15

ILC Draft Article	Votes on embedded clauses: first reading in favour	Votes on embedded clauses: first reading against	Votes on embedded clauses: first reading abstain	Votes on embedded clauses: second reading in favour	Votes on embedded clauses: second reading against	Votes on embedded clauses: second reading abstain	Article as a whole in favour	Article as a whole against	Article as a whole abstain
55 paragraph 3 (*7 paragraph 4*)**	38	1	18	32	22	8	–	–	–
55 as a whole	–	–	–	–	–	–	34	20	5
56 paragraph 2 (*8 paragraph 2*)**	46	7	4	(*not voted on separately*)	(*not voted on separately*)	(*not voted on separately*)	–	–	–
56 as a whole	–	–	–	–	–	–	42	8	6

Note: * Delegations present and voting at the meetings; absences not recorded. ** Corresponding article numbering under *Fishing & Conservation Convention* (1958).

even more pronounced, although the substance of the articles in most cases had been marginally revised. The only parameter that had changed in between the two stages, and which explains the rallying of opposition, is the final consolidation of the principle of compulsory adjudication in *ILC Draft Article* 57 (corresponding to article 9 FSS). The cast votes on the dispute settlement clause, which had preceded both times the taking of votes on the outstanding procedural paragraphs, were 38 in favour to 14, with 10 abstentions during the first reading,[59] and 39 in favour to 10, with 6 abstentions during the second reading.[60] This dynamic in the voting results illustrates that States not only acknowledged the functional rationale underlying the structure of embedded clauses but also, as reflected in the outcomes, largely endorsed their incorporation. Illustrative of this understanding is Romania's candid statement on this aspect during its vote against adopting *ILC Draft Article* 52, in stressing that:

> [It] had voted for paragraph 1 of article 52, and against article 52 as a whole, because although the Romanian Government endorsed the principles set forth in paragraph 1 of article 52, it was opposed to the establishment of compulsory arbitration. From the legal point of view, the reference to the principle of compulsory arbitration in several articles was to be deplored, as it would compel delegations which did not accept it to express their opposition time and again.[61]

In the same spirit, the USSR delegation explained its abstention from the vote on *ILC Draft Article* 54 as a whole, by stating that although it 'set great store by the provision in paragraph 1, it considered paragraph 5 unacceptable'.[62] Similarly, Peru abstained from voting on all the texts of the outstanding procedural paragraphs in *ILC Draft Articles* 52–56, on the grounds that the inclusion of procedural stipulations reinforced the arbitration system already adopted under *ILC Draft Article* 57; an arrangement to which it was opposed.[63] For the same reason, Chile, Ecuador, Mexico, and the USSR, during the second reading by the Third Committee, pressed motions for separate votes on the articles containing embedded clauses,[64] employing this as a tactical expedient to forestall the adoption of those specific paragraphs. It should be amply clear from the foregoing that the function of this textual morphology was to embed the compulsoriness principle within each substantive rule, thereby rendering the application of compulsory and binding third-party dispute settlement

---

59 Third Committee 30th meeting (12/4/1958) [*Geneva Off.Records* V:84].

60 Third Committee 38th meeting (18/4/1958) [*Geneva Off.Records* V:112].

61 *ibid.*, 113.

62 *ibid.*, 113–115; *note*: paragraph numbering reflects the renumbering adopted by the Third Committee, corresponding to article 54(5).

63 Third Committee 34th meeting (14/4/1958) [*Geneva Off.Records* V:95].

64 Third Committee 38th meeting (18/4/1958) [*Geneva Off.Records* V:112–114].

procedures an indispensable component of the rule itself. While recognizing the importance of consensus in establishing viable and effective regulatory regimes, this pattern of operational provisions simultaneously emphasizes the need for prior negotiation and agreement on common conservation measures. Crucially, it furnishes a legal avenue for addressing stalemates caused by a lack of cooperation.[65] To this end, the structural formation of the respective articles establishes an inseparable connection between the principle of compulsory adjudication and the duty to cooperate in circumstances requiring joint CMMs, irrespective of whether such measures arise within a permanent or *ad hoc* institutional framework. This approach was lauded by the majority of States in the Third Committee.[66]

International rules conceptualized and constructed in accordance with this perception fall into a *sui generis* legal category that academic commentary has appositely designated in their normative nature to be 'justiciable'. These are rules for which States have agreed in advance to submit disputes concerning their interpretation and application to compulsory and binding third-party settlement procedures. Such an *ex-ante* undertaking limits jurisdictional debate by preventing States from asserting the right to unilaterally interpret contested rules. This legal category notably includes progressively developed rules, often formulated in broad substantive terms, and whose effective application necessitates a corresponding *pari passu* development of the dispute settlement law.[67] On this account, the other *Geneva Conventions* have been rightly criticized for relegating dispute settlement provisions to the *Optional Protocol* (1958), rather than integrating them into the substantive body of the treaty text.[68] In contrast, the Convention has been praised for Part XV LOSC, which forms an essential component of the negotiated balance of interests under the package

---

[65] *see* Oda, 'New Trends in the Regime of the Seas: A consideration of the problems of conservation and distribution of marine resources', ZaöRV (1957) 275 and Dean, 'The Geneva Conference on the Law of the Sea: What was accomplished', AJIL (1958) 626.

[66] Comments on this specific aspect of the articles can be found in the statements made during the Third Committee's general debate, inter alios, by Norway at the 5th meeting (10/3/1958) [*Geneva Off.Records* V:3–4]; United Kingdom, The Netherlands and Sweden at the 7th meeting (13/3/1958) [*Geneva Off.Records* V:10–13]; Portugal and Greece at the 8th meeting (14/3/1958) [*Geneva Off.Records* V:14¶¶16–17 and 16, respectively]; and Spain at the 9th meeting (17/3/1958) [*Geneva Off.Records* V:17¶42].

[67] *see* Rosenne, 'Relations between Governments and the International Law Commission', *Year Book of World Affairs* (1965) 183*ff* and Cheng, 'On the Nature and Sources of International Law', in Cheng (Ed.) *International Law* (1982) 212.

[68] Darwin, 'General Introduction', in Waldock (Ed.) *International Disputes* (1972) 62–63.

deal.[69] Obviously, the *Fishing & Conservation Convention* (1958) should not be grouped with the other *Geneva Conventions* in this critique, as it not only uniquely incorporated dispute settlement provisions directly into its treaty text but moreover embedded them as inseparable elements of the substantive rules in articles reflecting progressive development. On this basis, it is appropriate to highlight two key observations. First, as discussed in the preceding chapter, the integration of dispute settlement procedures within the treaty text was a distinctive feature inherited by the Convention. Second, the law-making technique of linking substantive rules with dispute settlement provisions was specifically employed in drafting articles 74/83 LOSC, which merits closer examination in the remainder of this chapter.

4.3 THE 1982 UNITED NATIONS CONVENTION ON THE LAW OF THE SEA

The rule on maritime delimitation prescribed by articles 74/83 LOSC, which simply requires that delimitation be effected 'by agreement on the basis of international law', reflects perhaps the most extreme example of inherently evolutionary law in the Convention.[70] In contrast to article 15 LOSC governing the delimitation of territorial sea, which places primacy on the median/equidistant line by following the rule previously established in article 12(1) *Convention on the Territorial Sea and the Contiguous Zone* (1958), the Convention does not specify the methodology for EEZ/CS delimitation between States with opposite or adjacent coasts. The near-identical structure of articles 74/83 LOSC, coupled with the integration of substantive and procedural provisions to address disputes arising from overlapping maritime entitlements, represents a compromise intended to reconcile the major differences among States at UNCLOS/III regarding delimitation methods.[71] While the articles attained effortlessly the status of customary international law and are considered universally applicable,[72] the substance of the delimitation rule leaves ample scope for its interpretation and application, rendering negotiations susceptible to

---

69 Collier–Lowe, *The Settlement of Disputes in International Law* (1999) 86.

70 Boyle, 'Further Development of the Law of the Sea Convention: Mechanisms for change', ICLQ (2005) 568–569.

71 Oude-Elferink–Henriksen–Busch, 'The Judiciary and the Law of Maritime Delimitation', in Oude-Elferink–Henriksen–Busch (Eds) *Maritime Boundary Delimitation* (2018) 11.

72 *e.g. Delimitation of the Maritime Boundary in the Gulf of Maine Area* (Judgment) 1984, §81 *et seq*; *Maritime Delimitation and Territorial Questions* (Merits) 2001, §167; *Territorial and Maritime Dispute* (Merits) 2012, §139.

deadlock and frequently resulting in disputes between neighbouring littoral States with undelimited maritime boundaries.[73]

4.3.1 The Twin Articles on EEZ/CS Delimitation between States with Opposite or Adjacent Coasts

Articles 74/83 LOSC are identical in their statement of the delimitation terms (see Table 4.5). Paragraph (1) contains a neutral formula that prescribes as a fundamental standard the attainment of 'an equitable solution' through agreement based on general international law between the States concerned.[74] In doing so, UNCLOS/III deliberately sidestepped a central point of contention; namely, the unresolved debate between States supporting the median/equidistant line method and those favouring the application of equitable principles. The intentionally vague language of the delimitation provision preserves the flexibility for either approach to be adopted and does not foreclose the consideration of alternative criteria or relevant circumstances that may be appropriate in a given case. However, this deliberate ambiguity in the substantive rule was predicated on the prior acceptance of third-party dispute settlement as a means of last resort to resolve disagreements over the interpretation or application of the twin articles. This was a necessary component of the broader package deal, which established a dispute settlement regime permitting the unilateral initiation of compulsory procedures to which States consent in advance and *ipso facto* through adherence to the Convention. In that respect, the negotiations on the wording of the substantive rule in paragraph (1) were inextricably linked to, and made contingent upon, the outcome of negotiations on two corresponding procedural provisions.

The first of these provisions is contained in article 298 LOSC. Maritime delimitation disputes that are not resolved by agreement between the States concerned are, in principle, subject to the compulsory procedures entailing binding decisions provided for in section 2 Part XV LOSC. The jurisdiction of the court or tribunal to be seised under article 287 LOSC is limited only by the optional exception invoked through a duly submitted written declaration pursuant to article 298(1)(a) LOSC, whereby a State may exclude disputes relating to the delimitation of the territorial sea, EEZ and CS, or those involving historic bays or titles. Such declarations operate without prejudice to the obligations arising under section 1 Part XV LOSC. At the request of any party to the dispute, a matter falling within the scope of a valid declaration under

---

[73] *see* Tanja, *The Legal Determination of International Maritime Boundaries* (1990) 87; Prescott–Schofield, *The Maritime Political Boundaries of the World* (2005) 246; and Lando, *Maritime Delimitation as a Judicial Process* (2019) 17.

[74] 1993 *Virginia Commentary* II ¶74.1(800).

Table 4.5 Clausal embeddedness in Articles 74 and 83 of the 1982 UN Convention on the Law of the Sea

Article 74 Delimitation of the exclusive economic zone between States with opposite or adjacent coasts	Article 83 Delimitation of the continental shelf between States with opposite or adjacent coasts
1. The delimitation of the exclusive economic zone between States with opposite or adjacent coasts shall be effected by agreement on the basis of international law, as referred to in Article 38 of the Statute of the International Court of Justice, in order to achieve an equitable solution.	1. The delimitation of the continental shelf between States with opposite or adjacent coasts shall be effected by agreement on the basis of international law, as referred to in Article 38 of the Statute of the International Court of Justice, in order to achieve an equitable solution.
2. (P) *If no agreement can be reached within a reasonable period of time, the States concerned shall resort to the procedures provided for in Part XV.* (P)	2. (P) *If no agreement can be reached within a reasonable period of time, the States concerned shall resort to the procedures provided for in Part XV.* (P)
3. Pending agreement as provided for in paragraph 1, the States concerned, in a spirit of understanding and cooperation, shall make every effort to enter into provisional arrangements of a practical nature and, during this transitional period, not to jeopardize or hamper the reaching of the final agreement. Such arrangements shall be without prejudice to the final delimitation.	3. Pending agreement as provided for in paragraph 1, the States concerned, in a spirit of understanding and cooperation, shall make every effort to enter into provisional arrangements of a practical nature and, during this transitional period, not to jeopardize or hamper the reaching of the final agreement. Such arrangements shall be without prejudice to the final delimitation.
4. Where there is an agreement in force between the States concerned, questions relating to the delimitation of the exclusive economic zone shall be determined in accordance with the provisions of that agreement.	4. Where there is an agreement in force between the States concerned, questions relating to the delimitation of the continental shelf shall be determined in accordance with the provisions of that agreement.

Notes: The symbol '(P)' has been used to highlight the textual disposition of clausal embeddedness.

article 298(1)(a) LOSC, and arising after the Convention's entry into force,[75] may be submitted to compulsory conciliation under section 2 Annex V LOSC, where no agreement is otherwise reached through negotiation. If, following the conclusion of conciliation proceedings, the disagreement remains unresolved, the Convention obliquely provides that the parties 'shall, by mutual consent, submit the question to one of the procedures provided for in section 2 [Part XV LOSC], unless the parties otherwise agree'.[76] The formulation of this proviso is notably equivocal, as it appears to require submission to a procedure entailing binding decisions, while simultaneously conditioning such recourse on 'mutual consent'. Consequently, commentators differ in their views on the procedural avenue to be followed. Some argue that the unresolved disagreement creates practically a procedural renvoi that, in effect, leads to the re-engagement of the compulsory procedures in section 2 Part XV LOSC. Others maintain that the lack of agreement itself becomes a dispute, thus triggering the very procedures from which the delimitation matter was initially excluded.[77]

The other relevant procedural provision is found in the identical paragraph (2) of articles 74 and 83 LOSC. This paragraph constitutes an embedded clause, making express reference to the Convention's dispute settlement procedures by providing that 'If no agreement can be reached within a reasonable period of time, the States concerned shall resort to the procedures provided for in Part XV [LOSC]'. Academic commentary published in the period immediately following the adoption of the Convention generally maintained that, despite the ostensibly obligatory phrasing 'shall resort to', this reference 'does not itself necessarily amount to an acceptance of compulsory jurisdiction', given the qualification introduced by *lit.*(i) of article 298(1)(a) LOSC regarding the applicability of section 2 Part XV LOSC. On that basis, it has been argued that articles 74/83 LOSC 'do not contain a fully constraining third-party settlement system'.[78] However, more recent commentary—benefitting from interpretative

---

75 The relevant date to be considered is 16/11/1994, when the Convention entered into force generally, rather than the date of its entry into force *inter partes*, as affirmed in the *Timor Sea Conciliation* (Decision on Competence) 2016, §74*ff.*

76 LOSC 298(1)(a-ii).

77 *see* Highet 'Maritime Boundary Disputes and Settlement', in Kusuma-Atmadja–Mensah–Oxman (Eds) *Sustainable Development and Preservation of the Oceans* (1997) 764*ff*; Gautier, 'The International Tribunal for the Law of the Sea: Activities in 2006', CJIL (2007) 399; and Beckman–Sim, 'Maritime Boundary Disputes and Compulsory Dispute Settlement: Recent developments and unresolved issues', in Nordquist–Moore–Long (Eds) *Legal Order in the World's Oceans* (2017) 245.

78 Allott, 'Power Sharing in the Law of the Sea', AJIL (1983) 21–22.

guidance in subsequent jurisprudence—has rightly observed that 'it is not without significance' that paragraph (2) of these articles 'makes it obligatory for the States concerned to resort to the procedures under Part XV [LOSC]'.[79] Under paragraph (2), the lack of agreement 'within a reasonable period of time' gives rise to a positive obligation to submit the matter to adjudication, thus compelling recourse to the Convention's compulsory dispute settlement mechanisms. The principal significance of this embedded clause, therefore, is to reaffirm the compulsoriness principle in the context of maritime delimitation disputes. States must first attempt to resolve such disputes by exchanging views and engaging in negotiations under section 1 Part XV LOSC; if these efforts fail, either party may unilaterally invoke the compulsory procedures entailing binding decisions provided for in section 2 Part XV LOSC.[80]

Yet all these pertinent observations remain at the surface of the dispute settlement process, without truly grappling with the underlying *raison d'être* of paragraph (2) of articles 74/83 LOSC, particularly in instances where no optional declaration under article 298(1)(a) LOSC has been made. In such cases, should paragraph (2) be regarded as mere textual verbosity—floating *in vacuo*—with no legal consequence? Scholarly attention to this dimension of the rule has been both sparse and relatively modest, with limited inquiry into the drafting intention behind the textual pattern of clausal embeddedness in the twin articles. The analysis that follows, therefore, turns its focus to this second procedural stipulation.

4.3.2 Negotiating History of Paragraph (2) of Articles 74/83 of the 1982 United Nations Convention on the Law of the Sea

The guiding precept for the settlement of maritime boundary disputes—whether by special agreement or through compulsory recourse to binding third-party adjudication—finds its origins in the codification and progressive development of the law governing the legal nature and breadth of territorial waters by LoN in the 1920s. It was during this early formative period that

---

[79] Rao, 'Delimitation Disputes under the United Nations Convention on the Law of the Sea: Settlement procedures', in Ndiaye–Wolfrum (Eds) *Law of the Sea, Environmental Law, and Settlement of Disputes* (2007) 894.

[80] This point was expressly made in the statement delivered by the ITLOS President Rüdiger Wolfrum to the 'Informal Meeting of Legal Advisers of Ministries of Foreign Affairs' (23/10/2006); and *see* further in Treves, 'What Have the United Nations Convention and the International Tribunal for the Law of the Sea to Offer as Regards Maritime Delimitation Disputes?', in Lagoni–Vignes (Eds) *Maritime Delimitation* (2006) 73 and Fietta–Cleverly, *A Practitioner's Guide to Maritime Boundary Delimitation* (2016) 123.

the question of delimitation emerged as a matter of legal doctrine concerning the appropriate method of resolution. The view that ultimately prevailed, as reflected in the assiduous work of the Committee of Experts, recommended that 'it would be better to arrange for the conclusion of a special agreement between the States concerned, or for the settlement of the matter by arbitration or an ordinary tribunal, than to lay down an immutable principle'.[81] It is from this foundation that ILC derived its approach to CS delimitation. At its 1951 Session, during discussions on the draft articles on the *Regime of the High Seas*,[82] Commissioner Scelle proposed that the Commission adopt a rule stipulating that where 'two States could not reach agreement as to the partition of the CS, neither [of them] was entitled to exploit it', in which case they 'must either maintain the *status quo* or they would be under an obligation to refer the question to the [ICJ]'.[83] A refined version of the proposal was tabled for discussion at ILC's 1953 Session by Commissioner Fyodor Ivanovich Kozhevnikov (USSR), proposing a rule whereby 'the boundaries of a CS [...] of two or more States shall be established by agreement between those States. Failing such agreement, a dispute between them shall be resolved by one of the methods for the joint peaceful settlement of disputes'.[84] The core of this proposal gradually gained traction during the ILC deliberations, chiefly with regard to how its substance should be worded. The rule, as ultimately formulated in the Draft Articles on the Continental Shelf—and later adopted with only minor drafting modifications as *ILC Draft Article* 72—provided that, in the absence of agreement and unless justified by special circumstances, delimitation was to be effected by a median line in the case of States with opposite coasts, and by an equidistant line in cases involving adjacent coasts.[85] This delimitation rule was complemented by *ILC Draft Article* 73, which established a firm obligation for the settlement of disputes concerning the interpretation or application of these articles to be submitted to ICJ at the request of any party, unless the parties agreed upon another method of peaceful settlement. However, while the delimitation rule was ultimately adopted in article 6 CSC, UNCLOS/I refused to include a standing clause on compulsory dispute settlement. Instead, it opted

81 Committee of Experts for the Progressive Codification of International Law, 'List of Subjects the Regulation of which by International Agreement would seem to be most Desirable and Realizable' (8/4/1925) [(1926) AJIL *Suppl.* 20(3):12–16] and 'Memorandum by Rapporteur Mr Walther Schücking' (29/1/1926) [(1926) AJIL *Suppl.* 20(3):87–88], and *see* Rhee, 'Sea Boundary Delimitation Between States before World War II', AJIL (1982) 575.

82 A/CN.4/42 (1951), *supra cit.* (*fn*13).

83 ILC 116th meeting (3/7/1951) [1951*YbILC*(I) 288¶5 & 292¶65].

84 ILC 205th meeting (30/6/1953) [1953*YbILC*(I) 130¶4].

85 A/CN.4/76 (1953), *supra cit.* (*fn*10) 213, draft article 7.

to address such disputes through voluntary undertakings under the *Optional Protocol* (1958).[86]

Whereas CS delimitation had been previously addressed in considerable detail by ILC and UNCLOS/I, and had also been tested through adjudication,[87] the progressive development of the law governing the EEZ regime required UNCLOS/III to revisit the issue of delimitation *de novo*. This was necessary given that States increasingly sought, where feasible, to designate a single maritime boundary and therefore to deal with the applicable method in employing the same approach to both maritime areas.[88] It became apparent relatively early in the negotiations that, insofar as the content of the delimitation rule was inextricably linked to the regime of compulsory dispute settlement, the substantive and procedural aspects of the rule would need to be addressed as an integrated package.[89]

Crafting embedded clauses within the delimitation rule

A suitable starting point for examining the *travaux préparatoires* emanated from the UNCLOS/III negotiations is a submission made by The Netherlands during the Second Session (1974). In that proposal, it was envisaged that maritime delimitation disputes would be referred to a compulsory system of conciliation in cases where the States concerned either refused to enter into negotiations or failed to reach agreement within a specified period after negotiations had commenced. A commission—modelled in terms of its constitution and procedure on the conciliation mechanism established under the *Vienna Convention on the Law of Treaties* (1969)—was intended to act as an impartial body making recommendations to facilitate the parties in reaching agreement. However, given that such recommendations, in line with the conventional practice of conciliation in international law, are non-binding, this form of dispute

---

86 A wealth of insightful information regarding the wording of the compulsory jurisdiction clause and the background negotiations is offered by Whiteman, 'Conference on the Law of the Sea: Convention on the continental shelf', AJIL (1958) 629 and Verzijl, *International Law in Historical Perspective* (1976) 395–399.

87 Notably, the rule had already been considered in *North Sea Continental Shelf Cases* (1969) and its application was further examined, during the UNCLOS/III negotiations, in *Anglo-French Continental Shelf Arbitration* (Decision), 1977 and 1978.

88 1993 *Virginia Commentary* II ¶74.2(801).

89 *see* Oude-Elferink, *The Law of Maritime Boundary Delimitation* (1994) 27–28 and McRae, 'The Applicable Law: The Geneva Convention on the Continental Shelf, the LOSC, and customary international law', in Oude-Elferink–Henriksen–Busch (Eds) *Maritime Boundary Delimitation* (2018) 92*ff.*

settlement could not in itself produce a definitive resolution. Therefore, it was necessary to make this outcome explicitly clear by adding to the substantive delimitation rule a clause providing that:

> If agreement is not reached within [a period of time] ... after the Conciliation Commission has made its final recommendations, the question of delimitation may be submitted, by any of the States concerned, to the procedure for the compulsory judicial settlement of disputes, provided for in article...of the present Convention.[90]

Accordingly, when both direct negotiations and the conciliation procedure failed to produce an agreement, the clause in question would require the parties to resolve the dispute by judicial means entailing a binding and final decision on the delimitation question. The incorporation of this embedded clause into the EEZ/CS delimitation articles took place during negotiations on ISNT at the Geneva Session (1975). As it will be recalled, each of the main Committees was tasked with producing a tentative treaty text reflecting the progress achieved in the negotiations up to that point. During that session, an informal consultative group was established within the Second Committee,[91] prompted by the recognition that agreeing upon a rule governing EEZ/CS delimitation had become a markedly divisive issue. Following additional informal negotiations within this consultative group, the articles on EEZ/CS delimitation that were included in ISNT featured an identical paragraph, which drew upon the earlier proposal submitted by The Netherlands and stipulated that:

> If no agreement can be reached within a reasonable period of time, the States concerned shall resort to the procedures provided for in part ... (Settlement of disputes).[92]

Given that disputes involving undelimited or undefined maritime boundaries had been regarded as among the gravest threats to international peace and security, the embedded clauses within the EEZ/CS delimitation articles were seen as innovative procedural safeguards.[93] These clauses were designed to mitigate such dangers by providing a structured approach to dispute resolution.

---

90 A/Conf.62/C.2/L.14 'The Netherlands: Draft article on delimitation between States with opposite or adjacent coasts' (19/7/1974) [*UNCLOS/III Off.Records* III:190], paragraph 4.

91 A/Conf.62/C.2/L.89/Rev.1 'Statement on the work of the Second Committee' (15/7/1975) [*UNCLOS/III Off.Records* IV:195].

92 A/Conf.62/WP.8/Part-II (1975), *cit.* Chapter 2 (*fn*86), paragraph 2 of articles 61 and 70 [*UNCLOS/III Off. Records* IV:162–163].

93 Stevenson–Oxman, 'The Third United Nations Conference on the Law of the Sea: The 1975 Geneva session', AJIL (1975) 781.

The text of the identical paragraph remained unchanged in both delimitation articles of RSNT, with only renumbering adjustments.[94] Despite proposals from delegations sceptical of third-party settlement—for instance, Ireland—that sought a diluted reference to compulsory procedures,[95] the clauses retained their original wording. The persistent ambiguity in the substantive part of the delimitation rule underscored the necessity of incorporating dispute settlement provisions, even though the specific procedures under the Convention had yet to be finalized at that time.[96] Following the harmonization of the EEZ delimitation article with that of CS, the substance of the identical procedural stipulation contained in paragraph (2) solidified in ICNT.[97] Both paragraphs were eventually concretized into their current form within articles 74/83 LOSC, with only minor drafting updates to reflect the final numbering of Part XV LOSC in the treaty text.[98] The embedded clauses withstood the arduous negotiations that marked the concluding phase of UNCLOS/III, reflecting their critical function within the Convention's procedural architecture.

The interplay between the embedded clauses and the optional exception

Among the highly contentious issues that remained unresolved and showed little progress during the negotiations were the settlement of EEZ and CS disputes, including disagreements over their delimitation. Specifically, delegations were deeply divided during the Second Session (1974) and Third Session (1975) on whether the Convention should prioritize the median/equidistant line method or the application of equitable principles as the primary rule for effecting delimitation. Momentum in these negotiations was regained during the Fourth Session (1976) when the broader question of dispute settlement was introduced to the UNCLOS/III Plenary for general debate. Prior to this, each of the three main Committees addressed dispute settlement only insofar as it pertained to their respective mandates. In preparation for the resumption of the negotiations in New York, President Amerasinghe drafted the dispute settlement section of ISNT. As discussed in Chapter 3, article 18 ISNT incorporated limited exceptions aimed at resolving the stalemate that had hindered

---

94 A/Conf.62/WP.8/Rev.1/Part-II (1976), *cit.* Chapter 2 (*fn*87), paragraph 2 of articles 62 and 71, respectively [*UNCLOS/III Off.Records* V:164–165].

95 1993 *Virginia Commentary* II ¶74.6(808) & ¶83.7(961).

96 Brown, 'The Continental Shelf and the Exclusive Economic Zone: The problem of delimitation at UNCLOS/III', *Maritime Policy and Management* (1977) 395–398.

97 A/Conf.62/WP.10 (1977), *cit.* Chapter 2 (*fn*127), paragraph 2 of articles 74 and 83, respectively [*UNCLOS/III Off.Records* VIII:16–17].

98 A/Conf.62/WP.10/Rev.1 (1979), *loc. cit.* Chapter 3 (*fn*138).

progress on EEZ/CS disputes.[99] To specifically address the persistent impasse over EEZ/CS delimitation, article 18(2) ISNT provided:

> When ratifying the present Convention, or otherwise expressing its consent to be bound by it, a Contracting Party may declare that it does not accept some or all of the procedures for the settlement of disputes specified in the present Convention with respect to one or more of the following categories of disputes: [...] (b) Disputes concerning sea boundary delimitations between adjacent States, or those involving historic bays or titles, provided that the State making such a declaration shall indicate therein a regional or other third-party procedure, entailing a binding decision, which it accepts for the settlement of these disputes.

It should be noted that ISNT was intended as a procedural tool to advance the negotiations by incorporating the outcomes of both formal and informal discussions held up to that point. However, it did not constitute a negotiated text nor reflect any accepted compromise. As an informal document, its purpose was to avoid prejudicing the position of any delegation. Nonetheless, the President's intention was to reinforce the broader dispute settlement process. Accordingly, while it affirmed the general obligation to resolve disputes by peaceful means, the text did not constrain the choice of dispute settlement methods available to States, with the expectation that consensus would eventually be reached on the exact corresponding provisions.[100] At the same time, the proposed treaty text was drafted with deliberate flexibility, as its acceptance remained contingent on the outcomes of ongoing negotiations in other parts of the Convention. It thus served as a basis for parallel discussions on dispute settlement alongside the development of RSNT as a whole.[101]

The negotiating group sessions

The Sixth Session (23/5 to 15/8/1977, New York) witnessed the continuation of a 'tilting battle' over the delimitation formula. While primary focus remained on the negotiations concerning the substantive content of the rule—with delegations maintaining irreconcilable positions between supporters of the median/equidistant line method and proponents of equitable principles—the issue of compulsory third-party settlement in maritime delimitation disputes persisted as an essential, yet equally contentious, element of the articles.[102]

---

[99] A/Conf.62/WP.9 (1975), *loc. cit.* Chapter 3 (*fn*120).

[100] A/Conf.62/WP.9/Add.1 (1976), *cit.* Chapter 1 (*fn*69), [*UNCLOS/III Off. Records* V:122–123].

[101] A/Conf.62/WP.9/Rev.2 'Revised Single Negotiating Text – Part Four' (23/11/1976) [*UNCLOS/III Off. Records* VI:2–3].

[102] Oxman, 'The Third United Nations Conference on the Law of the Sea: The 1977 New York session', AJIL (1978) 79.

Overall, the slower-than-anticipated progress across the agenda items revealed the growing need to transition to negotiations framed by package deal compromises. In the Second Committee, this approach unfolded on two distinct levels. First, exploratory discussions were conducted privately during intersessional meetings between like-minded delegations and across interest groups, aimed at drafting compromise texts capable of securing broader consensus. However, unlike other areas where identifiable common interests facilitated the formation of negotiation blocs, the issue of delimitation lacked such cohesion. Second, negotiations proceeded more openly within Committee sessions which, despite their formal designation under the UNCLOS/III process, were treated as informal in character.[103]

At the Geneva Session (1978), the issue of 'Delimitation of maritime boundaries between adjacent and opposite States and settlement of disputes thereon' was identified as one of the seven core contentious issues and assigned to Negotiating Group 7 ('NG7').[104] Significantly, both the identification and elaboration of the substantive rule and the corresponding mechanisms for the settlement of disputes concerning its interpretation and application were treated as interdependent components of the broader delimitation process. The President explicitly acknowledged this interrelation, while clarifying that NG7's mandate on dispute settlement was confined strictly to delimitation matters, with the broader dispute settlement agenda remaining under the purview of the Plenary.[105] It is worth noting, however, a revealing comment made by the President in justification of this structural division, which simultaneously reflected the essence of such disputes: maritime boundary delimitation issues, he noted, could be effectively resolved only through the prior establishment of suitable procedures for their settlement. This law-making approach was also deemed germane to disputes involving the exercise of sovereign rights by coastal States in their EEZs. Accordingly, NG7 was afforded a degree of discretion, given that the Plenary was not well-positioned to address the technical and legal complexities of maritime delimitation in detail. Against this background, the embedded clauses in the EEZ/CS delimitation articles assumed particular importance, as they accentuated the inherent functional connection between the substantive and procedural components of the rule. The President, in his explanatory remarks, meaningfully pointed to paragraph

103 *see* Miles, 'The Structure and Effects of the Decision Process in the Seabed Committee and the Third United Nations Conference on the Law of the Sea', *International Organization* (1977) 185*ff* and Buzan, '"*United We Stand* ...": Informal negotiating groups at UNCLOS/III', *Marine Policy* (1980) 188.

104 A/Conf.62/62 (1978), *cit.* Chapter 1 (*fn*56), 8.

105 1993 *Virginia Commentary* II ¶74.9(812) & ¶83.10(965).

(2) of articles 74/83 ICNT as already providing for the applicable dispute settlement procedures.[106]

Under the chairmanship of Eero Johannes Manner (Finland), the negotiations within NG7 strove for a rule on delimitation that would offer a substantially improved prospect of consensus. From the outset, discussions on paragraph (1) of articles 74/83 ICNT were marked by concerted efforts to bridge the yawning gap between delegations supporting the median/equidistant line method and those emphasizing the application of equitable principles in maritime delimitation. In equal measure there were also the opposing arguments on the nature of settlement procedures for maritime boundary disputes, as reflected in two principal informal proposals. One such proposal, sponsored by a coalition of 20 States, provided in paragraph (2) of the articles an embedded clause that stipulated:

> If no agreement can be reached, within a period of … from the time when one of the interested parties asks for the opening of negotiations on delimitation, the States concerns shall resort to the procedures provided for in Part … (Settlement of disputes) or by other third-party procedure entailing a binding decision which is applicable to them.[107]

The other informal proposal, backed by a group of 22 States emphasizing the application of equitable principles, also retained paragraph (2) as an embedded clause. However, it revised the provision in a manner that diluted the compulsoriness principle by deferring the settlement of disputes to procedures under the Charter, subject to mutual consent and accordingly read:

> If no agreement can be reached within a reasonable period of time, the States concerned shall resort to the procedures of settlement of disputes provided for in Part XV of this Convention or such other procedures agreed upon in accordance with Article 33 [UNC].[108]

Despite intensive efforts within NG7, no compromise emerged between the two opposing camps on a formula capable of offering a more promising prospect for consensus. In an effort to reconcile the differences, Chairman Manner established a subgroup composed of experts from 14 delegations, led by Louis Bruno Sohn (United States), tasked with examining and proposing potential

---

[106] UNCLOS/III Plenary 90th meeting (12/4/1978) [*UNCLOS/III Off.Records* IX:11–12].

[107] NG7/2 'Informal Suggestions Relating to Paragraphs 1, 2 and 3 of Articles 74 and 83 Article ICNT' (20/4/1978) [*Platzöder* IX:392].

[108] NG7/4 'Informal Suggestions Relating to Articles 74 and 83 Article ICNT' (21/4/1978) [*Platzöder* IX:397].

mechanisms for resolving maritime boundary disputes.[109] The subgroup, following private consultations, prepared an informal paper entitled 'A survey of possible conciliatory approaches to the question of settlement of sea boundary disputes'.[110] This document outlined a range of procedural combinations intended to inform negotiations towards a compromise on compulsory recourse to binding third-party procedures during the resumed Seventh Session in New York. Sohn, reportedly, had recognized that a viable dispute settlement system required unavoidably complex compromises, and accordingly his approach within the subgroup emphasized flexibility to facilitate consensus on formal third-party procedures.[111] Notwithstanding the Chairman's appeal for meaningful engagement with the survey's conciliatory options, the two sides were not willing to converge in a compromise formula.

This predicament was most clearly exhibited in a submission circulated by the Israeli delegation, which presented an informal working paper advocating for the replacement of compulsory adjudication with compulsory conciliation. The proposal was intended to provide a gateway to consensus for those delegations opposed to binding third-party procedures in the settlement of maritime boundary disputes. From Israel's perspective, there existed no general obligation under international law to subject such disputes to procedures culminating in final and binding decisions by a third party. In so arguing, the associated draft specifically dismantled the embedded clause in paragraph (2) of article 74 ICNT, with the same logic extended to article 83 ICNT.[112] The draft was introduced mainly as a compromise basis for paragraph (1) of articles 74/83 ICNT, proposing a residual rule that would operate only in the absence of agreement.[113] However, it unequivocally stated that the embedded clause in paragraph (2) should be excised. The rationale, as explained by Israel's representative, Shabtai Rosenne, was that the inclusion of dispute settlement provisions was exacerbating an already complex issue. While the proposal

109 1989 *Virginia Commentary* V ¶298.2(118).

110 NG7/20 (15/5/1978) [Platzöder IX:408] as later elaborated in NG.7/20/Rev.1 'Supplemental Working Paper' (25/8/1978) [Platzöder IX:412] and refined in NG7/27 'Maritime Boundary Disputes' (27/3/1979) [Platzöder IX:438] and NG7/37 'A Selection from the Basic Choices for Treatment of Maritime Boundary Disputes' (12/4/1979) [Platzöder IX:457].

111 Noyes, 'Louis B. Sohn and the Law of the Sea', *Willamette Journal of International Law and Dispute Resolution* (2008) 248–249.

112 NG7/28 'Article 74 "Title: reserved"'; the text was reproduced, following Israel's request, in the transcript of the summary record of the UNCLOS/III Second Committee 57th meeting (24/4/1979) [*UNCLOS/III Off.Records* XI:61(*fn*8)].

113 NG7/30 'Informal Working Paper by Israel; dated 6 February 1979' (2/4/1979) [*Platzöder* IX:451] reissued for circulation at the Eighth Session.

reaffirmed the primacy of delimitation by agreement, it rejected the necessity of applying Part XV to such disputes. On that basis, he concluded that 'paragraph 2 of the two articles seemed incorrect and unacceptable'.[114]

As Chairman Manner noted, the discussions on paragraph (2) remained 'characterized by opposing arguments on the desirability of compulsory dispute settlement procedures'.[115] On one side, the group of States supporting the median/equidistant line as the primary method of delimitation generally promoted the inclusion of compulsory dispute settlement. On the other, those favouring the application of equitable principles expressed a preference for excluding all disputes concerning maritime boundary delimitation from the Convention's compulsory dispute settlement system. In an effort to find common ground, Chairman Manner informally drafted and proposed a compromise text for further negotiations. This text consolidated the core elements of both groups' proposals. Seeking a neutral solution, paragraph (2) of the EEZ/CS delimitation articles provided:

> If no agreement can be reached within a reasonable period of time, the States concerned shall, subject to the provisions of Article 297 [ICNT], resort to the procedures of settlement of disputes provided for in Part XV of [the present]/[this] Convention [or to any other procedures agreed]/[unless any other procedure] is agreed upon in accordance with Article 33 [UNC].[116]

This proposal presented a compromise, adopting a 'middle-of-the-road' position by providing that maritime delimitation disputes would be subject to the compulsory procedures of the Convention only with the mutual consent of the parties concerned. The embedded clause in the proposed EEZ/CS delimitation articles largely reflected the substance of the corresponding paragraphs in the informal proposals submitted by delegations, incorporating only minor drafting refinements. Notably, it included the addition of the phrase 'subject to the provisions of Article 297 [ICNT]', which at that time set forth the provisions concerning optional exceptions.

---

114 UNCLOS/III Second Committee 57th meeting (24/4/1979) [*UNCLOS/III Off.Records* XI:61¶¶50–51].

115 A/Conf.62/RCNG/1 'Report by the NG7 Chairman on the work of the group' (19/5/1978) [*UNCLOS/III Off.Records* X:125]; and *see* 1993 *Virginia Commentary* II ¶83.10(965–968).

116 NG7/9 'Untitled' (27/4/1978) [*Platzöder* IX:401] and NG.7/11 'Informal Suggestions by the Chairman' (2/5/1978) [*Platzöder* IX:405] with the drafting iterations to reflect minor improvements made by the Drafting Committee to align internal cross-references more consistently in ICNT; and *see* further the comments recorded at the UNCLOS/III Plenary 116th meeting (27/4/1979) [*UNCLOS/III Off.Records* XI:38(¶¶145–146)].

As far as paragraph (2) is concerned, several observations can be made concerning its treatment in the informal proposals submitted by delegations and by Chairman Manner. Firstly, Chairman Manner expressed a preference for the clause to provide directly for submission to a conciliation commission, of which the report would serve as a basis for further negotiations between the parties. This option could be variously cascaded within the whole dispute settlement process, given the expectation that parties would have to continue their efforts on the basis of the report for reaching a final agreement. He suggested that, if conducive to a final compromise, the provision could further require the parties—should negotiations not result in an agreed delimitation within a fixed period—to submit the question of delimitation, 'by mutual consent', to the procedures provided for in section 2 Part XV ICNT, unless they agreed otherwise.[117] While it was acknowledged that among the various permutations of dispute settlement procedures, compulsory conciliation appeared more likely to enhance prospects for consensus, Chairman Manner noted that conciliation did not entail a binding decision, and parties would still need to be obligated to reach a final agreement. Consequently, proposals centred on the method of conciliation lacked consensus within NG7.[118] A second approach involved proposals to delete paragraph (2) of article 74 ICNT entirely, with some also suggesting the deletion of the corresponding paragraph in article 83 ICNT. Yet no solid conclusion could be drawn from these suggestions. In contrast, a third approach garnered the most substantial support, being endorsed by 30 States in NG7. This approach was to retain paragraph (2) in the EEZ/CS delimitation articles, with minor drafting improvements,[119] and add a phrase rendering its application subject to the corresponding provisions on optional exceptions.[120] It was this perspective that ultimately encouraged Chairman Manner to propose the compromise text. Nevertheless, the employed phrasing did not alleviate the disagreement between the two groups. States supporting the median/equidistant line, and generally promoting compulsory dispute settlement, remained dissatisfied with the reference to dispute settlement. Specifically, paragraph (2) was considered confusingly worded, invoking both article 297 ICNT and the compulsory settlement procedures provided for in Part XV ICNT. This concern extended beyond mere drafting issues, touching upon the substantive formulation of the delimitation rule, and importantly of its declarative construction *vis-à-vis* the extent to which it would integrally incorporate the principle of compulsory adjudication. As article 297(1) ICNT

117 UNCLOS/III Second Committee 57th meeting (24/4/1979) [*UNCLOS/III Off.Records* XI:60¶41].

118 1993 *Virginia Commentary* II ¶83.13(970–973).

119 NG7/10/Rev.2 'Articles 74 and 84 ICNT' (28/3/1980) [*Platzöder* IX:402].

120 As informally suggested by the Chairman in NG.7/11 (1978), *loc. cit.* (*fn*116).

stood at that time, it allowed States the option to exclude disputes concerning delimitations between opposite or adjacent littoral States, or those involving historic bays or titles, from the compulsory settlement procedures.

Consequently, Chairman Manner could not confidently propose any revisions to ICNT based on the discussions in NG7.[121] Three interrelated aspects in the articles 74/83 ICNT remained open: the substantive delimitation rule in paragraph (1), the settlement of maritime boundary disputes in paragraph (2), and the possibility of provisional arrangements pending final delimitation in paragraph (3).[122] These issues were inevitably returned to the Second Committee during the Eighth Session (19/3 to 27/4, held in Geneva, and 19/7 to 24/8, resumed in New York, 1979), when UNCLOS/III was scheduled to undertake the revision of ICNT. It shall be borne in mind that, in parallel negotiations, NG5 had effectively fulfilled its mandate by presenting the Second Committee with a potential compromise on disputes concerning disputes related to the exercise of the coastal State's sovereign rights in its EEZ. Despite the seemingly inconclusive outcome of NG7, which necessitated the continuation of negotiations at a later stage of UNCLOS/III, a strong 'general feeling' emerged that both aspects of the EEZ/CS delimitation articles were inseparable and, therefore, had to be considered in tandem. This sentiment was encapsulated in the guiding precept that 'the rules of delimitation and the settlement of disputes thereon should not be separated from each other'.[123] This principle in constructing the delimitation rule was distinctly emphasized. As Chairman Manner reported back to the Second Committee 'there did not seem to be much prospect of finding the sought-after compromise on the basis of a rule in which one form or another provides for the acceptance of an alternative procedure entailing a binding decision'.[124] NG7 discussions on these matters appeared more polarized than before. The division between the two sides seemed to have deepened owing to the interrelationship between the EEZ/

---

121 NG7/39 'Report of the Chairman on the Work of NG7' (20/4/1979) [*Platzöder* IX:459–463].

122 A convenient analysis of paragraphs (1) and (3) at that stage of the UNCLOS/III negotiations is offered in Tanaka, *Predictability and Flexibility in the Law of Maritime Delimitation* (2019) 36 *et seq*; yet, given the focus on the substance of the delimitation rule, the author understandably omits matters relating to paragraph (2).

123 A/Conf.62/RCNG/1 (1978), *loc. cit.* (*fn*115); and *see* Adede, 'Toward the Formulation of the Rule of Delimitation of Sea Boundaries Between States with Adjacent or Opposite Coasts', VJIL (1978) 219–232.

124 NG7/26 'Statement by the Chairman: Made at the 28th meeting of NG7 prepared for the last series of negotiations of the group' (26/3/1979) [*Platzöder* IX:435].

CS delimitation articles and the binding procedures to form Part XV ICNT, with the former failing to inspire the necessary confidence for an *ex-ante* acceptance of compulsory jurisdiction. This lack of confidence stemmed from the presence of numerous provisions that either remained vague or exhibited more apparent shortcomings at the time, principally concerning the procedures entailing binding decisions subject to the limitations contained in article 296 ICNT and the optional exceptions addressed in article 297 ICNT. Despite the extensive discussions and the 'abundantly clear' support among most delegations for compulsory and binding procedures, it remained 'equally clear' that consensus on such a solution could not be reached.[125]

During the resumed Eighth Session in New York, discussions in NG7 only served to confirm the unresolved controversy over the formulation to be employed for the delimitation of EEZ/CS maritime areas, without recording any further progress. Regarding the question of dispute settlement, the prevailing view was that compulsory conciliation on certain aspects, rather than arbitration or the complete elimination of compulsory third-party procedures on all aspects, was to remain a key consideration in the EEZ/CS delimitation articles, especially given the unlikelihood of the negotiations yielding a precise or definitive substantive rule. However, in the absence of consensus on this question matter, all related issues remained subject to further negotiations.[126]

The final wording

At the Ninth Session (3/3 to 4/4 and 28/7 to 29/8 in New York, 1980), when UNCLOS/III undertook the second revision of ICNT,[127] discussions on the EEZ/CS delimitation not only stalled but in fact regressed to positions previously held during the Seventh Session (1978). The two principal unresolved issues concerned the substantive rule for maritime delimitation and the applicability of compulsory procedures, including conciliation, for the settlement of such disputes. The group in support of the median/equidistant line as the primary method of delimitation, represented by Spain, insisted on compulsory dispute settlement through judicial means.[128] Conversely, the group favouring the application of equitable principles maintained its earlier stance. For

---

125 UNCLOS/III Second Committee 57th meeting (24/4/1979) [*UNCLOS/III Off.Records* XI:59(¶27) & 60(¶40)].

126 NG7/45 'Report of the Chairman of NG7' (22/8/1979), which is contained in A/Conf.62/91 'Reports to the plenary Conference' (19/11/1979) [*UNCLOS/III Off. Records* XII:107*ff*]; and *see* Oxman, 'The Third United Nations Conference on the Law of the Sea: The eight session (1979)', AJIL (1980) 4–5.

127 A/Conf.62/WP.10/Rev.2 'Informal Composite Negotiating Text; revision 2' (11/4/1980).

128 1993 *Virginia Commentary* II ¶83.15(977).

instance, Argentina outrightly stated that it would accept '[o]nly a formula which clearly established that no compulsory judicial procedure is to apply to disputes regarding delimitation unless the parties to the dispute specifically agree to it'.[129] Despite these entrenched positions, some conciliatory views emerged within the equitable principles group. For instance, Ireland, having moderated its original stance on dispute settlement, notably responded to the Argentinean statement by seizing the opportunity to stress that it would regard it as a '[r]egrettable outcome … if the Conference were not to adopt a provision making delimitation disputes subject to compulsory and binding settlement procedures'. This, however, as stated by the Irish representative Mahon Francis Hayes, was not the result of the 'relationship between substantive and procedural provisions' forming part of an 'artificial package' regarding the delimitation rule, but rather reflected a decision grounded in the evolving development of international law on the subject.[130] An important counterstatement regarding the declarative construction of the delimitation rule—reflecting the disagreement shared by numerous delegations, regardless of their alignment with either negotiating group—was delivered by the Colombian representative, Charry Hector Samper. He argued that 'it was inaccurate to speak of an "artificial package" since NG7 had been established on the understanding that delimitation criteria, provisional rules, and settlement procedures were inseparable and required a common solution'. This, he emphasized, captured the essence of the challenge in the negotiations: the principle of compulsory dispute settlement could only be treated as part of a 'package deal' once balanced and equitable provisions were agreed upon for the delimitation criteria, and provided that the resulting solution would apply uniformly to all categories of disputes.[131]

Notwithstanding the entrenched divergence of views between the two principal groups, the EEZ/CS delimitation articles, as adapted under Chairman Manner's initiative, were incorporated into the revisions of ICNT, leading to the *Draft Convention on the Law of the Sea*.[132] It was at this critical point that paragraph (2) acquired its final wording in the treaty text. Yet the remainder of article 83 ICNT continued to lack consensus, placing a solution at the resumed

---

129 A/Conf.62/WS/5 'Statement by the Delegation of Argentina; dated 2 April 1980' (10/4/1980) [*UNCLOS/III Off.Records* XIII:106(¶24)].

130 UNCLOS/III Plenary 126th meeting (2/4/1980) [*UNCLOS/III Off.Records* XIII:15(¶49)].

131 UNCLOS/III Plenary 127th meeting (3/4/1980) [*UNCLOS/III Off.Records* XIII:27(¶¶42–43)].

132 *compare* A/Conf.62/WP.10/Rev.3 (1980), *loc. cit.* Chapter 3 (*fn*139), *with* A/Conf.62/L.78 'Draft Convention on the Law of the Sea' (28/8/1981) [*UNCLOS/III Off.Records* XV:187&189].

Ninth Session beyond reach. Beyond the 'deceptive simplicity' that paragraph (2) may convey, this provision endured arduous negotiations and encapsulates extensive debates concerning the application of compulsory procedures entailing binding decisions under Part XV ICNT to maritime delimitation disputes. Despite the complexity of these negotiations, the rationale for embedding this provision within the EEZ/CS delimitation articles was fundamentally straightforward: the more imprecise and uncertain the substantive delimitation rule, the greater the necessity for some form of compulsory settlement.[133]

By the opening of the Tenth Session (9/3 to 16/4, held in New York, and 3–28/8, resumed in Geneva, 1981), EEZ/CS delimitation was widely recognized as one of the most intractable issues in the entire UNCLOS/III negotiations. Under these challenging circumstances, the package deal principle had to be stringently applied. A breakthrough occurred during the preparation of the final revision of the treaty text, which became the official *Draft Convention on the Law of the Sea*, subject only to specific conditions recorded therein.[134] The newly elected President of the Conference, Tommy Koh, with the assistance of Satya Nandan, engaged in direct negotiations with the delegations of Spain and Ireland, and crafted the compromise formula now embodied in paragraph (1) of articles 74/83 LOSC.[135] In short, the substantive rule on delimitation was simply freed of any express normative stipulation, instead deferring to general international law sources. While lauded for diplomatic ingenuity, the provision was also vehemently criticized as an 'anodyne text' that lacked substantive significance, having sidestepped engagement with developments in jurisprudence and scholarly discourse on delimitation doctrine.[136]

4.3.3 Jurisprudential Pronouncements on the Embedded Clauses

Whereas the substantive rule is not explicitly defined in the formula of the twin articles, instead being guided by the overarching principle of achieving an equitable solution,[137] the procedural tenet mandating that maritime delimitation disputes be resolved through compulsory third-party settlement in case of disagreement between the concerned States is firmly established. Consequently, the evolving body of case law constitutes a significant subsidiary

133 Brown, 'Delimitation of Offshore Areas: Hard labour and bitter fruits at UNCLOS/III', *Marine Policy* (1981) 180.
134 A/Conf.62/114 'Recommendations of the General Committee' (25/8/1981) [UNCLOS/III *Off.Records* XV:100].
135 1993 *Virginia Commentary* II ¶¶83.16-17(979).
136 Oxman, 'The Third United Nations Conference on the Law of the Sea: The tenth session (1981)', AJIL (1982) 14–15.
137 *Continental Shelf* (Judgment) 1985, §28.

means for determining the applicable substantive law to maritime delimitation, effectively informing the interpretation of articles 74/83 LOSC.[138] While courts and tribunals are not generally tasked with the progressive development of international law, the application of the formula within these articles has positioned the judiciary in a role that closely aligns with this function.[139] The applicable substantive principles, the relevance of various factors and circumstances considered during the delimitation process, and the overall methodology have largely been shaped through judicial decisions, contributing to the progressive development of the *acquis judiciaire* in maritime delimitation.[140]

It should be noted that several delimitation cases heard by ICJ have offered insightful *obiter dicta* concerning the interpretation of paragraph (2) of articles 74/83 LOSC. However, these are not examined in detail here, as the Court was not seised under article 36(1) ICJ Statute and, pursuant thereto, in accordance with section 2 Part XV LOSC. For instance, in the case concerning the *Land and Maritime Boundary between Cameroon and Nigeria*, the respective arguments of the parties on Nigeria's seventh preliminary objection would otherwise merit consideration.[141] In brief, Nigeria contended that issues of maritime delimitation were inadmissible in the absence of sufficient action by the parties to effect a delimitation 'by agreement on the basis of international law', as stipulated in paragraph (1) of articles 74/83 LOSC.[142] Cameroon, in its counter-submissions, responded that paragraph (2) of articles 74/83 LOSC should not be interpreted as barring recourse to third-party settlement but, on the contrary, as establishing an obligation for such recourse to avoid unilateral delimitation.[143]

Turning to jurisprudence directly relevant due to the procedures followed, there exists a growing body of case law wherein competent courts and tribunals have identified, considered, and consistently pronounced on three significant

---

138 ICJ Statute 38(1)(d); and *see* in *Delimitation of the Maritime Boundary in the Bay of Bengal* (Judgment) 2012, §184 and *Bay of Bengal Maritime Boundary Arbitration* (Award) 2014, §339.

139 Vidas, 'The Delimitation of the Territorial Sea, the Continental Shelf, and the EEZ: A comparative perspective', in Oude-Elferink–Henriksen–Busch (Eds) *Maritime Boundary Delimitation* (2018) 41.

140 This is echoed in *Delimitation of the Maritime Boundary in the Bay of Bengal* (Judgment) 2012, *Declaration of Judge Wolfrum*, 137–138.

141 *Land and Maritime Boundary between Cameroon and Nigeria* (Preliminary Objections) 1998, §§108–109.

142 *Land and Maritime Boundary between Cameroon and Nigeria*; *Counter-Memorial of the Federal Republic of Nigeria* (21/5/1999), Volume II:570*ff*.

143 *Land and Maritime Boundary between Cameroon and Nigeria*; *Replique de Law Republique du Cameroun* (4/4/2000), Livre I:392*ff*.

and interrelated questions concerning the interpretation and application of the procedural stipulations embedded in the twin articles. These questions are instructive for a close examination of the construction of embedded clauses. In summary, they pertain to the undertaking to settle disputes through compulsory third-party procedures entailing binding decisions, the right of referral to such procedures in cases of unsuccessful negotiations and whether a unilateral referral constitutes an abuse of the dispute settlement process.

The undertaking to adjudicate by third-party compulsory and binding procedures

As to the first question, the relevant pronouncements address the principle enshrined in articles 74/83 LOSC, which provides that delimitation shall be effected by agreement or, failing that, through the compulsory third-party settlement procedures of Part XV LOSC, unless a declaration under article 298 LOSC has been duly made. The structure of clausal embeddedness was examined in the *Barbados v. Trinidad and Tobago Arbitration*, where the tribunal's findings have generally been viewed as affirming and strengthening the compulsory jurisdiction under Part XV LOSC.[144] The parties had been entangled in a protracted disagreement over the use of marine natural resources, including living resources, in maritime areas subject to overlapping claims dating back to the late 1970s. Following nine rounds of negotiations over five years, which addressed both boundary delimitation and fisheries cooperation in potentially affected waters, Trinidad and Tobago reportedly declared during a bilateral meeting that the boundary issue was 'intractable' and 'challenged Barbados to take it to arbitration, if it so wished'.[145] In response, Barbados promptly initiated proceedings before an Annex VII LOSC tribunal, requesting EEZ/CS delimitation on the basis of a single unified boundary line.[146]

The significance of this case, in the context of the present book, lies in the tribunal's reasoning regarding the construction of embedded clauses, which supports the compulsory settlement of disputes under the Agreement, given that the latter applies *mutatis mutandis* the procedures of the Convention.[147] One of the preliminary objections raised was whether the tribunal had jurisdiction and, if so, whether its jurisdiction was subject to limitations under

144 Tanaka, 'Arbitral Tribunal Award, Annex VII LOSC Barbados v Trinidad and Tobago', IJMCL (2006) 523.

145 *Barbados v. Trinidad and Tobago Arbitration*; *Reply of Barbados* (9/6/2005), Volume 1: 45¶80 & 54¶93.

146 *Barbados v. Trinidad and Tobago Arbitration* (Decision) 2006, §§56–68.

147 A comparable parallel has been drawn by Kwiatkowska, 'The 2006 *Barbados/Trinidad and Tobago Award*: A landmark in compulsory jurisdiction and equitable maritime boundary delimitation', IJMCL (2007) 23.

Part XV LOSC. Barbados argued that, in the absence of a valid declaration under article 298(1) LOSC, jurisdiction was grounded in articles 286, 287, and 288 LOSC, in conjunction with Annex VII LOSC. Of particular importance was Barbados's emphasis on the operation of paragraph (2) of articles 74/83 LOSC. It is important to note that Barbados elaborated its arguments in favour of compulsory jurisdiction by stressing in particular the operation of paragraph (2) of articles 74/83 LOSC. It underscored the necessity of preserving the textual integrity of the twin articles by drawing attention to the procedural stipulations embedded therein, contending that any contrary interpretation would undermine the object and purpose of Part XV LOSC.[148] The tribunal, exercising the *compétence de la compétence* to adjudge this question, upheld its jurisdiction in the matter.[149]

The tribunal's *ratio decidendi* directly reflected the drafting rationale underpinning the textual pattern of clausal embeddedness. Accordingly, its reasoning can be applied by analogy to support the argument advanced in this book regarding the functionality of the embedded clauses in article 7 FSA. Specifically, the tribunal began by emphasizing that the substantive content of articles 74/83 LOSC imposes a general obligation on the States concerned to reach agreement on delimitation based on international law with a view to achieving an equitable solution. It then turned to the procedural dimension of the rule, observing that the parties had engaged in negotiations for a reasonable period, as required by paragraph (2) of articles 74/83 LOSC. The tribunal found it crystal clear that by 'the very fact of their failure to reach agreement within a reasonable time … and by their failure even to agree upon the applicable legal rules … there was a dispute between them'.[150] In elaborating its reasoning, the tribunal invoked the settled *jurisprudence constante* developed by PCIJ and ICJ to confirm that the requirement of meaningful negotiations does not oblige parties to persist in talks that clearly appear unproductive from the outset.[151]

Having determined that the requirement for negotiations had been satisfied, the tribunal proceeded—by invoking the procedural stipulation contained in

---

148 *Barbados v. Trinidad and Tobago Arbitration*; *Statement of the Claim and the Grounds on which it is Based by Barbados* (16/2/2004), 1¶7 and *Memorial of Barbados* (30/10/2004), 1¶5*ff* & 34¶62.

149 *Barbados v. Trinidad and Tobago Arbitration* (Decision) 2006, §384 *dispositif.*

150 *ibid.*, §§193–197.

151 *see* in *Mavrommatis Palestine Concessions* (Judgment) 1924, 13; *South West Africa Cases* (Preliminary Objections) 1962, 345–346; and *Applicability of the Obligation to Arbitrate under Section 21 of the United Nations Headquarters Agreement of 26 June 1947* (Advisory Opinion) 1988, §55.

paragraph (2) of articles 74/83 LOSC—to initiate the procedures set forth in Part XV LOSC.[152] This approach underscores that the embedded procedural clauses are integrally linked with the principle of compulsory adjudication. The tribunal found that, given the existence of a dispute that had 'not been settled within a reasonable period of time', the States were placed under 'the obligation to resort to the procedures of Part XV [LOSC]'. It further emphasized that the obligation to submit the dispute to binding procedures is embedded specifically in these clauses, affirming Barbados's position that 'since … negotiations failed to result in a settlement of [the] dispute, then both by way of [articles 74(2)/83(2)] and 281(1) LOSC the procedures of Part XV [LOSC] were applicable'.[153] The legal effect of the embedded clauses was specifically reiterated by the tribunal in viewing that:

> In practice the only relevant obligation upon the parties under section 1 of Part XV [LOSC] is to seek to settle their dispute by recourse to negotiations, an obligation which in the case of delimitation disputes overlaps with the obligation to reach agreement upon delimitation imposed by articles 74/83 [LOSC]. Upon the failure of the parties to settle their dispute by recourse to section 1, *ie.* to settle it by negotiations, article 287 [LOSC] entitles one of the parties unilaterally to refer the dispute to arbitration.[154]

This dictum addresses a critical question regarding the *ratione temporis* jurisdiction of a competent court or tribunal under section 2 Part XV LOSC. While paragraph (2) of articles 74/83 LOSC provides that where States are unable to reach agreement 'within a reasonable period of time', recourse shall be had to the dispute settlement procedures of Part XV LOSC, the procedural implications of this stipulation—specifically its interaction with the prerequisite of negotiations under section 1 Part XV LOSC—was a question to which the Convention had not given a clear answer. Would it be possible for a party engaged in unproductive negotiations, to argue that the elapsed time already satisfies both the 'reasonable period of time' requirement under paragraph (2) and the negotiation phase required under section 1 Part XV LOSC, thereby justifying the invocation of compulsory procedures entailing binding decisions under section 2 Part XV LOSC?[155] In the *Barbados v. Trinidad and Tobago Arbitration*, the tribunal interpreted paragraph (2) as indicating that

152 *Barbados v. Trinidad and Tobago Arbitration* (Decision) 2006, §199.
153 *ibid.*, §200(ii).
154 *ibid.*, §206.
155 Prior to the instant case, this issue was raised in the literature, including by Churchill, 'The Role of the International Court of Justice in Maritime Boundary Delimitation', in Oude-Elferink–Rothwell (Eds) *Ocean Management in the 21st Century* (2004) 136–137.

no additional negotiations are required beyond those undertaken in the pursuit of agreement 'within a reasonable period of time' to meet the condition set by section 1 Part XV LOSC. However, this approach may not extend to disputes concerning provisions that lack an embedded procedural stipulation of this kind, for which the negotiation requirement under section 1 Part XV LOSC remains distinct and may, as a result, necessitate meeting a different procedural threshold.[156]

The procedural right to refer a dispute following unsuccessful negotiations

Turning to the question regarding the referral of a dispute to compulsory third-party settlement procedures, significant are the pronouncements on its timing against unsuccessful negotiations. This was contained in the last sentence of the dictum in the passage quoted earlier. The question, which featured prominently in the *Barbados v. Trinidad and Tobago Arbitration*, was whether resort to the compulsory dispute settlement procedures of section 2 Part XV LOSC could be unilaterally initiated by one of the disputing parties. As this question flows directly from the issue of conducting negotiations 'within a reasonable period of time' under the embedded clause, the arguments presented by the parties and the tribunal's reasoning are closely connected with the first pronouncement. Analyzing the issue from this perspective allows for a deeper exploration of the interpretation—particularly in identifying grammatical nuances—of paragraph (2) of articles 74/83 LOSC.

Trinidad and Tobago maintained that paragraph (2) of articles 74/83 LOSC comprises three key parts. First, the use of the modal verb 'shall' signifies a mandatory action to resort to the procedures of Part XV LOSC. Second, this undertaking is reciprocal, implying that unilateral initiation of the procedures under section 2 Part XV LOSC, without prior notice, is not permissible. Third, the procedures to be followed encompass those outlined in section 1 Part XV, notably article 283 LOSC, which mandates an exchange of views as a requisite initial step in the settlement process. Consequently, Trinidad and Tobago argued that the reference made to 'States' in paragraph (2) of articles 74/83 LOSC—occurring in the plural form—does not leave open to one of the parties in the dispute to act unilaterally by having resort directly to the procedures of section 2 Part XV LOSC and, acting in that manner, to bypass the procedures of section 1. It was therefore alleged that the ordinary meaning of paragraph

---

[156] As alluded in *Obligations concerning Negotiations relating to Cessation of the Nuclear Arms Race and to Nuclear Disarmament* (Preliminary Objections) 2016, §38.

(2) of articles 74/83 LOSC presupposes a joint decision by the States concerned to proceed with the dispute under the procedures of section 2 Part XV LOSC.[157]

In response, Barbados contended that the procedures of section 2 Part XV LOSC would be rendered pointless if the applicant party were not entitled to unilaterally initiate them after concluding that negotiations had failed to produce a positive outcome. This, in practice, would allow the respondent party to trigger article 298(1)(a) LOSC, without having previously formally declaring it as an optional exception, merely by asserting that the requisite exchange of views had not occurred, thereby circumventing the compulsory procedures entailing binding decisions.[158] The tribunal upheld Barbados's position, distinguishing the obligations under paragraph (2) of articles 74/83 LOSC from those under article 283 LOSC. It observed that the duty to exchange views expeditiously under article 283 LOSC does not align with the circumstances addressed by articles 74/83 LOSC. Article 283 LOSC operates as a general provision applicable to all disputes arising under the Convention, initiating the dispute settlement process upon the mere existence of a dispute. In contrast, articles 74/83 LOSC impose a specific obligation on States to reach an agreement on delimitation through negotiations. Paragraph (2) of articles 74/83 LOSC then directs the parties to the procedures of Part XV LOSC when such negotiations fail. The tribunal reasoned that, in situations where negotiations have already been conducted over a 'reasonable period of time' without resolution, article 283 LOSC cannot logically be interpreted as requiring the parties either to engage in further exchanges of views, whether on settlement by negotiation or on resorting to other peaceful means. The negotiation process mandated by articles 74/83 LOSC effectively satisfies the requirements of article 283 LOSC, rendering additional procedural steps unnecessary in this context.[159]

The tribunal carefully examined the issue of unilateral recourse to compulsory dispute settlement procedures, offering two principal observations on the interpretation of the procedural stipulations contained in paragraph (2) of articles 74/83 LOSC. These pronouncements lend substantial support to the central argument advanced in this book: that rules constructed through the textual pattern of clausal embeddedness constitute a *sui generis* legal category,

---

[157] The development of this argument can be traced in *Barbados v. Trinidad and Tobago Arbitration*; *Counter-Memorial of Trinidad and Tobago* (30/3/2005), Volume 1(1):37¶112, and *Rejoinder of Trinidad and Tobago* (18/8/2005), Volume 1:19(¶41).

[158] *Barbados v. Trinidad and Tobago Arbitration*; *Reply of Barbados* (9/6/2005), Volume 1:61–62.

[159] *Barbados v. Trinidad and Tobago Arbitration* (Decision) 2006, §§201–203.

one that inherently precludes self-judging interpretations by the States concerned when a disagreement escalates into a legal dispute. As discussed previously in connection with the *Fishing & Conservation Convention* (1958), where the parties fail to agree on the interpretation or application of such rules, the embedded clauses function as textual mechanisms that mandate recourse to compulsory third-party settlement procedures at the request of either party. In its first observation on this point, the tribunal affirmed that Part XV LOSC confers upon each party the unilateral right to initiate a compulsory, binding third-party procedure under section 2 Part XV LOSC, such as arbitration pursuant to Annex VII LOSC; as in the instant case. In this respect, it further noted that '[the] unilateral right would be negated if the States concerned had first to discuss the possibility of having recourse to that procedure'.[160] The second observation relates to the strict and proper grammatical construction of the procedural stipulations embedded in the relevant clauses. In elaborating its pronouncement on the unilateral procedural right to refer a dispute under the applicable law, the tribunal stressed that:

> [Paragraph (2) of articles 74/83 LOSC], which refers to 'the States concerned' (in the plural) resorting to the procedures (stated generally) provided for in Part XV [LOSC], must be understood as referring to those procedures in the terms in which they are set out in Part XV [LOSC]: where the procedures require joint action by the States in dispute they must be operated jointly, but where they are expressly stated to be unilateral their invocation on a unilateral basis cannot be regarded as inconsistent with any implied requirement for joint action which might be read into articles 74(2) or 83(2) [LOSC].[161]

The tribunal acknowledged, in the passage above, that the phraseology 'the States concerned' does not preclude unilateral recourse to compulsory procedures. Rather, the permissibility of such a submission must be assessed in light of whether the procedures in question require mutual consent. On this specific point, scholarly commentary has noted the recurrence of similar wording across numerous bilateral and multilateral treaties. As has been noted, a provision establishing a compulsory process may, for example, be phrased as follows: 'Any dispute … shall … be submitted, at the request of any party to the dispute, to the [specified settlement procedure]',[162] or similar expressions employing singular subject syntax. Under the terms of reference established by articles 74/83 LOSC, the *ex ante* acceptance of compulsory jurisdiction is sufficient to unilaterally invoke the procedures of section 2 Part XV LOSC,

---

160 *ibid.*, §204.

161 *ibid.*, §207.

162 Irwin, 'Settlement of Maritime Boundary Disputes: An analysis of the law of the sea negotiations', ODIL (1980) 111.

unless a State has entered an optional exception pursuant to article 298 LOSC. In its analysis, the tribunal concluded that, under the specific circumstances, both article 286 LOSC and article 1 of the Annex VII LOSC permit such unilateral invocation.[163] As a matter of plainly reading the text of the Convention, the unilateral nature of the action contemplated is evident in both provisions. Article 286 LOSC, which governs the application of all the procedures set out in Part XV LOSC, provides that:

> Subject to section 3, any dispute concerning the interpretation or application of this Convention shall, where no settlement has been reached by recourse to Section 1, be submitted at the request of any party to the dispute to the court or tribunal having jurisdiction under this section.

Similarly, article 1 of the Annex VII LOSC, which prescribes the procedure for the institution of arbitral proceedings, provides that:

> Subject to the provisions of Part XV [LOSC], any party to a dispute may submit the dispute to the arbitral procedure provided for in this Annex by written notification addressed to the other party or parties to the dispute. The notification shall be accompanied by a statement of the claim and the grounds on which it is based.

This constitutes a 'certainly welcome interpretation' of the relationship between articles 74/83 and 283 LOSC,[164] through which the tribunal affirmed its assumption of jurisdiction. It follows that, in the context of maritime delimitation disputes, the obligation to resolve the dispute through negotiations—as mandated under section 1 Part XV LOSC—is subsumed by the procedural stipulation embedded in paragraph (2) of articles 74/83 LOSC. Where negotiations fail, either party is consequently entitled to refer the matter unilaterally and without delay to the compulsory and binding dispute settlement procedures of section 2 Part XV LOSC.

A year later, a comparable interpretative approach was endorsed in another arbitration, although the legal effect of the embedded clause in paragraph (2) had not been explicitly contested by the respondent party. In the *Guyana v. Suriname* arbitration, the Annex VII LOSC tribunal was seised with the delimitation of the maritime boundary between the two States, along with allegations of breaches of international law in the disputed maritime areas. Guyana initiated proceedings pursuant to articles 286 and 287 LOSC, submitting, inter alia, the question of the delimitation of overlapping EEZ/CS entitlements. The case touched upon several critical aspects of articles 74/83 LOSC, including disputes over the substantive rule in paragraph (1) and alleged violations of the

---

[163] *Barbados v. Trinidad and Tobago Arbitration* (Decision) 2006, §209 *et seq.*

[164] Karaman, *Dispute Resolution in the Law of the Sea* (2012) 195.

obligation under paragraph (3) to enter into provisional arrangements of a practical nature pending final delimitation. Guyana argued that, having exhausted all avenues for settlement through direct negotiations, it was entitled to invoke the compulsory procedures under Part XV LOSC.[165] Suriname's jurisdictional objections were framed primarily around a so-called 'mixed dispute', questioning the tribunal's competence insofar as it might involve the determination of territorial sovereignty.[166] The tribunal, however, had little difficulty in asserting jurisdiction, noting that neither party had made a declaration under article 298 LOSC to exclude such disputes. It further upheld Guyana's position on the obligation to exchange views and the procedural right to invoke section 2 Part XV LOSC once negotiations had been exhausted and no agreement was forthcoming.[167]

In the useful light shed by the above pronouncements on the textual construction of articles 74/83 LOSC, the reader is invited to revisit article 7(4) FSA, particularly for a comparative assessment of the language employed in their respective embedded clauses. Whereas paragraph (2) of articles 74/83 LOSC provides that '[i]f no agreement can be reached within a reasonable period of time, the States concerned shall resort to the procedures provided for in Part XV', paragraph (4) of article 7 FSA stipulates that 'any of the States concerned may invoke the procedures for the settlement of disputes provided for in Part VIII'. This nuanced grammatical difference—most notably, the shift from the plural 'the States concerned' to the singular 'any of the States concerned'—bears important legal implications. It reflects a clearer and more direct articulation of the compulsoriness principle, particularly in affirming the unilateral right of the applicant party to initiate the procedures set out in Part VIII FSA when no agreement is reached within a reasonable period of time. The treaty language of article 7(4) FSA reinforces, *expressis verbis*, the availability of unilateral recourse to binding third-party dispute settlement. As a general remark on this point, academic commentary has consistently interpreted the use of unpluralized nouns or equivalent periphrastic legal formulations—for example, the phrase 'at the request of any party'—as indicative of compulsory jurisdiction, permitting the unilateral seising of a court or tribunal under the procedures of section 2 Part XV LOSC.[168]

---

165 *Guyana v. Suriname Arbitration*; Memorial of Guyana (22/2/2005), Volume I:74, 125–128.

166 *Guyana v. Suriname Arbitration*; Rejoinder by Suriname (1/11/2006), Volume I:34–37.

167 *Guyana v. Suriname Arbitration* (Award) 2007, §§167, 407–416, 446 & 484.

168 Treves, 'The Law of the Sea Tribunal: Its status and scope of jurisdiction after November 16, 1994', ZaöRV (1995) 428–429.

Unilateral referral and abuse of the dispute settlement process

Finally, pronouncements of particular relevance concern the extent to which the unilateral referral of a dispute to section 2 Part XV LOSC, pursuant to the procedural stipulations contained in the embedded clauses, may amount to abuse of rights or the dispute settlement process. On this point, the *Barbados v. Trinidad and Tobago Arbitration* is once again instructive, having unequivocally stated that the unilateral invocation of arbitration procedures—and, by extension, other adjudicative procedures available under section 2 Part XV LOSC—'cannot by itself be regarded as an abuse of right contrary to article 300 [LOSC], or an abuse of right contrary to general international law'. Article 286 LOSC, as the *chapeau* article of section 2 Part XV LOSC, 'confers a unilateral right', and its exercise—unilaterally and without discussion or agreement with the other party—constitutes a straightforward invocation of the right granted by the treaty, in the manner therein envisaged.[169] This interpretation has been affirmed in subsequent jurisprudence. Notably, the Annex VII LOSC tribunal in the *South China Sea Arbitration* held that the mere unilateral initiation of proceedings under Part XV LOSC does not, in itself, amount to an abuse of rights.[170]

This question was examined at greater length in the *Delimitation of the Maritime Boundary in the Indian Ocean* between Mauritius and the Maldives. The arbitral proceedings under Annex VII LOSC were, in this instance, transferred to a Special Chamber of ITLOS pursuant to a *compromis* concluded between the parties. The case addressed issues of considerable legal and political significance, particularly in the aftermath of the rulings on the legal status of the Chagos Archipelago in the 2015 *Chagos MPA* arbitration and the 2019 *Legal Consequences of the Separation of the Chagos Archipelago from Mauritius in 1965* advisory opinion by ICJ. The Maldives raised several preliminary objections concerning both the jurisdiction of the Chamber and the admissibility of the claims. After thorough consideration, the Chamber rejected all objections, affirming its jurisdiction over the maritime delimitation dispute and declaring the claims admissible.[171] Turning now to the objections of direct relevance, the following two had a direct bearing on the interpretation of the embedded clauses in articles 74/83 LOSC.

The Maldives argued that the parties had not engaged in meaningful negotiations, which are a prerequisite to the exercise of jurisdiction in disputes concerning the interpretation or application of articles 74/83 LOSC. Only once such negotiations had been undertaken and failed to resolve the disagreement

---

[169] *Barbados v. Trinidad and Tobago Arbitration* (Decision) 2006, §208.

[170] *South China Sea Arbitration* (Jurisdiction and Admissibility) 2015, §126.

[171] *Delimitation of the Maritime Boundary in the Indian Ocean* (Preliminary Objections) 2021, §351.

could any of the States resort to the procedures in Part XV LOSC. The Maldives asserted that this step in the dispute settlement process extended beyond the negotiations envisaged in article 283 LOSC.[172] In response, Mauritius contended that paragraph (2) of articles 74/83 LOSC imposed no additional jurisdictional requirement beyond those set out in section 1 Part XV LOSC, and that the parties had, in any event, engaged in the requisite negotiations. In this regard, it was emphasized that the precondition of negotiations in articles 74/83 'appears outside of but before' Part XV LOSC, and that this structural arrangement renders the procedures subject only to the exhaustion of negotiations as outlined in the provisions of Parts V and VI LOSC, relating to the EEZ and CS, respectively. While acknowledging that parties should not hasten to adjudication, Mauritius further argued that the drafters of the Convention clearly intended to preserve a State's right to invoke Part XV LOSC—particularly the compulsory procedures entailing binding decisions under section 2—once negotiations had failed.[173]

The Chamber observed that, while articles 74/83 LOSC impose an obligation on the States concerned to conduct negotiations in good faith, this obligation does not require that the negotiations be successful. Paragraph (1) of these articles provides that delimitation shall be effected by agreement, clearly distinguishing this obligation from unilateral action.[174] This closely follows the established jurisprudence regarding the scope of the obligation to negotiate.[175] The record before the Chamber indicated that Mauritius had made repeated attempts to initiate negotiations on the delimitation of the overlapping EEZ and CS, whereas the Maldives largely refrained from engaging.[176] Notably, the Chamber expanded upon its analysis in a manner consistent with this chapter's argument concerning the inseparability of the substantive and procedural constituent elements in *sui generis* rules formulated through the textual

[172] *Delimitation of the Maritime Boundary in the Indian Ocean*; *Written observations of the Republic of Maldives* (15/4/2020) 32–34 and ITLOS, 'Minutes of public hearings in accordance with article 86 of the Rules of the Tribunal' (13–19/10/2020), 44–46.

[173] *Delimitation of the Maritime Boundary in the Indian Ocean*; *Written observations of the Republic of Mauritius on the preliminary objections raised by the Republic of Maldives* (17/2/2020), 36–38.

[174] *Delimitation of the Maritime Boundary in the Indian Ocean* (Preliminary Objections) 2021, §§270–273.

[175] *Maritime Delimitation in the Indian Ocean* (Preliminary Objections) 2017, §90 and *Delimitation of the Maritime Boundary in the Atlantic Ocean* (Merits) 2017, §604.

[176] *Delimitation of the Maritime Boundary in the Indian Ocean* (Preliminary Objections) 2021, §§288–289.

pattern of clausal embeddedness. To begin with, the Chamber clearly endorsed Mauritius's argument that articles 74/83 LOSC set out 'two interrelated substantive obligations'. Namely, that while a State may not unilaterally delimit its EEZ/CS, since delimitation must initially be effected by agreement with the other State(s) concerned, if such agreement cannot be reached, the States concerned are obliged to resort to the procedures provided for in Part XV LOSC. More specifically, the Chamber interpreted paragraphs (1) and (2) of articles 74/83 LOSC as operating 'in a mutually reinforcing way', establishing substantive obligations for States to delimit their EEZ/CS either by agreement or, failing such agreement within a reasonable period of time, by recourse to the dispute settlement procedures of Part XV LOSC.[177] In addressing the effect of the embedded clauses, the Chamber unequivocally stated that 'in situations in which "no agreement can be reached", to resort to the procedures of Part XV [LOSC], as set out in paragraph (2) of each of articles 74 and 83 [LOSC], is not only justified but also an obligation of the States concerned'.[178] As a result, the Chamber rejected the preliminary objection to its jurisdiction by a vote of eight to one, with the sole dissenting vote cast by Bernard Oxman, the *ad hoc* Judge appointed by the Maldives.[179]

The other preliminary objection, closely linked to and directly relevant to the first, was the Maldives' submission that the claims brought by Mauritius constituted an abuse of process and should therefore be dismissed as inadmissible. The Maldives contended that Mauritius was improperly invoking the Convention's compulsory dispute settlement procedures to obtain a ruling on a territorial sovereignty dispute to which a third State was a party.[180] Conversely, Mauritius maintained that the objection lacked merit, asserting that it did not request, nor was it necessary to render, a ruling on sovereignty over the Chagos Archipelago.[181] The Chamber referred to its earlier finding that the obligation under paragraph (1) of articles 74/83 LOSC had been fulfilled, and that a dispute concerning the delimitation of the maritime boundary existed between the parties at the time the proceedings were initiated.[182] It further noted that 'Mauritius resorted to the dispute settlement procedures provided for in Part XV [LOSC], in accordance with paragraph (2) of articles 74/83 LOSC',

177 *ibid.*, §§262, 267, 271–275.

178 *ibid.*, §292.

179 *ibid.*, §354 *dispositf.*

180 *Written observations of the Republic of Maldives* (2020), *supra cit.* (*fn*172), 37–38.

181 *Written observations of the Republic of Mauritius on the preliminary objections raised by the Republic of Maldives* (2020), *supra cit.* (*fn*173), 41–43.

182 *Delimitation of the Maritime Boundary in the Indian Ocean* (Preliminary Objections) 2021, §345.

and that the claims were 'confined to articles 74/83 [LOSC]'.[183] Accordingly, the Chamber found no abuse of process and unanimously rejected this preliminary objection.[184]

4.4 CONCLUSION

The rationale underlying the textual pattern of clausal embeddedness lies in the necessity of establishing an inseparable link between the substantive rule and the provisions governing the settlement of disputes concerning its interpretation or application. This textual morphology indicates the presence of a *sui generis* rule, whose distinctiveness derives from its legal normativity as a product of progressive development, contingent upon the proviso of compulsory third-party procedures entailing binding decisions. Embedded clauses function as affirmative declarations for invoking the treaty's dispute settlement procedures and, accordingly, express *ex abundante cautela* the principle of compulsory adjudication as an integral component of the substantive rule itself. Where States fail to reach agreement on the interpretation or application of such rules, these clauses operate as textual triggers for compulsory third-party procedures at the request of either party, thereby precluding self-judging interpretations by the States concerned once a disagreement crystallizes into a legal dispute.

---

183 *ibid.*, §§347–348.
184 *ibid.*, §§349 & 354 *dispositf.*

5. The drafting of the compatibility rule in the 1995 United Nations Fish Stocks Agreement

5.1 INTRODUCTION

While much attention has been devoted to the ambiguity inherent in the compatibility rule with respect to its spatial scope, uncertainty remains as to the extent to which compulsory settlement procedures entailing binding decisions apply to related disputes. This chapter probes and clarifies the interpretative relationship between the substantive and procedural constituent elements of the rule as integrated within the structure of article 7 FSA, and, in light of this analysis, further examines the legal relationship between the rule as a whole and the fisheries limitation in article 32 FSA, with a view to determining its effect on the remit of the dispute settlement procedures contained in Part VIII FSA. In doing so, the analysis adopts a strict *litera legis* approach, testing the interpretative flexibility of the Agreement's treaty text by delving into historical, legislative, and technical aspects of the drafting process at the *Fish Stocks Conference*. It is argued that significant doctrinal considerations have shaped the textual morphology of the compatibility rule in article 7 FSA, echoing the rationale behind similar patterns previously employed by ILC, UNCLOS/I, and UNCLOS/III in formulating provisions that embody progressive developments of international law.

5.2 THE UNITED NATIONS FISH STOCKS CONFERENCE DOCUMENTS

During the *Fish Stocks Conference*, Chairman Nandan personally led the drafting of the Agreement, adhering firmly to his preferred 'leading with a text' method of negotiation.[1] Regrettably, no official *procès-verbal* of the

[1] Nandan, 'Revisiting the Law of the Sea Negotiations', in Nordquist–Moore–Long (Eds) *The Marine Environment and United Nations Sustainable Development Goal 14* (2018) 19.

proceedings exists, as insufficient funding precluded the Secretariat from producing verbatim records.[2] The absence of an official record poses a challenge to definitively clarifying the Agreement's provisions, as the scope for methodological recourse to supplementary means of interpretation—in order to confirm the meaning resulting from the application of the general rule under article 31 VCLT[3]—is confined to the formal documents, without the benefit of the accompanying oral accounts. An unendorsed source of the major points debated during the conference is the Earth Negotiations Bulletin ('ENB'), an independent reporting service covering multilateral negotiations on environment and sustainable development.[4] However, the paucity of this otherwise significant legal material has little practical consequence for the interpretation of the Agreement, given that the conference's *modus operandi* centred on extensive informal consultations to address contentious matters raised in the plenary sessions. Intersessional consultations, whether bilateral or in small groups of interested States, along with other exchanges, formal or otherwise, conducted while the conference was not in session, were in fact encouraged by the Chairman as an effective means of advancing negotiations by identifying areas requiring further time and effort to achieve consensus. Nevertheless, in the interest of transparency—and because these diplomatic interactions were not intended to pre-empt the decisions of the conference—their outcomes were communicated to all delegations at the next plenary session and were reflected in the Chairman's reports.[5] While additional information that may have transpired from this process could shed light on the legitimate expectations and shared understanding between DWFS and coastal States, it should be accorded persuasive rather than binding effect. This is because, at the Chairman's request, statements made during informal negotiations—regardless of how they were recorded or relayed—were not to be given formal attribution.[6]

By implication of these methodological constraints, the analysis that follows concentrates on the officially tabled treaty drafts that functioned as the principal instruments guiding the negotiations. A number of preparatory and progress-reporting documents used to inform delegations—along with a

---

2 Lévy–Schram (Eds) *United Nations Conference on Straddling Fish Stocks and Highly Migratory Fish Stocks* (1996) 13.

3 VCLT 32.

4 The conference is documented in ENB Volume 7 across 55 issues, available from the *International Institute for Sustainable Development* {enb.iisd.org/negotiations/un-conference-straddling-fish-stocks-and-highly-migratory-fish-stocks-fsa}.

5 A/Conf.164/17 'Statement made by the Chairman of the Conference at the opening of the Third Session, held on 14 March 1994' (16/3/1994), 3.

6 ENB:7(30) 'Summary report, 14–31 March 1994'.

considerable number of national proposals, formal texts prepared and revised by the Chairman himself, and documents produced by the Secretariat—are accessible through the dedicated UN webpage listing all publicly registered and numbered documents.[7] A limited selection of these materials appears in an edited compilation.[8] Other key materials consist of internally circulated 'conference room papers' issued during the negotiations. These papers primarily contain original versions and proposed revisions of the core drafts, the study of which, even by the simple literary technique of juxtaposition, offers valuable insights into the formal negotiation texts and contextualizes the national submissions of delegations. Within this category of texts, the redrafted iterations of the compatibility rule and its components, as proposed by the Chairman, are particularly significant. They provide critical insights into the structural evolution of the embedded clause in article 7 FSA, revealing a discernible trend towards the principle of compulsory adjudication.[9] The core drafts that constituted the basis of negotiations and will be examined below are the *Negotiating Text*, which was introduced in the Second Session (12–30/7/1993);[10] the *Revised Negotiating Text*, which was circulated at the

---

[7] *Division for Ocean Affairs and the Law of the Sea* {un.org/Depts/los/index .htm}.

[8] The compilation of documents by former DOALOS Director Jean-Pierre Lévy and Gunnar Schram, *supra cit.* (*fn*2), preserves essential information and offers a valuable attempt at providing a comprehensive record of the materials discussed during the conference proceedings.

[9] A/Conf.164/CRP.1 'Redraft of Sections IV, V, VI and VII of the Negotiating Text (A/Conf.164/13 of 23/11/1993) (Prepared by the Chairman)' (23/3/1994); A/Conf.164/CRP.2 'Redraft of Sections II and III of the Negotiating Text (1993) (Prepared by the Chairman)' (25/3/1994); A/Conf.164/CRP.3 'Chairman's Revision of the Negotiating Text (1993) (Section I)' (28/3/1994); A/Conf.164/CRP.4 'Chairman's Revision of the Negotiating Text (1993) Section VIII and Annex 2)' (29/3/1994); A/Conf.164/CRP.5 'Chairman's Revision of the Negotiating Text (1993) (Sections X and XI)' (29/3/1994); A/Conf.164/CRP.6 (6/4/1995) and A/Conf.164/CRP.6/Add.1 (7/4/1995) containing the 'to be completed' versions of the 'Revised Text of Draft Agreement for the Implementation of the Provisions of the United Nations Convention on the Law of the Sea of 10 December 1982 relating to the Conservation and Management of Straddling Fish Stocks and Highly Migratory Fish Stocks (Prepared by the Chairman)'; and A/Conf.164/CRP.7 'Draft Agreement for the Implementation of the Provisions of the United Nations Convention on the Law of the Sea of 10 December 1982 relating to the Conservation and Management of Straddling Fish Stocks and Highly Migratory Fish Stocks (Prepared by the Chairman)' (19/7/1995).

[10] A/Conf.164/13 'Negotiating Text (Prepared by the Chairman)' (23/11/1993), as reissued for technical reasons.

end of the Third Session (14–31/3/1994);[11] the amended and restructured treaty text, in the form of the *Draft Agreement*, which was presented three days before the close of the Fourth Session (15–26/8/1994);[12] and the complete *Revised Draft Agreement*, which was released at the end of the Fifth Session (27/3 to 12/4/1995).[13] The final text of the Agreement[14] was approved by consensus during the Sixth, and final, Session of the conference (24/7 to 4/8/1995), incorporating important last-minute changes to the treaty text that bear on the interpretation of the procedural stipulations contained in articles 7(4) and 32 FSA.

5.3 THE NEGOTIATING TEXT OF THE UNITED NATIONS FISH STOCKS CONFERENCE

Substantive proceedings at the *Fish Stocks Conference* began on the basis of the *Negotiating Text*, which Chairman Nandan prepared after reviewing the national proposals and position papers submitted by delegations. This text was divided into 11 sections, namely: Section I (Nature of CMMs to be established through cooperation); Section II (Mechanisms for international cooperation); Section III (RFMO/As); Section IV (Duties of the flag State); Section V (Compliance and enforcement of high seas CMMs); Section VI (Port States); Section VII (Non-parties to subregional or RFMO/As); Section VIII (Dispute settlement); Section IX (Compatibility and coherence between national and international conservation measures for the same stock); Section X (Special requirements of developing countries); and Section XI (Review of the implementation of CMMs). The text further contained two draft annexes:

---

[11] A/Conf.164/13/Rev.1 'Revised Negotiating Text (Prepared by the Chairman)' (30/3/1994).

[12] A/Conf.164/22 'Draft Agreement for the Implementation of the Provisions of the United Nations Convention on the Law of the Sea of 10 December 1982 relating to the Conservation and Management of Straddling Fish Stocks and Highly Migratory Fish Stocks (Prepared by the Chairman)' (23/8/1994).

[13] A/Conf.164/22/Rev.1 'Revised Draft Agreement for the Implementation of the Provisions of the United Nations Convention on the Law of the Sea of 10 December 1982 relating to the Conservation and Management of Straddling Fish Stocks and Highly Migratory Fish Stocks (Prepared by the Chairman)' (11/4/1995).

[14] A/Conf.164/33 'Draft Agreement for the Implementation of the Provisions of the United Nations Convention on the Law of the Sea of 10 December 1982 relating to the Conservation and Management of Straddling Fish Stocks and Highly Migratory Fish Stocks (Proposal by the Chairman)' (3/8/1995).

the first specified minimum data requirements for the conservation and management of fish stocks, while the second proposed provisional arrangements for arbitration.

As the primary purpose of the document was to serve as a starting point for negotiations aimed at reaching consensus, key issues, both within and across sections, were initially examined through informal consultations based on a series of working papers separately prepared by the Chairman to gauge the views of delegations on the following topics: 'A precautionary approach to fisheries management'; 'Compatibility and coherence between national and international conservation measures for the same stock'; 'Minimum data requirements for the conservation and management of straddling fish stocks and highly migratory fish stocks'; 'Procedures for the settlement of high seas fisheries disputes'; 'Compliance and enforcement'; 'The nature of conservation and management measures to be established through cooperation'; 'The mechanisms for international cooperation'; 'Regional fisheries management organizations or arrangements'; 'Port State enforcement'; 'Non-parties to a subregional or regional agreement or arrangement'; and 'Special requirements of developing countries'.[15]

From the outset of the negotiations, the Chairman emphasized the significance of achieving compatible and coherent CMMs. It was noted that the biological integrity of fish stocks necessitated the implementation of measures spanning their full distribution and migratory range. In this context, a central challenge for the conference was to agree on arrangements that would ensure the harmonization of regimes applicable to the stocks across jurisdictional sea areas, without prejudice to the sovereign rights of coastal States in their EEZs. Anticipating severe differences of opinion between DWFS and coastal States on this matter, the Chairman called for constructive negotiations, acknowledging it as 'a difficult and sensitive issue', yet stressing that 'the gravity of the problems facing not only high seas fisheries but marine fisheries as a whole' made it imperative 'to find acceptable solutions'.[16] Accordingly, the key provision introducing Section IX adopted a balanced approach, stating that 'coastal States and States fishing on the high seas have a duty to cooperate and achieve compatible, coherent and coordinated measures for the conservation and management of straddling fish stocks and highly migratory fish stocks'.[17] Nevertheless, the remainder of the section largely favoured coastal

[15] A/Conf.164/16 'Report on the Second Session of the United Nations Conference on Straddling Fish Stocks and Highly Migratory Fish Stocks (Prepared by the Secretariat)' (17/8/1993).

[16] A/Conf.164/11 'Statement made by the Chairman of the Conference at the opening of the Second Session, held on 12 July 1993' (16/7/1993), 3.

[17] A/Conf.164/13 (1993), *supra cit.* (*fn*10), Section IX (¶47).

State interests, as reflected in provisions assessing the consistency between in-zone national measures and international measures applied on the high seas. Among other stipulations, it was provided that, in determining compatible CMMs, States should give due regard to the measures prescribed by the coastal State in areas under national jurisdiction. Additionally, it was suggested that measures established for high seas areas be no less stringent than those adopted by the coastal State in its EEZ for the same stock.[18] In the event of disagreement concerning the high seas components of CMMs, the section proposed:

> States shall nevertheless continue their efforts to reach agreement and States fishing on the high seas shall observe, provisionally and voluntarily, conservation and management measures equivalent in effect to those applying in respect of the same stock(s) in areas under national jurisdiction and, in the absence of such measures, observe minimum international standards or otherwise act in a manner consistent with the duties imposed on States under the Convention, until agreement is reached.[19]

Contrariwise, the 'no less stringency' requirement was not intended to apply reciprocally to coastal States. Where agreed CMMs applicable to high seas areas were more stringent than those adopted in areas under national jurisdiction for the same stock, coastal States were merely expected to apply such measures—or other equivalent in effect—on a voluntary basis, without attributing to them any provisional or binding legal force.[20]

Despite the advantageous position afforded to coastal States under these provisions, the *Negotiating Text* crucially refrained from any reference to the concept of special interest in high seas areas. During the Second Session, several coastal States had extensively invoked the concept of special interest or preferential rights in high seas areas adjacent to and beyond EEZ. A joint draft submitted by Argentina, Canada, Chile, Iceland, and New Zealand proposed that CMMs should 'recognize and give effect to the special interest of coastal States' in relevant fish stocks.[21] Canada's individual submission went

---

[18] *ibid.*, Section IX (¶49) *litt.*(b)–(c) & (d)(i); this requirement had appeared in the submissions of coastal States, notably in A/Conf.164/L.14 'Elements of an International Agreement on the Conservation and Management of Straddling Fish Stocks and Highly Migratory Fish Stocks on the High Seas (Working Paper Submitted by the Delegations of Chile, Colombia, Ecuador and Peru)' (16/7/1993), article X(3).

[19] A/Conf.164/13 (1993), *supra cit.* (*fn*10), Section IX (¶51).

[20] *ibid.*, Section IX (¶50).

[21] A/Conf.164/L.11/Rev.1 'Draft Convention on the Conservation and Management of Straddling Fish Stocks and Highly Migratory Fish Stocks on

further, suggesting that beyond the recognition of 'special interest', the treaty text should also emphasize that CMMs must 'be consistent with [the measures] applied by the relevant coastal State or States within their [EEZs]'.[22] In contrast, DWFS—notably South Korea, China, including the European Union—advocated for the principle of common and equal responsibilities and obligations among all States concerned.[23] This position was clearly expounded in the Japanese statement, which asserted that the conservation of fish stocks is a matter of common interest for all States. It was rightly observed that neither customary international law nor the Convention provides for special or preferential rights for coastal States on the high seas. Therefore, it was essential to maintain a balanced and equitable distribution of rights and duties between DWFS and coastal States. As CMMs applicable to the respective jurisdictional sea areas were required to be complementary, their adoption had to be science-based and pursued through cooperative efforts among the States concerned.[24] This position was reiterated in response to the Chairman's informal working paper on 'Compatibility and coherence between national and international conservation measures for the same stock', which emphasized that the rights of coastal States must not infringe upon the freedom of fishing on the high seas.[25]

Turning now to consider Section IX through the lens of dispute settlement, it is noteworthy that there was a catholic recognition among States of the necessity of incorporating compulsory procedures as an integral part of the Agreement. This position was reflected in several national submissions as early as the Secretariat's initial call for statements and comments. The position papers of Australia[26] and the United States[27] indicated a willingness to consider institutional mechanisms within RFMO/As, while Argentina explicitly supported the

---

the High Seas' (14/7/1993), article 4(a)(iii); and *see* Hayashi, 'United Nations Conference on Straddling Fish Stocks and Highly Migratory Fish Stocks: An analysis of the 1993 sessions', *Ocean Yearbook* (1994) 42.

22 A/Conf.164/L.5 'Letter dated 28 May 1993 from the Chairman of the delegation of Canada addressed to the Chairman' (4/6/1993), ¶III(a)(iv).

23 ENB:7(16) 'Summary report, 12–30 July 1993'.

24 A/Conf.164/L.6 'Organization of work (List of issues), Japan' (8/6/1993), ¶¶I(2) & II6(4).

25 A/Conf.164/L.28 'Comments on compatibility and coherence between national and international conservation measures for the same stock, Japan' (27/7/1993), ¶3.

26 A/Conf.164/L.9 'Comments on issues before the Conference, Australia' (1/7/1993), ¶11.

27 A/Conf.164/L.15, 'Principles on straddling fish stocks and highly migratory fish stocks, United States' (16/7/1993), ¶I(8).

inclusion of an 'obligatory procedure' of some kind.[28] The Russian Federation maintained that States should settle disputes concerning the conservation and rational utilization of straddling stocks through the peaceful means listed in article 33(1) UNC. However, if no resolution was attained within a reasonable period, the Russian position proposed that such disputes should, at the request of either party, be promptly submitted to a special arbitral tribunal for a binding and final decision, in accordance with the relevant provisions of the Convention.[29] Even at that early stage of the conference, several delegations moved quickly to submit detailed proposals on procedural mechanisms. Notable influences included Canada's individual submission,[30] the joint draft convention proposed by Argentina, Canada, Chile, Iceland, and New Zealand,[31] and a working paper tabled by the CPPS member States.[32] A degree of caution towards compulsory recourse to binding procedures was openly expressed in the Japanese position statement, which emphasized the importance of dispute prevention and, in cases of unresolved disagreement, identified consensual arbitration as the more appropriate mechanism.[33]

The Chairman repeatedly underscored the quintessential role of dispute settlement procedures in addressing disagreements over the adoption of compatible CMMs, beginning from the early stages of the Conference. At the opening of the Second Session, he noted that, owing to the frequent occurrence of disputes in fisheries matters, effective high seas fisheries management necessitated a dispute settlement mechanism that was both efficient and readily accessible for the expeditious resolution of such disputes.[34] Likewise, in his remarks concluding the general debate, he reiterated this position before the Plenary, stressing that effective CMMs must be underpinned by robust dispute settlement mechanisms that take into account the specific nature of fisheries

---

28 A/Conf.164/L.10, 'Letter from the permanent representative of Argentina to the UN addressed to the Chairman' (12/6/1993), preamble of the annex.

29 A/Conf.164/L.25 'Letter dated 26 July 1993 from the alternate chairman of the delegation of the Russian Federation addressed to the Chairman' (26/7/1993), ¶I(5) of the annex.

30 A/Conf.164/L.5 (1993), *supra cit.* (*fn*22), ¶VI & 'Appendix on Arbitration'.

31 A/Conf.164/L.11/Rev.1 (1993), *supra cit.* (*fn*21), article 26 & 'Annex III: Arbitration'.

32 A/Conf.164/L.14 (1993), *supra cit.* (*fn*18), articles I(5) & ¶IX.

33 A/Conf.164/L.6 (1993), *supra cit.* (*fn*24), ¶II(7(7)) pointing to the procedures employed under article XXV(2) of the *Convention on the Conservation of Antarctic Marine Living Resources* (1980); an exposition of the Japanese view on this matter is offered in Watanabe, 'Current Fisheries Issues: The position of a "fishing nation" and current cases', in Nordquist–Moore (Eds) *Entry into Force of the Law of the Sea Convention* (1995) 307.

34 A/Conf.164/11 (1993), *supra loc. cit.* (*fn*16).

disputes and the need to ensure their prompt resolution.[35] It is worth noting that, at the closing of the Second Session, the Chairman unequivocally stated that:

> In cases where there are disagreements over conservation and management measures, it is essential that provisions exist for the speedy and binding settlement of disputes. If such expeditious procedures are not established, stocks may be progressively depleted while we await the outcome of a more prolonged procedure. Such a situation would be contrary to the fundamental goal of this Conference.[36]

In response to these considerations, the *Negotiating Text* made general provision in the treaty text for dispute settlement procedures by means of compulsory arbitration.[37] At this point, it is important to note the absence of any procedural stipulation within the compatibility rule as it was originally formulated in Section IX (see Box 5.1). That is to say, the compatibility rule was not originally formatted as a treaty article containing an embedded clause; rather, this structural feature was incorporated later in the course of negotiations, culminating in the *Revised Negotiating Text* at the end of the Third Session. Nevertheless, where States failed to agree on the adoption of compatible CMMs, it was clearly anticipated from the outset of negotiations that dispute settlement procedures would be invoked.[38]

BOX 5.1 THE COMPATIBILITY RULE IN THE NEGOTIATING TEXT (1993) OF THE UN FISH STOCKS CONFERENCE

SECTION IX. Compatibility and Coherence between National and International Conservation Measure for the Same Stock

47. Coastal States and States fishing on the high seas have a duty to cooperate and achieve compatible, coherent and coordinated measures for the conservation and management of straddling fish stocks and highly migratory fish stocks.

---

35 A/Conf.164/12 (1993), *cit.* Chapter 1 (*fn*113).

36 A/Conf.164/15 'Statement made by the Chairman of the Conference at the closing of the Second Session, held on 30 July 1993' (10/8/1993), 4.

37 A/Conf.164/13 (1993), *supra cit.* (*fn*10), annex II.

38 Jacobson, 'Managing Marine Living Resources in the Twenty-First Century: The next level of ocean governance?', in Nordquist–Moore (Eds) *Entry into Force of the Law of the Sea Convention* (1995) 316–317.

48. In developing conservation and management measures for straddling fish stocks and highly migratory fish stocks, States shall recognize the interdependence of stock components harvested in areas under national jurisdiction and on the high seas. States regulating fisheries in areas under their national jurisdiction, and subregional or regional organizations or arrangements establishing conservation and management measures for the same stock(s) on the high seas, should achieve compatible and coherent conservation and management measures. Establishment of such measures shall be without prejudice to the sovereign rights of the coastal State(s) for the purpose of exploring, exploiting, managing, and conserving living marine resources within areas under national jurisdiction, exercised in accordance with the United Nations Convention on the Law of the Sea.
49. In determining conservation and management measures for straddling fish stocks and highly migratory fish stocks on the high seas, States, either directly or through subregional or regional organizations or arrangements, shall:
   (a) Ensure that the measures do not result in transferring, directly or indirectly, a disproportionate burden of the need for conservation action onto the coastal State(s);
   (b) Ensure that the measures do not result in undue harmful impact on the living marine resources within the areas of national jurisdiction;
   (c) Ensure that the measures established in respect of the high seas are no less stringent than those established in areas under national jurisdiction in respect of the same stock(s);
   (d) Give due regard to the interests of all States concerned, and to: (i) the measures taken or proposed by the coastal State(s) within areas under national jurisdiction; (ii) the relative dependence of the coastal State(s) and States fishing on the high seas on the stock(s) concerned; (iii) the impact of high seas fishing on the stock(s) and on associated and dependent species within areas under national jurisdiction; (iv) the particularities of the region and the biological characteristics of the stock(s) concerned.
50. Where agreement is reached on conservation and management measures for the high seas that are more stringent than those applied in areas under national jurisdiction, in respect of the same stock(s), the coastal State(s) concerned shall voluntarily apply conservation and management measures equivalent in effect to the relevant

measures applicable on the high seas in areas under their national jurisdiction.

51. If, in spite of the processes outlined above, States are unable to agree on conservation and management measures for the high seas, States shall nevertheless continue their efforts to reach agreement and States fishing on the high seas shall observe, provisionally and voluntarily, conservation and management measures equivalent in effect to those applying in respect of the same stock(s) in their national jurisdiction and, in the absence of such measures, observe minimum international standards or otherwise act in a manner consistent with the duties imposed on States under the Convention, until agreement is reached.

5.4 FROM THE REVISED NEGOTIATING TEXT OF THE UNITED NATIONS FISH STOCKS CONFERENCE TO THE DRAFT FISH STOCKS AGREEMENT

The *Fish Stocks Conference* resumed its Third Session with the convening of the Plenary to afford delegations the opportunity to make statements on the *Negotiating Text*. Delegations had benefited from a number of technical papers, which were considered in two open-ended working groups addressing the application of the precautionary approach to fisheries management.[39] The Conference also drew on a report prepared under the auspices of IATTC concerning the role of regional regimes in gathering scientific and statistical data, which had been produced prior to the resumption of its work.[40] In parallel with the plenary deliberations, the Chairman held informal consultations to resolve outstanding points of difference and to gain further insights necessary for revising the treaty text.[41] These consultations contributed to greater textual

[39] A/Conf.164/INF/8 'The Precautionary approach to fisheries with reference to straddling fish stocks and highly migratory fish stocks' (26/1/1994) and A/Conf.164/INF/9 'Reference points for fisheries management: Their potential application to straddling and highly migratory resources' (26/1/1994).

[40] A/Conf.164/INF/10 'Ad hoc consultation on the role of regional fishery agencies in relation to high seas fishery statistics' (27/1/1994).

[41] A/Conf.164/20 'Report on the Third Session of the United Nations Conference on Straddling and Highly Migratory Fish Stocks (Prepared by the

clarity, evidencing the significant progress made in building consensus on key issues, most notably the complex matter of ensuring compatibility and coherence in CMMs. Significant understanding was also reached on related issues, such as the general principles underpinning these measures, the importance of cooperation for effective conservation and management, and the need for compulsory and expeditious procedures for dispute settlement.[42] The *Revised Negotiating Text*, which was circulated at the end of the Third Session, contained two important and interrelated elements in the formulation of the compatibility rule. The first was reflected in the drafting of provisions designed to ensure that CMMs adopted in EEZs and on the high seas are compatible and coherent, with a view to addressing the biological unity of the stocks, as required by the ecosystem approach. The second took form in the incorporation of compulsory and binding dispute settlement procedures, consistent with the Convention, while allowing States the flexibility to choose the specific means of dispute resolution.

5.4.1 The Principle of Ecosystem Approach and the Compatibility Rule

The treaty text was revised to include a new Section I, which set out the objective of the Agreement. This proposed objective was closely interwoven with the compatibility rule and was formulated as follows:

> States have a duty to ensure the long-term sustainability of straddling fish stocks and highly migratory fish stocks. The biological unity of stocks which occur both in the high seas and in areas under national jurisdiction requires that measures taken on the high seas and those taken in areas under national jurisdiction be compatible in order to ensure conservation and management of the stocks overall. To this end coastal States and States fishing on the high seas have a duty to cooperate for the purpose of achieving compatible measures so as to effectively conserve and manage straddling fish stocks and highly migratory fish stocks.[43]

A comparison with the *Negotiating Text* suggests that the above formulation of the objective was intended to reflect a more refined and expansive version of the previously neutral framing of the compatibility rule. In the earlier text, this rule had been presented as a key proposition concerning the balance of

Secretariat)' (26/5/1994), ¶12 & ¶16 and *see* the revisions contained in the conference room papers A/Conf.164/CRP.1 (1994) to A/Conf.164/CRP.5 (1994), *supra cit.* (*fn*9).

42 A/Conf.164/19 'Statement made by the Chairman of the Conference at the closing of the Third Session, held on 31 March 1994' (9/5/1994), 2–3.

43 A/Conf.164/13/Rev.1 (1994), *supra cit.* (*fn*11), Section I (¶1).

interests between DWFS and coastal States, as presented in the introductory provision of Section IX.[44] Accordingly, the *Revised Negotiating Text* contemplated the application of compatible CMMs throughout the full geographic extent of the stocks' distribution or migratory range, including areas under the national jurisdiction of coastal States. This approach was further reinforced by corresponding revisions to Section II on the scope of application, which stipulated that:

> Except as provided for in parts B [*viz.*, 'Precautionary approaches to fisheries management'] and C [*viz.*, 'Compatibility'] of Section III, the provisions set out in this document shall apply to conservation and management on the high seas of straddling fish stocks and highly migratory fish stocks. In accordance with the 1982 United Nations Convention on the Law of the Sea, the coastal State has responsibility for conservation and management of such stocks in areas under national jurisdiction.[45]

By adopting the ecosystem approach without any preassigned jurisdictional qualification, the text of both Sections I and II deliberately avoided the formulations proposed in submissions by certain coastal States, which sought to confine the coordination of CMMs to exceptional cases within national jurisdiction and to limit the Agreement's objective exclusively to high seas areas adjacent to or beyond a coastal State's EEZ. These submissions further proposed that international measures applicable on the high seas should be compatible and coherent with those adopted in areas under national jurisdiction, but not the reverse, leaving no room for adjusting in-zone national measures on the basis of international measures adopted for the high seas.[46] The insertion of the phrase 'on the high seas' in Sections I and II prompted extended debate over the *Negotiating Text*, revealing the sensitivity and interpretative implications of spatially suggestive terminology.[47] DWFS advocated for the complete omission of such references, seeking to preserve the concept of biological unity unburdened by any spatial correlation. In this vein, Poland—supported by South Korea—proposed that the provision concerning the scope of the Agreement be revised to explicitly affirm the concept of biological unity by

---

[44] *see* Section IX (¶47) in A/Conf.164/13 (1993), *supra cit.* (*fn*10), and accompanying text.

[45] A/Conf.164/13/Rev.1 (1994), *supra cit.* (*fn*11), Section II (¶2).

[46] *e.g.* A/Conf.164/L.44 'Presentation of the working paper for a Draft Convention on the Conservation and Management of Straddling and Highly Migratory Fish Stocks on the High Seas, Ecuador' (28/3/1994), articles 2, 4 & 38, which, on the issue of spatial scope and particularly the application of the compatibility rule, followed the relevant provisions previously set out in the joint proposal of the five coastal States under A/Conf.164/L.11/Rev.1 (1993), *supra cit.* (*fn*21).

[47] As reported in ENB:7(30) 'Summary report, 24 March 1994'.

referencing the full distribution or migratory range of stocks, both within and beyond areas of national jurisdiction. Coastal States, however, viewed such proposals as an encroachment on their sovereign rights within EEZ. Chile, expressing concern that the objective had already blurred the distinction between EEZ and adjacent high seas areas, called for the inclusion of a clause explicitly safeguarding coastal State sovereignty. Similarly, Ecuador proposed that the objective be expressly limited in its application in areas of national jurisdiction—and then, only by exception—and called for language that would reinforce this characterization.[48]

Despite the neutral predisposition of the compatibility rule in its association with the objective and the spatial scope of the Agreement, the *Revised Negotiating Text* retained a position favourable to coastal States—although framed in far more temperate language—by privileging their prescriptive authority in certain instances. Notably, DWFS, in determining compatible CMMs, were expected to 'ensure that the measures established in respect of the high seas areas are no less stringent than those established, in accordance with the Convention, in sea areas under national jurisdiction in respect of the same stock(s)'.[49] Furthermore, in the event of a compatibility dispute with a coastal State, and pending its final settlement, DWFS were required to 'observe [CMMs] equivalent in effect to the measures applying in the area(s) under national jurisdiction'.[50] As part of the exploratory revisions intended to induce consensus on this issue, the rule was substantially reformulated through the incorporation of articles 63(2) and 64 LOSC. This reformulation reiterated that compatible CMMs 'shall be without prejudice to the sovereign rights of coastal States … within areas under national jurisdiction, exercised in accordance with the [Convention]'.[51]

Nonetheless, this incorporation ultimately intensified uncertainty regarding the application of the compatibility rule and its implications for the sovereign rights of coastal States, particularly in relation to the regime governing straddling fish stocks.[52] The jurisdictional complexities associated with the regulation of such stocks were further highlighted in the special case of high seas enclaves. In this regard, submissions by the Russian Federation—acting in that context on the basis of its coastal, rather than distant-water fishing, interests—are particularly illustrative of coastal State perspectives. It was argued that the

---

48 ENB:7(39) 'Summary report, 15–26 August 1994'.

49 A/Conf.164/13/Rev.1 (1994), *supra cit.* (*fn*11), Section III/C (¶7(d)).

50 *ibid.*, Section III/C (¶8).

51 *compare* A/Conf.164/13 (1993), *supra cit.* (*fn*10), Section IX (¶48) *with* A/Conf.164/13/Rev.1 (1994), *supra cit.* (*fn*11), Section III/C (¶5).

52 Barston, 'United Nations Conference on Straddling and Highly Migratory Fish Stocks', *Marine Policy* (1995) 163–164.

conservation and management of living marine resources remained imprecise in international fisheries law, and that this lack of clarity was a principal factor contributing to the persistent difficulties in reaching consensus on the provisions relating to the compatibility rule. Elaborating on what it described as the 'correct interpretation of the rights and duties of the coastal State under article 61 LOSC', Russia asserted that this article provided a legal basis for the extension of in-zone national measures beyond the EEZ in respect of specific categories of straddling stocks, viewing that:

> The Convention obliges the coastal State to ensure the appropriate maintenance of the living resources in the zone, but it does not provide that the measures taken by the coastal State for this purpose should be confined solely to that part of the stock which is located within the 200-mile economic zone at the time in question. The issue of the legality of the application of such measures by the coastal State to the part of the straddling stock which is located outside the zone is related to the correct interpretation of the concept of management of living marine resources.[53]

Even though the proffered interpretation was factually tied to 'straddling-out' stocks in an 'adjacent region' presenting specific geographical circumstances,[54] a special case that was ultimately accommodated in article 16 FSA, the proposed reading does not align with the clear distinction between articles 61 and 63(2) LOSC, nor does it satisfy the requirement of interpretative consistency between the Convention and the Agreement. The UNCLOS/III negotiations on record, as already discussed in Chapter 2, confirm that the issue of CMMs for straddling stocks in high seas areas adjacent to EEZ was addressed under article 63(2) LOSC, not article 61 LOSC. Whereas the Russian interpretation attempted to establish a link between these two articles, it failed to offer a satisfactory explanation of how the duty of States fishing on the high seas to comply with coastal State measures under article 61 LOSC would interact with their obligation to cooperate with DWFS in adopting agreed measures under article 63(2) LOSC.[55] This reading, which would in effect condone the seaward extension of coastal State interests into high seas areas, was not endorsed by other delegations. As a key example, Ukraine distinguished its position from such broad interpretation of the compatibility rule by stating that it could not

---

[53] A/Conf.164/L.38 'Conceptual approach to the conservation of straddling fish stocks by improving their management, Russian Federation' (2/3/1994), ¶2.

[54] *see* the submissions by the Russian Federation in A/Conf.164/L.18 'Definition of straddling stocks of marine life, and List of their main species' (20/7/1993) and A/Conf.164/L.47, 'Letter from the permanent representative of the Russian Federation to the UN addressed to the Chairman' (24/3/1995).

[55] Oude-Elferink, 'Fisheries in the Sea of Okhotsk High Seas Enclave – The Russian Federation's Attempts at Coastal State Control', IJMCL (1995) 9–15.

support the efforts of some coastal States to confer upon the EEZ the legal status of a territorial sea or to regulate fishing in adjacent high seas areas based on national measures inconsistent with the Convention.[56] Sweden advanced a contrasting position, favouring a more flexible interpretation of the compatibility rule. In particular, it was argued that compatible and coherent measures could only be ensured within the framework of regional cooperation, either through RFMO/As that institutionally incorporate EEZs into the regulatory areas for the purpose of conservation and management, or through *ad hoc* regulatory schemes achieving the same effect.[57] This approach had, in fact, already been implemented prior to the conclusion of the *Fish Stocks Conference*, with the adoption of the *Convention on the Conservation and Management of Pollock Resources in the Central Bering Sea* (1994). That instrument avoided the jurisdictional impasse in the debate over the compatibility rule by establishing agreed arrangements among the contracting parties within the convention area, thereby discouraging any tendency towards creeping jurisdiction.[58]

5.4.2 The Compatibility Rule and the Principle of Compulsory Adjudication

The *Revised Negotiating Text* envisaged a dispute settlement process premised on the principle of free choice of means, while also reflecting a principled commitment to securing settlement through compulsory procedures entailing binding and final decisions. Section VIII of the document reiterated the imperative of cooperation to avoid or resolve disagreements in the first instance. However, where such disagreements could not be effectively and promptly resolved through diplomatic channels, parties were given the option of resorting either to the procedures of Part XV LOSC or to the compulsory procedures provided by RFMO/As to which they were members. As a practical outcome, this option obligated regional fisheries regimes to establish dispute settlement procedures, consistent with the Convention, to facilitate the prompt resolution of compatibility disputes. Where the parties were unable to agree on the applicable procedure within 30 days of the notification of a dispute, the matter was

---

[56] A/Conf.164/L.40 'Conservation and rational utilization of straddling and highly migratory fish species, Ukraine' (17/3/1994), ¶6.

[57] A/Conf.164/L.39 'Elements of a draft instrument on conservation and management of straddling fish stocks and highly migratory fish stocks compatible with sustainable development, Sweden' (16/3/1994), ¶3(2).

[58] Dunlap, 'Bering Sea', IJMCL (1995) 117–126.

to be unilaterally submitted to compulsory arbitration, in accordance with the terms proposed in an appended annex.[59]

A limitation to the applicability of compulsory procedures was also introduced, reflecting the approach previously adopted in the context of the Convention, by stipulating that:

> The dispute settlement provisions herein shall not apply to disputes with coastal States relating to the sovereign rights of coastal States with respect to the living resources in their exclusive economic zones or the exercise of those rights and they do not affect in any way the provisions of Article 297 of the Convention.[60]

It is readily apparent that the limitation clause in the *Revised Negotiating Text* was significantly more expository in its phrasing than the final formulation now found in article 32 FSA, which tersely provides that 'Article 297, paragraph 3, of the Convention applies also to this Agreement'. The wording of the provision introducing the fisheries limitation was redrafted during the Fourth Session, following deliberations on the *Draft Agreement*, which, by employing the technique of incorporation by reference, stated that 'the provisions for the settlement of disputes contained in this Agreement do not affect in any way the provisions of the article 297 of the Convention'.[61] The development of the treaty text across the core drafts does not furnish any evidence suggesting that the limitation was intended to exclude, in principle, compatibility disputes from the remit of compulsory dispute settlement procedures. On the contrary, it is important to note that the phrasing of the limitation in the *Revised Negotiating Text* purposely avoided formulations that would restrict the application of compulsory adjudication to only the high seas components of a fisheries dispute. For instance, Ecuador proposed that the application of the procedures to be adopted for the Agreement should not prejudice the rights and duties of coastal States under the Convention, and recommended replacing the existing provision with the following:

> In no case shall those procedures be applied to disputes concerning the exercise of the sovereign rights of coastal States with regard to the exploration, exploitation, conservation or administration of the living resources within their exclusive economic zones. Such disputes may be submitted only to a conciliation commission as provided for in annex V to that Convention.[62]

---

[59] A/Conf.164/13/Rev.1 (1994), *supra cit.* (*fn*11), Section VIII (¶¶44–49) & annex III.

[60] *ibid.*, Section VIII (¶50).

[61] A/Conf.164/22 (1994), *supra cit.* (*fn*12), draft article 31.

[62] A/Conf.164/L.44 (1994), *supra cit.* (*fn*46), article 47.

The above wording was not only rejected, but, conversely, the understanding expressed by other States—including those with coastal interests—regarding the functional role of dispute settlement in this type of dispute underscored the need for a qualified approach to the concept of the 'right to fish', a right that is, in principle, equally enjoyed by all States. For example, although Russia had, in its submissions, expounded a conceptual formulation of the compatibility rule that favoured coastal States, it nevertheless made significant concessions in support of the application of compulsory and binding procedures. It emphasized that the compatibility rule should not prejudice the right of DWFS to participate in the adoption of CMMs in agreement with the coastal State, in accordance with article 63(2) LOSC. Moreover, Russia held the view that, where a coastal State refused to agree on such measures, or where a State failed to comply with non-discriminatory high seas measures adopted by the coastal State, the application of compulsory dispute settlement procedures was necessary.[63] The Chairman himself alluded to this ineluctable modality during the negotiations on the treaty text, reminding delegations of the 'stewardship' character of the right to fish. He observed that States often appeared to 'ignore the fact that the right to fish in the oceans is a conditional right'—whether in EEZ or on the high seas—as that right is necessarily accompanied by the duty to manage and conserve fishery resources.[64]

The clausal construction of embeddedness in the compatibility rule

In view of the two significant normative conditions grafted onto the compatibility rule—namely, the neutral predisposition of the requirement to achieve compatible CMMs in accordance with the principle of the ecosystem approach, and the principle of compulsory adjudication—it is not surprising that the most conspicuous revision in the structural formation of the rule in the *Revised Negotiating Text* was its clausal construction, as it appeared in Section III/C(8) (see Box 5.2). Consistent with the development of the objective to ensure effective CMMs spanning the biological range of stocks, in line with the ecosystem approach, the treaty text introduced a procedural clause to address the eventuality of disagreements, stipulating that:

> If, in spite of having made every effort to cooperate for the purposes specified in paragraph 1, States are unable to agree on compatible and coordinated conservation and management measures, they shall resolve their differences in accordance with the dispute settlement provisions set out in section VIII.

---

63 A/Conf.164/L.38 (1994), *supra cit.* (*fn*53), ¶4(4)–(5).

64 A/Conf.164/21 'Statement made by the Chairman of the Conference at the opening of the Fourth Session, held on 15 August 1994' (17/8/1994), 2.

BOX 5.2 THE STRUCTURE OF CLAUSAL EMBEDDEDNESS IN THE COMPATIBILITY RULE OF THE REVISED NEGOTIATING TEXT (1994) OF THE UN FISH STOCKS CONFERENCE

SECTION III. GENERAL PRINCIPLES

C. Compatibility

5. Without prejudice to the sovereign rights of coastal States for the purpose of exploring and exploiting, conserving and managing the living marine resources within areas under national jurisdiction as provided for in the Convention, and the right of all States for their nationals to engage in fishing on the high seas in accordance with the Convention:
   (a) with respect to straddling fish stocks, the relevant coastal State(s) and the States whose nationals fish for such stocks in the adjacent high seas area shall seek, either directly or through the appropriate mechanisms for cooperation provided for in section IV, to agree upon the measures necessary for the conservation of these stocks in the adjacent high seas area;
   (b) with respect to highly migratory fish stocks, the relevant coastal State(s) and other States whose nationals fish for such stocks in the region shall cooperate, either directly or through the appropriate mechanisms for cooperation provided for in section IV, with a view to ensuring conservation and promoting the objective of optimum utilization of such stocks throughout the region, both within and beyond the exclusive economic zone.
6. Coastal States shall regularly notify States fishing on the high seas in the subregion or region, either directly or through appropriate subregional or regional fisheries management organization or arrangement, or through other appropriate means, of the measures adopted by such coastal States for straddling fish stocks and highly migratory fish stocks in areas under national jurisdiction.
7. In determining the manner in which compatible conservation and management measures are to be achieved for straddling fish stocks and highly migratory fish stocks, and the nature and extent of those measures, States shall respect any measures and arrangements adopted by relevant coastal States in accordance with the Convention in areas under national jurisdiction and shall:

(a) take into account the biological characteristics of the stock(s), the relationship between the distribution of the stock(s), and the fisheries and the geographical particularities of the region, including the extent to which the stock(s) occur and are fished in areas under national jurisdiction;
(b) take into account the respective dependence of the coastal State(s) and States fishing on the high seas on the stock(s) concerned;
(c) ensure that the measures do not result in undue harmful impact on the living marine resources;
(d) ensure that the measures established in respect of the high seas are no less stringent than those established, in accordance with the Convention, in areas under national jurisdiction in respect of the same stock(s).

8. (P) *If, in spite of having made every effort to cooperate for the purposes specified in paragraph 1, States are unable to agree on compatible and coordinated conservation and management measures, they shall resolve their differences in accordance with the dispute settlement provisions set out in section VIII.* (P) In the meantime, until the dispute settlement process is ended, States shall continue to observe the provisions herein, and relevant minimum international standards and otherwise act in a manner consistent with the duties imposed on States under the Convention, and:
(a) where coordinated measures for conservation and management of the stock(s) have been adopted by the relevant coastal States; or
(b) where there is only one coastal State involved, and that coastal State has adopted measures for conservation and management of the stock(s);

States fishing on the high seas shall observe conservation and management measures equivalent in effect to the measures applying in the area(s) under national jurisdiction. If measures have been agreed in respect of the high seas, in the absence of conservation and management measures as described in (a) or (b) above, the relevant coastal State(s) shall observe measures equivalent in effect to those agreed of the same stocks(s) in the high seas.

Note: The symbol '(P)' has been used to highlight the textual disposition of clausal embeddedness.

Subsection (8) of Section III/C constitutes an embedded clause, given that the *Revised Negotiating Text* had already adopted Section VIII on 'Dispute Settlement' to which it was explicitly cross-referenced. Although it may be immaterial who initially proposed the revision, the textual development of the compatibility rule—viewed in light of the available negotiating record—gives rise to two plausible conjectures. The first is that the Chairman drafted the clause in the exercise of his *ex officio* mandate to develop the core treaty texts concurrently with the deliberative process. Just as plausible, however, is the possibility that the clause originated from the initiative of the Polish delegation, was presented to the Chairman, and, following his consideration, became the subject of discussion during informal consultations. The ENB records of the proceedings chronicled that the plenary session on 17/8/1994 was devoted to the first reading of the *Revised Negotiating Text*, following an informal session that morning focused primarily on the 'General Principles' in Section III, including Part C, which contained the compatibility rule. At that time, some delegations also raised questions concerning the provisions in Sections I and II, specifically those addressing the 'Objective' and the 'Application' of the principles across different jurisdictional sea areas. Upon the resumption of plenary deliberations, the Polish representative stated, on behalf of his delegation, that 'any extension of the rights of coastal States' was not acceptable, and that, for this reason, they could not support the concluding sentence of the *chapeau* in Section III/C(7), which essentially required DWFS to respect measures and arrangements adopted by the relevant coastal States. Following this intervention, Poland circulated an alternative text for subsection (8), addressing dispute settlement procedures and proposing that States should resolve disputes 'over incompatible and uncoordinated' CMMs.[65]

Given, however, that the *Revised Negotiating Text* was released by the Secretariat on 30/3/1994, it is reasonable to infer that the embedded clause had already been incorporated into the compatibility rule by Chairman Nandan at that stage. As an unofficial document not listed in the index subsequently compiled by the Secretariat,[66] the Polish submission most likely furnished a refined version of the clause as it ultimately appeared in paragraph (4) of article 7 *Draft Agreement* (see Box 5.3), which was released by the Chairman three days before closing the Fourth Session. The *Draft Agreement* introduced extensive structural revisions to the embedded clause previously found in Section III/C(8), setting out three procedural phases for the application of the compatibility rule. To begin with, paragraph (3) required States, in giving effect to their duty to cooperate, to make every effort to agree on CMMs

65 ENB:7(34) 'Summary report, 17 August 1994'.
66 A/Conf.164/INF/16 'List of documents' (8/12/1995).

'within a reasonable period of time'. Where that was not possible, the second procedural step—under paragraph (4)—required States to resort to the dispute settlement procedures contained in Part VIII *Draft Agreement.*

BOX 5.3 CONSOLIDATION OF THE EMBEDDED CLAUSE WITHIN THE COMPATIBILITY RULE IN THE DRAFT AGREEMENT (1994) OF THE UN FISH STOCKS CONFERENCE

Article 7

Compatibility of Conservation and Management Measures

1. Without prejudice to the sovereign rights of coastal States for the purpose of exploring and exploiting, conserving and managing the living marine resources within areas under national jurisdiction as provided for in the Convention, and the right of all States for their nationals to engage in fishing on the high seas in accordance with the Convention:
   (a) with respect to straddling fish stocks, the relevant coastal State(s) and the States whose nationals fish for such stocks in the adjacent high seas area shall seek, either directly or through the appropriate mechanisms for cooperation provided for in Part III, to agree upon the measures necessary for the conservation of these stocks in the adjacent high seas area;
   (b) with respect to highly migratory fish stocks, the relevant coastal State(s) and other States whose nationals fish for such stocks in the region shall cooperate, either directly or through the appropriate mechanisms for cooperation provided for in Part III, with a view to ensuring conservation and promoting the objective of optimum utilization of such stocks throughout the region, both within and beyond the exclusive economic zone.
2. Conservation and management measures taken on the high seas and those taken in areas national jurisdiction shall be compatible in order to ensure conservation and management of the stocks overall. To this end, coastal States and States fishing on the high seas have a duty to cooperate for the purpose of achieving compatible measures in respect of straddling fish stocks and highly migratory fish stocks. In determining compatible conservation and management measures, States shall:

(a) take into account the conservation and management measures established in respect of the same stock(s) by coastal States in areas under national jurisdiction and ensure that measures established in respect of the high seas do not undermine the effectiveness of those measures established in respect of the same stock(s) by coastal States in areas under national jurisdiction;
(b) take into account the biological unity and other characteristics of the stock(s) and the relationship between the distribution of the stock(s), the fisheries and the geographical particularities of the region, including the extent to which the stock(s) occur and are fished in areas under national jurisdiction;
(c) take into account the respective dependence of the coastal State(s) and the State(s) fishing on the high seas on the stocks concerned; and
(d) ensure that the measures do not result in undue harmful impact on the living marine resources as a whole.

3. In giving effect to their duty to cooperate, as provided for in paragraph 1, States shall make every effort to agree on compatible conservation and management measures within a reasonable period of time.
4. (P) *If no agreement can be reached within a reasonable period of time, the States concerned shall resort to the procedures for the settlement of disputes provided for in Part VIII of this Agreement without prejudice to the provisions of article 31.* (P)
5. Pending agreement on compatible conservation and management measures, the States concerned, in a spirit of understanding and cooperation, shall make every effort to enter into provisional arrangements of a practical nature. In the event that they are unable to agree on such arrangements, the States concerned shall resort to the procedure for the determination of provisional measures provided for in article 30 of this Agreement.
6. The provisional arrangements or measures entered into or prescribed pursuant to paragraph 5 shall take into account any conservation and management measures agreed by relevant subregional or regional fisheries management organizations or arrangements, shall have due regard to the rights and obligations of all States concerned, and shall be without prejudice to the final outcome of the dispute settlement procedure.

7. During the interim period, States shall not jeopardize or hamper the reaching of the final settlement of the dispute or undermine the objective of this Agreement.

Note: The symbol '(P)' has been used to highlight the textual disposition of clausal embeddedness.

Two aspects of these revisions merit particular attention. On the one hand, the use of the modal verb 'shall' signalled an imperative obligation for the States concerned to follow the procedure. On the other hand, recourse to those procedures was expressly made without prejudice to article 31 *Draft Agreement*, which contained limitations on the applicability of dispute settlement procedures and provided that such procedures '[did] not affect in any way the provisions of article 297 [LOSC]'. Pending agreement on CMMs, the third step, outlined in paragraphs (5)–(7), provided for the adoption of provisional arrangements. Where agreement on such arrangements could not be reached between the States concerned, the text envisaged the prescription of provisional measures, provided such measures would not jeopardize or hamper the prospects of reaching a final settlement, nor undermine the objective of the Agreement. Regardless of whether the embedded clause originated with the Chairman or the Polish delegation, Part VIII *Draft Agreement*—which is substantively identical to Part VIII FSA—was premised on the principle of compulsoriness, envisioning arbitration by default unless the parties to the dispute agreed upon an alternative procedure under Part XV LOSC. This was without prejudice to the limitations set out in article 297 LOSC. Such an understanding is consistent with the positions expressed by several delegations, suggesting that the textual pattern of clausal embeddedness drew upon established treaty practice. Reportedly, India supported the use of established conventional language with respect to dispute settlement.[67]

The temporal element of the embedded clause

Indeed, a clear thread of conceptual continuity can be traced from the structure of clausal embeddedness in paragraph (2) of articles 74/83 LOSC, through their precursors in articles 6(5), 7(4), 8(2) FCC, to its reaffirmation in paragraph (4) of article 7 *Draft Agreement*. This continuity is especially apparent in the recurring stipulation within the embedded clause that States must act 'within a reasonable period of time'. The inclusion of a temporal element in each of these provisions reflects a core aspect of the compulsoriness

[67] ENB:7(34) 'Summary report, 17 August 1994'.

principle, that time acquires a distinct procedural significance by functioning as a mechanism to compel parties towards action. In this respect, an analysis of the drafting rationale behind the prescribed time limits under the *Fishing & Conservation Convention* (1958) is instructive in understanding the logic and function of such provisions. It is worth recalling, in this context, that the Chairman initially had proposed that, if the parties were unable to agree on the applicable procedure within 30 days from the notification of the existence of a dispute, the matter was to be submitted unilaterally to compulsory arbitration, in accordance with the terms proposed in the annex to the *Revised Negotiating Text*.

As previously discussed in Chapter 4, the textual pattern of clausal embeddedness came under scrutiny at multiple stages of its conventional development. From its early origins in the *ILC Draft Articles* to its eventual consolidation in the treaty text of the *Fishing & Conservation Convention* (1958), this drafting technique provoked extensive debate concerning its role within substantive treaty provisions. Opposition to its inclusion intensified during consideration by the Third Committee of UNCLOS/I, where the embedded clauses—grouped as 'outstanding paragraphs in articles 52 to 56 [*ILC Draft Articles*]'—were subjected to detailed scrutiny. These clauses, though phrased with minor variations, uniformly stipulated that any of the States concerned could initiate dispute settlement procedures if agreement could not be reached 'within a reasonable period of time' (see Table 4.3). The Third Committee adopted amendments to replace this phrase with specific time limits. The negotiations once again revealed a division among States, with two opposing groups forming around the structural use of clausal embeddedness. The first group, representing the majority in UNCLOS/I, supported amendments that preserved the embedded clauses and, furthermore, introduced concrete time limits.[68] The second group, consisting of States that had persistently resisted this drafting technique, proposed amendments aimed at removing all such paragraphs from

68 *see* indicatively A/Conf.13/C.3/L.3 'France: Proposal (articles 51 to 58)' (13/3/1958) [*Geneva Off.Records* V:134]; A/Conf.13/C.3/L.31 'Uruguay: Proposal (articles 52, 53 and 56)' (26/3/1958) [*Geneva Off.Records* V:143]; A/Conf.13/C.3/L.33 'Japan: Proposal (articles 54 to 56)' (26/3/1958) [*Geneva Off.Records* V:144]; A/Conf.13/C.3/L.36 'Sweden: Proposal (articles 52 to 57)' (26/3/1958) [*Geneva Off.Records* V:144]; A/Conf.13/C.3/L.37 'Spain: Proposal (articles 51 to 56)' (26/3/1958) [*Geneva Off.Records* V:145]; and A/Conf.13/C.3/L.43 'The Netherlands, Portugal, United Kingdom of Great Britain and Northern Ireland, and United States of America: Proposal (article 54)' (28/3/1958) [*Geneva Off. Records* V:148].

the substantive articles.[69] Ultimately, the Third Committee not only retained the structure of clausal embeddedness but also reinforced the commitment to compulsory and binding procedures by introducing explicit time limits. This amendment was welcomed by States, who regarded the replacement of the vague 'reasonable period of time' with strict deadlines as a means of enhancing procedural certainty. Uruguay, among other States in favour of the unilateral referral of disputes to binding and compulsory procedures,[70] emphasized that time limits were a matter of essence, particularly given the need for the speedy and effective resolution of fisheries disputes.[71]

5.5 FROM THE REVISED DRAFT FISH STOCKS AGREEMENT TO THE FINAL TEXT OF THE 1995 UNITED NATIONS FISH STOCKS AGREEMENT

In presenting the *Draft Agreement* at the conclusion of the Fourth Session, Chairman Nandan emphasized the normative neutrality of the compatibility rule, stating that it '[ensured] that the measures taken for the conservation and management of those stocks in areas under national jurisdiction and in the adjacent high seas areas are compatible and coherent'. He also underscored the application of the principle of compulsory adjudication, noting that the text 'provides for the peaceful settlement of disputes relating to fisheries matters through compulsory binding dispute settlement mechanisms, which also ensures flexibility for the parties to use the procedure of their choice for the settlement of such disputes'.[72] This reflected a principled approach intended to encompass even the special case of compatibility disputes arising in high seas enclaves. In such situations—although the coastal State's prescriptive authority

[69] *e.g.* A/Conf.13/C.3/L.30 (1958), *cit.* Chapter 4 (*fn*49), which was jointly submitted by Poland and the USSR regarding *ILC Draft Article* 56.

[70] *see* comments at Third Committee 9th meeting (17/3/1958) [*Geneva Off. Records* V:19¶18] and 28th meeting (11/4/1958) [*Geneva Off.Records* V:73–74].

[71] Third Committee 29th meeting (12/4/1958) [*Geneva Off.Records* V:81].

[72] A/Conf.164/24 'Statement made by the Chairman of the Conference at the closing of the Fourth Session, held on 26 August 1994' (8/9/1994), 2; the neutral legal character of the rule has also been affirmed in the scholarly writings of Satya Nandan, *see* Nandan, 'The Draft Agreement on the Conservation and Management of Straddling Fish Stocks and Highly Migratory Fish Stocks', in Nordquist–Moore (Eds) *Entry Into Force of the Law of the Sea Convention* (1995) 296.

would take precedence in establishing the regulatory regime—it would not possess an 'unqualifiedly superior right' to the straddling fish stocks.[73]

The compatibility rule acquired a neutrally articulated formulation by granting *vice-versa* status to the compatibility requirements applicable to both DWFS and coastal States. Notably, a new paragraph was added to place the measures adopted in their respective jurisdictional sea areas on equal footing with respect to the factors to be considered when determining agreed compatible CMMs. More specifically, the text had previously provided that, in making such determinations, States 'shall ensure that measures established in respect of the high seas do not undermine the effectiveness of those measures established in respect of the same stocks by coastal States in sea areas under national jurisdiction'.[74] Particular attention should be drawn to the fact that, in seeking to secure the balance of interests between DWFS and coastal States, the *Draft Agreement* removed from the embedded clause—and, the compatibility rule as a whole—the obligation previously imposed on DWFS to observe CMMs equivalent in effect to those applied in areas under national jurisdiction, pending agreement. This earlier obligation was intended to operate as a negotiating lever for coastal States in the process of reaching agreed measures.[75] The corresponding obligation for coastal States applied only insofar as they had not adopted any CMMs in the sea areas under their national jurisdiction.[76] The *Draft Agreement* replaced all such provisions that had previously favoured coastal States by establishing a shared obligation—common to both DWFS and coastal States—to have recourse to third-party compulsory dispute settlement procedures for the prescription of provisional measures, in the event that they were unable to reach agreement on provisional arrangements.[77]

The *Fish Stocks Conference* was reconvened for its Fifth Session to consider the *Draft Agreement*. At the opening of the session, the Chairman noted that the compatibility rule *vis-à-vis* the dispute settlement procedures under the Convention had generated significant debate during the preceding deliberations. In this context—and, undoubtedly alluding to the escalating *Turbot war* between Canada and the European Union during the negotiations—he urged delegations to remain focused and committed to achieving consensus,

---

[73] Goltz, 'The Sea of Okhotsk Peanut Hole: How the United Nations draft agreement on straddling stocks might preserve the pollack fishery', PRL&PJ (1995) 465–467.

[74] A/Conf.164/22, *supra cit.* (*fn*12), as iterated in draft article 7(2)(a).

[75] Van Dyke, 'Modifying the 1982 Law of the Sea Convention, New Initiatives on Governance of High Seas Fisheries Resources: The straddling stocks negotiations', IJMCL (1995) 221–223.

[76] A/Conf.164/13/Rev.1 (1994), *supra cit.* (*fn*11), Section III/C (¶8).

[77] A/Conf.164/22, *supra cit.* (*fn*12), as iterated in draft article 7(5).

emphasizing that the treaty text, in its final form, must 'reflect the balance of interests that States have in matters relating to fisheries'.[78] In light of the general comments made by delegations in the Plenary, as well as the outcome of informal meetings and consultations on unresolved issues,[79] the Chairman circulated his proposed revisions to the text of the *Draft Agreement* in the form of conference room papers. These revisions included suggested drafting changes and editorial improvements.[80] A prompt review of the proposed revisions was not possible, and although the Plenary was initially scheduled to reconvene during the second week of negotiations, it was postponed, and all other informal meetings were likewise suspended. The Plenary resumed on the final day of the scheduled proceedings, during which the Chairman invited comments on his latest revisions—distributed the previous day[81] and allowed for the delivery of general statements before formally closing the session.

The compatibility rule in article 7 *Revised Draft Agreement* was supplemented with a new provision in paragraph (2), which counterbalanced the obligation previously imposed solely on DWFS by requiring coastal States to take into account previously agreed measures established and applied in accordance with the Convention for the same stocks on the high seas.[82] The revisions placed both categories of States on equal footing by subjecting them to a reciprocal obligation to keep one another regularly informed—either directly or through the appropriate RFMO/As—of CMMs adopted in the respective sea areas.[83] Taken as a whole, the revised formulation of the compatibility rule addressed a longstanding issue in the negotiations: how best to harmonize the exercise of rights and the fulfilment of obligations with respect to fish stocks of common interest. It successfully reconciled the competing interests of DWFS and coastal States by grounding CMMs in the principle of biological unity, which holds that such measures must be applied across the full range of the stock and must not be obstructed by artificial legal boundaries.[84] With regard

---

[78] A/Conf.164/26 'Statement made by the Chairman of the Conference at the opening of the Fifth Session, held on 27 March 1995' (31/3/1995), 3–4.

[79] A/Conf.164/29 'Report on the Fifth Session of the United Nations Conference on Straddling Fish Stocks and Highly Migratory Fish Stocks (Prepared by the Secretariat)' (18/5/1995).

[80] A/Conf.164/CRP.6 (6/4/1995) and A/Conf.164/CRP.6/Add.1 (7/4/1995), *supra cit.* (*fn*9).

[81] A/Conf.164/22/Rev.1 (1995), *supra cit.* (*fn*13).

[82] *ibid.*, iterated as revised draft article 7(2)(b).

[83] *ibid.*, paragraphs (7) and (8) of the revised draft article 7.

[84] de Fontaubert, 'The United Nations Conference on Straddling Fish Stocks and Highly Migratory Fish Stocks: Another step in the implementation of the Law of the Sea Convention', *Ocean Yearbook* (1996) 85.

to the stipulation in paragraph (4) of article 7 *Draft Agreement*, there was only one—but nonetheless a pivotal—revision concerning the manner in which dispute settlement procedures were to be invoked when agreement could not be reached within a reasonable period of time. The *Revised Draft Agreement* replaced the phraseology 'the States concerned shall resort to the procedures' with the unilateral procedural right, reflected in the wording 'any of the States concerned may invoke the procedures'.

An event reported by ENB further supports the argument regarding the functional rationale underlying the textual pattern of clausal embeddedness. During the plenary debate of the Fifth Session on dispute settlement, the United States proposed a set of amendments aimed at revising the embedded clause of article 7 *Draft Agreement*. The proposal also called for the deletion of article 28(1) *Draft Agreement*, which served as the general jurisdiction clause for the settlement of disputes, along with the corresponding deletion of article 29 *Draft Agreement*, which extended jurisdiction to disputes within RFMO/As, and article 30 *Draft Agreement*, which provided for the prescription of provisional measures *pendente lite*.[85] Canada also tabled a similar amendment proposing the deletion of article 28(1) *Draft Agreement*. The precise rationale concerning paragraph (4) of article 7 *Draft Agreement* remains unspecified, as no further record exists on this specific point. A prior exchange of views between several DWFS and coastal States did include some reference to articles 297 and 298 LOSC, although no conclusive discussion took place on the matter.[86] The subsequent official transmittal of the Agreement by the United States government to the Congress may provide a useful perspective on its anticipated legal impact in disputes over compatible CMMs. The accompanying explanatory notes clearly stated that 'should parties be unable to achieve the compatibility of such measures within a reasonable period of time, any party could bring the matter to compulsory, binding dispute settlement in accordance with Part VIII [FSA]'.[87]

The *Revised Draft Agreement* was further scrutinized during the Sixth Session. Prior to its opening, the Chairman circulated to delegations an updated version of the suggested drafting changes and editorial improvements.[88] At the Plenary, several delegations observed that the document contained more than

85 ENB:7(41) 'Summary report, 3–7 April 1995'.

86 ENB:7(41) 'Summary report, 27–31 March 1995'.

87 *see* United States, 'Letter of Submittal from the Secretary of State Warren Christopher to the President of the United States, 24 January 1996' in the 'Message of President Clinton Transmitting the 1995 Fish Stocks Agreement to the US Senate for Ratification in 1996', Senate Treaty Document No.104, 24 Congress 2nd Session (1996), viii–ix.

88 A/Conf.164/CRP.7 (1995), *supra cit.* (*fn*9).

mere editorial or stylistic revisions, noting that some changes could potentially affect the substance of the articles. At that point, the conference was under mounting pressure to complete its work by the session's end. This clearly prompted the Chairman to adopt a more active and direct approach than in previous sessions, and in doing so, to press for compromises that were inevitably framed in the consensus-driven language of the treaty text. Following a round of general statements by delegations, the Chairman authorized direct negotiations and personally undertook informal consultations as part of his renowned 'shuttle diplomacy'. This strategy proved effective in resolving all outstanding issues, including the special case of a high seas enclave entirely surrounded by sea areas under the national jurisdiction of a single State, as well as matters related to dispute settlement.[89] At the close of the negotiations, the final treaty text presented two drafting changes of paramount importance to the present analysis.

Prior to its revision, paragraph (4) of article 7 *Revised Draft Agreement* had provided that:

> If no agreement can be reached within a reasonable period of time, the States concerned shall resort to the procedures for the settlement of disputes provided for in Part VIII of this Agreement without prejudice to the provisions of article 31.

This paragraph was ultimately revised by deleting the proviso 'without prejudice to the provisions of article 31'. As a result, the procedural stipulation now makes the dispute settlement procedures in Part VIII FSA directly applicable at the request of any party to a compatibility dispute. Should the matter proceed to a court or tribunal seised under section 2 Part XV LOSC, the limitations on the applicability of dispute settlement procedures are no longer accorded the legal privilege that the 'without prejudice to' clause would have otherwise preserved, had it remained in the text of the provision. Following this, the other notable revision appears in article 31 *Revised Draft Agreement*; renumbered as article 32 FSA to correspond with the drafting updates in Part VIII FSA. The earlier version had provided that 'The provisions of Article 297, paragraph 3, of the Convention shall apply also to this Agreement'. In the final text of article 32 FSA, aside from a slight improvement in the opening phrasing, the present-tense conjugation of the verb 'apply' replaced the modal 'shall'. It is far from clear whether this change was introduced for technical or stylistic reasons. The shift in grammatical form may nonetheless have substantive interpretative implications. In conventional treaty language, the choice between the modal verb 'shall' and the present tense is not treated as stylistically neutral. While 'shall' is generally understood to carry a normative imperative, the UNCLOS/

89 ENB:7(54) 'Summary report, 24 July to 4 August 1995'.

III Drafting Committee had expressly recommended that its use be avoided where the present tense sufficiently conveys the meaning, describes the operation of one rule as opposed to another, or clarifies its relationship to other provisions.[90] Whether this drafting principle was applied consistently—and, whether the grammatical revision was consciously made—in the finalization of the treaty text remains unclear. Nonetheless, it may reasonably be presumed to have informed the relevant drafting choices, particularly in view of the requirement of interpretative consistency between the Convention and the Agreement.

The Chairman's remarks in the final year of negotiations are particularly significant in relation to the progressive development of conservation and management law through the adoption of the compatibility rule. In a notably pointed observation, he stated that, in examining the treaty text, 'one should look for balance in its substantive content, and not on how often certain preferred words or phrases have been used'.[91] In light of the balanced approach taken between the interests of DWFS and coastal States, the Chairman presented the final text of the *Draft Agreement* as being built upon 'three essential pillars': the conservation and management principles, the collective responsibility of all States, and the dispute settlement procedures. Each of these pillars underpins the critical *à l'espèce role* of compulsory dispute settlement in achieving the overarching objective of ensuring the long-term conservation and sustainable use of fish stocks. Under the first pillar, all States are required to pursue CMMs both within and beyond areas of national jurisdiction, thereby addressing the full distribution or migration range of the stocks concerned. The compatibility rule, accordingly, stands as one of the 'cornerstones' of the Agreement, inclusive in its scope to encompass each stock as a whole, while also fully respecting the distinct jurisdictional responsibilities of States. As he aptly remarked, 'no one should escape from the conservation and management principles of the Agreement'. The second pillar upholds the common responsibility of all States, as the 'present-day custodians' of fishery resources, to secure the objective of sustainable regulation. The Chairman ultimately concluded that the treaty text not only achieves a balanced approach between the interests of DWFS and coastal States but also reflects the collective interests of the international community in these resources, for the benefit of present and future generations alike. In support of this fundamental purpose, the third

[90] A/Conf.62/L.40 'Report of the Chairman of the Drafting Committee' (22/8/1979) [*UNCLOS/III Off.Records* XI:104] and *see* further 1995 *Virginia Commentary* III (xliii–xliv).

[91] A/Conf.164/28 'Statement made by the Chairman of the Conference at the closing of the Fifth Session, held on 12 April 1995' (1/5/1995), 2.

pillar, i.e. dispute settlement procedures, serves to promote the sustainable use of marine resources through enhanced cooperation among States.[92]

The *Fish Stocks Conference* might well have remained deadlocked were it not for the eventual recognition by both DWFS and coastal States that they shared a common interest in reducing conflicts and ending the overexploitation of fish stocks. The negotiations reflected a *non-zero-sum* dynamic, in which the parties could either cooperate to achieve mutual success or fail to do so, with losses incurred on both sides.[93] The substantive work of the conference was brought to a close with the adoption, by consensus, of the *ne varietur* treaty text on 4/8/1995, during the final meeting of the Sixth Session.[94] The conference then entrusted the Secretariat with the task of consolidating the text of the Agreement—making any remaining editorial and drafting improvements, ensuring linguistic consistency across all official UN languages[95]—and preparing the final act for signature,[96] which took place on 4/12/1995.

5.6 CONCLUSION

The textual evolution of the compatibility rule, as reflected in the core drafts of the *Fish Stocks Conference*, reveals a clear drafting intent to establish a connection between disputes over its interpretation or application and the principle of compulsory adjudication. The rule was initially framed in favour of coastal State interests and notably lacked any provision for dispute settlement procedures. However, the rule's development over the course of the negotiations makes it evident that, as its normative content assumed a more neutral tone, the compulsoriness principle became increasingly pronounced in the structure of article 7 FSA, thereby serving to counterbalance the inherent constructive ambiguity surrounding its spatial scope. There is no documentary evidence to support an argument that the fisheries limitation under article 32 FSA

---

92 *see* the statements recorded in A/Conf.164/28 (1995), *supra cit.* (*fn*91), 3–4 and A/Conf.164/35 (1995), *cit.* Chapter 2 (*fn*166), 3.

93 de Fontaubert, 'The Politics of Negotiation at the United Nations Conference on Straddling Fish Stocks and Highly Migratory Fish Stocks', OCM (1995) 81.

94 A/Conf.164/36 'Report on the Sixth Session of the United Nations Conference on Straddling Fish Stocks and Highly Migratory Fish Stocks (Prepared by the Secretariat)' (31/8/1995), VI¶5.

95 A/Conf.164/37 'Agreement for the Implementation of the Provisions of the United Nations Convention on the Law of the Sea of 10 December 1982 relating to the Conservation and Management of Straddling Fish Stocks and Highly Migratory Fish Stocks' (8/9/1995).

96 A/Conf.164/32 'Draft Final Act of the United Nations Conference on Straddling Fish Stocks and Highly Migratory Fish Stocks' (2/8/1995).

excludes, *ipso facto* and *ab initio*, the EEZ components of compatibility disputes from the remit of compulsory procedures entailing binding decisions. On the contrary, the progressive development of the conservation and management law embodied in article 7 FSA was guided by two interrelated normative conditions. First, the requirement to achieve compatible CMMs, in accordance with the principle of the ecosystem approach, is neutrally predisposed with respect to the jurisdictional regime of fish stocks. Second, in the event that no agreement is reached within a reasonable period of time, any of the parties concerned may unilaterally invoke the dispute settlement procedures of Part VIII FSA. The legislative and historical study in Chapter 4 justifies the longer route taken to examine the rulemaking logic under the *Fishing & Conservation Convention* (1958) and the Convention, as a means of tracing the progressive development of fisheries law leading to the Agreement. The analysis across Chapters 4 and 5 highlights the striking conceptual continuity in the pattern of clausal embeddedness linking articles 6(5), 7(4), and 8(2) FCC, their drafting derivative in paragraph (2) of articles 74/83 LOSC, and the doctrinal redeployment of that pattern in paragraph (4) article 7 FSA. The procedural stipulation contained in the embedded clause is neither meaningless verbosity in the substantive formulation of the compatibility rule nor a mere textual embellishment within the structure of article 7 FSA. Rather, it constitutes a *consensus ad idem* on the principle of compulsory adjudication and, as such, reinforces the textual basis of the forum's compulsory jurisdiction over compatibility disputes.

6. Subsequent practice and approaches in regional regimes

6.1 INTRODUCTION

Building on the momentum generated by the *Fish Stocks Conference*, States re-engaged with renewed interest in regional negotiations aimed at overhauling existing RFMO/As and establishing new ones where necessary. Most of these bodies have since assumed an evolving role, taking on expanded functions in the wake of the Agreement, which significantly enhanced their regulatory authority over the conservation and management of fish stocks. While article 7 FSA reflects a consensus that CMMs must be compatible across jurisdictional sea areas, it does not define specific substantive measures. This flexibility accommodates the biological characteristics of fisheries, the geographic particularities, and the ecosystemic circumstances unique to each ocean region. Regional approaches to implementing the compatibility rule may therefore vary, provided that measures are adopted by agreement or, in cases of disagreement, referred to third-party dispute settlement procedures. This chapter examines the relevant legal provisions in the constitutive instruments of specialized multilateral organizations and arrangements, wherein the interpretation and application of the compatibility rule reveal broader legal input and normative reach than is typically found in generic or bilateral arrangements, without disregarding the contribution of the schemes in the latter category for the regulation of subregional fisheries.

6.2 INTERNATIONAL COOPERATION FOR THE SUSTAINABLE REGULATION OF FISHERIES

The duty of States to cooperate inheres in general international law, where it is consistently upheld as an indispensable legal, economic, and social norm for the orderly progression of the global community.[1] Its significance has

[1] *see* UNC 55–56, and A/Res.2625/25 'Declaration on Principles of International Law Concerning Friendly Relations and Cooperation among States

been specifically enunciated in the field of environmental protection by the *Declaration on the Human Environment* (1972), which calls upon all States to cooperate—while respecting their sovereignty and interests—through bilateral or multilateral treaties and other appropriate means, to effectively control, prevent, reduce, and eliminate adverse environmental effects.[2] Since then, the scope of the duty has been broadened by numerous specialized legal instruments aimed at advancing sustainable development. Notably, the need to cooperate 'in good faith and in a spirit of partnership' for environmental matters was universally reaffirmed in the *Declaration on Environment and Development* (1992).[3] Substantial emphasis has been placed on cooperation as a means to adopt concrete steps and establish quantifiable targets for implementing the sustainable development agenda. In particular, SDG 14(4) calls on States to work together to regulate harvesting, end overfishing and IUU fishing, eliminate destructive practices, and implement science-based plans to restore fish stocks in the shortest time feasible.[4]

It is unsurprising that this duty derives its normative foundation from the practice of States engaging in coordinated actions for the regulation of marine living resources. The 1893 *Rights of Jurisdiction in the Bering's Sea and the Preservation of Fur Seals* arbitration affirmed the principle of freedom of fishing on the high seas, while also urging the States concerned to cooperate in conserving marine living resources. This led to the eventual adoption of the *Convention Between Japan, Great Britain, Russia and the United States for the Protection and Preservation of Fur Seals and Sea Otters in the North Pacific Ocean* (1911). The duty has since evolved to reflect increased expectations in light of the biological, scientific, technological, economic, social, and political dimensions of fisheries conservation, all of which necessitate cooperation.[5] It has been highlighted as an imperative need in special circumstances involving the management of marine living resources across maritime boundaries, both in judicial pronouncements[6] and in the treaty practice of States[7]

in Accordance with the Charter of the United Nations' (24/10/1970).

[2] UNCED, 'Report of the United Nations Conference on the Human Environment' (16/6/1972), seventh proclamatory recital & Principle 24.

[3] *cit.* Chapter 1 (*fn*85), Principle 27.

[4] *cit.* Chapter 1 (*fn*18).

[5] Hey, *The Regime for the Exploitation of Transboundary Marine Fisheries Resources* (1989) 15–41.

[6] *see* in *North Sea Continental Shelf Cases* (Judgment) 1969, *Separate Opinion of Judge Jessup*, pp.83–84 and *Delimitation of the Maritime Boundary in the Gulf of Maine Area* (Judgment) 1984, §240.

[7] *e.g.* the *Treaty between Norway and the Russian Federation concerning the Maritime Delimitation and Cooperation in the Barents Sea and the Arctic Ocean*

regarding maritime delimitation. Among States at the international plane, cooperation concerning straddling and highly migratory fish stocks is considered essential owing to their distribution and migration range. This necessity is expressly recognized by articles 63(2) and 64 LOSC, which, together with articles 116–119 LOSC, form the legal foundation for the conservation and management of these two categories of fish stocks that the Agreement was designed to operationalize.[8] However, the terms set out in these provisions for fulfilling the duty of cooperation are somewhat imprecise. The occurrence of the *pactum de negotiando* in article 63(2) LOSC heightens the uncertainty regarding the outcome of cooperation in the event of disagreement over CMMs for straddling stocks. Nor does the *pactum de contrahendo* concerning stocks of highly migratory species offer greater certainty, at least as addressed in the *Southern Bluefin Tuna Cases*. On one hand, ITLOS, interpreting the duties under articles 64 and 116–119 LOSC, imposed on the parties, through provisional measures, an obligation to resume negotiations without delay for the adoption of CMMs, and to intensify cooperation with other participants in the fishery areas to ensure conservation and promote the objective of optimum utilization of the stocks.[9] On the other hand, the arbitral tribunal, in dismissing the case at the jurisdictional stage and revoking all provisional measures, ultimately observed that the entry into force of the Agreement will resolve many of the shortcomings of the earlier conventional regime, given that its provisions are more detailed and far-reaching.[10]

The conservation and management principles now enshrined in the Agreement are essentially cooperative in nature.[11] The commitment to cooperation is not only made *expressis verbis* with respect to straddling and highly migratory fish stocks, but is also integrally established as a normative condition within the compatibility rule, which unequivocally provides that DWFS and coastal States 'have a duty to cooperate for the purpose of achieving compatible measures in respect of such stocks'.[12] States are required to discharge this duty by entering into negotiations in good faith and without delay, either

(2010), 4 and Annex I; *qv.* Vylegzhanin–Young–Berkman, 'Governing the Barents Sea Region: Current status, emerging issues, and future options', ODIL (2018) 57–58.

8 A/Conf.164/11 (1993), *cit.* Chapter 5 (*fn*16), 3.

9 *Southern Bluefin Tuna Cases* (Provisional Measures) 1999, §§48, 78 & *dispositif* 1(e).

10 *Southern Bluefin Tuna Cases* (Jurisdiction) 2000, §71.

11 Munro, 'The United Nations Fish Stocks Agreement of 1995: History and problems of implementation', *Marine Resource Economics* (1995) 274*ff.*

12 FSA 7(2), *chapeaux*.

directly or, where available, through RFMO/As.[13] To this end, the Agreement elaborates institutional and procedural mechanisms to facilitate cooperation, significantly employing the precept of collective governance by integrating the role of RFMO/As into the implementation of its provisions.[14] Regional regimes have now become the primary mechanism through which States fulfil their duty to cooperate in the interest of the sustainable use of fish stocks.[15] The general principles of the Agreement—including the ecosystem and precautionary approaches, as well as the compatibility rule—are gradually being integrated into regional regulatory frameworks, with growing emphasis on uniform application. This includes efforts to identify and coordinate best practices in the development and implementation of CMMs.[16] The critical role of RFMO/As is further underscored by the expanding range of issues on which States are directed to cooperate,[17] their outreach to non-members and non-participants,[18] and the growing recognition of the need to provide assistance to developing States, including through capacity-building and financial support.[19] Although controversial, it may ultimately be the case that access to high seas fisheries is conditioned upon membership or participation in RFMO/As, with such membership limited to 'States having a real interest in the fisheries concerned'.[20] In the absence of a defined meaning of 'real interest' under the Agreement, growing global demand for access to fishery resources is likely to exacerbate tensions, especially if regional practice moves away from the current framework premised on mutual interest.[21]

---

[13] FSA 8(2) & (5).

[14] FSA 8–16.

[15] Molenaar, 'International Cooperation through Regional Fisheries Management Organizations', in Stokke–Østhagen–Raspotnik (Eds) *Marine Resources, Climate Change and International Management Regimes* (2022) 23*ff.*

[16] Refer to the positive remarks of the UN Secretary-General in 'Report Submitted to the Resumed Review Conference, 2016' (1/3/2016), ¶63 and the 'Report Submitted to the Resumed Review Conference, 2023' (13/2/2023), ¶152.

[17] FSA 10.

[18] FSA 17.

[19] FSA 24–26.

[20] FSA 8(3).

[21] *see* Molenaar, 'The Concept of "Real Interest" and Other Aspects of Co-operation through Regional Fisheries Management Mechanisms', IJMCL (2000) 475; Serdy, *The New Entrants Problem in International Fisheries Law* (2016); and Fuller–Schleit, 'The Future of Managing Fisheries and the Global Commons through Regional Fisheries Management Organizations: Steps toward global stewardship', in Werle *et al.* (Eds) *The Future of Ocean Governance and Capacity Development* (2018) 361.

The Agreement has also strengthened the compliance mechanisms available to RFMO/As to ensure effective implementation of agreed CMMs within their regulatory areas. In addition to reinforcing the duties of flag States[22] and enhancing the role of port States,[23] the Agreement introduces innovative provisions on MCS measures, including a scheme for the boarding and inspection of fishing vessels as part of cooperative enforcement efforts.[24] While a robust enforcement regime is now in place to support compliance with agreed CMMs, the duty to cooperate is most critically tested at an earlier stage, namely, during the adoption of the substantive measures themselves. How can this duty be meaningfully discharged when parties, although presumed to be negotiating *bona fide*, maintain irreconcilable positions that jeopardize the objective of achieving compatible CMMs? After all, the duty entails a positive obligation to act, and where no further constructive steps are taken, genuine cooperation between the States concerned cannot be said to exist.[25] A definitive resolution to this dilemma would probably remain unattainable under the Convention, arguably intended to preserve flexibility and avoid political controversy. As examined in Chapter 2, articles 63(2) and 64 LOSC remain silent on the critical issue of how to ensure cooperation when disagreements arise.

General international law has interpreted the duty to cooperate as comprising specific procedural elements, such as obligations to notify, consult, exchange information, and negotiate in a manner that inspires mutual trust and confidence. However, it has not been interpreted to require States to reach agreement.[26] In practice, the duty is most often expressed as an abstract principle, lacking mechanisms of redress and rarely structured to impose obligations to agree or to accept limitations on sovereignty.[27] Nonetheless, evolving jurisprudence increasingly reflects a normative shift towards enhanced coordination and progressively deeper forms of cooperation in the governance of common-pool marine living resources.[28] As regards highly migratory species,

---

[22] FSA 18 & 19.

[23] FSA 23.

[24] FSA 20–22.

[25] Pinto, 'The Duty of Co-Operation and the United Nations Convention on the Law of the Sea', in Bos–Siblesz (Eds) *Realism in Law-Making* (1985) 154.

[26] *Pulp Mills on the River Uruguay* (Judgment) 2010, §§145–150.

[27] *see* Boyle, 'The Principle of Co-operation: The environment', in Lowe–Warbrick (Eds) *The United Nations and the Principles of International Law* (1994) 120 and Craik, 'The Duty to Cooperate in International Environmental Law: Constraining State discretion through due respect', YbIEL (2019) 28.

[28] *Whaling in the Antarctic* (Judgment) 2014, §§83 & 137–144; and *see* Hagiwara, 'Sustainable Development before International Courts and Tribunals:

it is worth recalling that ITLOS expressed the view that States must make a substantial effort to adopt effective measures to ensure the sustainable regulation of these stocks.[29] Academic commentary has likewise supported the imposition of a more stringent duty to cooperate in situations involving the serious depletion of fish stocks. Some scholars have suggested that the interpretation of the Convention's fisheries provisions could evolve to render untenable any position that condones disagreement in good faith, thereby imposing on States a positive obligation to reach agreement on necessary CMMs.[30] By the same reasoning, it has been persuasively argued that in the aftermath of unresolved disagreements, the obligation to cooperate—or at a minimum, to coordinate measures—with respect to both straddling and highly migratory fish stocks remains within the jurisdiction of a court or tribunal seised under the procedures of section 2 Part XV LOSC, even if article 297(3) LOSC was relied upon to exclude disputes concerning these fish stocks in EEZ.[31]

The uncertainty surrounding the questions raised above is now squarely addressed by the Agreement. The possibility of a genuine disagreement obstructing the adoption of compatible CMMs is managed in a structured and technical manner within the rule itself. By expressly including the duty to cooperate in the *chapeau* of article 7(2) FSA, the Agreement affirms that this obligation constitutes a core substantive element of the compatibility rule. Beyond affirming the duty to cooperate, it also introduces a remedial mechanism applicable where such disagreements remain unresolved within a reasonable period of time. More precisely, paragraph (3) of article 7 FSA reiterates that 'States shall make every effort to agree on compatible conservation and management measures within a reasonable period of time', framing this obligation as an action required 'in giving effect to their duty to cooperate'. The remedial course of action in such circumstances is set out in the embedded clause of article 7(4) FSA, which explicitly provides that 'if no agreement can be reached within a reasonable period of time, any of the States concerned may invoke the procedures for the settlement of disputes provided for in Part VIII'.

Duty to cooperate and States' good faith', in Voigt (Ed.) *International Judicial Practice on the Environment* (2019) 167*ff.*

29 *SRFC Request* (Advisory Opinion) 2015, §210.

30 Edeson, 'A Brief Introduction to the Principal Provisions of the International Legal Regime Governing Fisheries in the EEZ', in Ebbin–Hoel–Sydnes (Eds) *A Sea Change: The Exclusive Economic Zone and governance institutions for living marine resources* (2005) 26.

31 Klein–Parlett, *Judging the Law of the Sea* (2022) 331.

6.3 THE COMPATIBILITY RULE IN REGIONAL REGIMES ESTABLISHED AFTER THE 1995 UNITED NATIONS FISH STOCKS AGREEMENT

The emphasis placed by the Agreement on strengthening the role of regional regimes prompted States to reconsider their duty to cooperate, particularly in relation to the requirement of membership or participation in the work of RFMO/As. While the regulatory authority of existing competent bodies—both in terms of geographic coverage and types of stocks—remained unaffected, ensuing State practice tended primarily towards the establishment of new organizations or arrangements. As a result of subsequent negotiations that accounted for regional needs and legal gaps, these newly established bodies were, from the outset, able to integrate advances achieved in the progressive development of fisheries law under the Agreement and to address the challenges associated with promoting sustainable fisheries regulation.[32] All RFMO/As established since the conclusion of the Agreement have introduced the rule of compatible CMMs into their applicable law, framing it in a deliberately neutral and balanced manner so as to avoid privileging either DWFS or coastal States. These regimes are examined in chronological order below, beginning with WCPFC, whose constitutive convention was adopted in 2000. CPPS is addressed last, as its convention has not yet entered into force, perhaps not coincidentally, given that it represents the only instance in subsequent State practice that diverges markedly from the neutral and balanced approach reflected in article 7 FSA.

6.3.1 Western and Central Pacific Fisheries Commission

The world's largest tuna fisheries—targeting commercially significant species such as *Thunnus alalunga* (Albacore), *Thunnus albacares* (Yellowfin), *Thunnus obesus* (Bigeye) and *Katsuwonus pelamis* (Skipjack)—are located in the western and central Pacific Ocean. Unsurprisingly, the region has historically witnessed tensions between SIDS, whose coastal fishing interests are coordinated through SPFF based on scientific advice from FFA, and DWFS.[33] Prolonged and intensive international negotiations, held under the

---

[32] *see* Barston, 'The Law of the Sea and Regional Fisheries Organisations', IJMCL (1999) 348*ff* and Harrison, 'Key Challenges Relating to the Governance of Regional Fisheries', in Caddell–Molenaar (Eds) *Strengthening International Fisheries Law in an Era of Changing Oceans* (2019) 79.

[33] Tsamenyi–Manarangi-Trott, 'Role of Regional Organizations in Meeting LOS Convention Challenges: The Western and Central Pacific Experience', in

Multilateral High-Level Conference on the Conservation and Management of Highly Migratory Fish Stocks in the Western and Central Pacific ('WCPFC MHLC'), culminated in the adoption of the Convention on the Conservation and Management of Highly Migratory Fish Stocks in the Western and Central Pacific Ocean (2000) ('WCPFC *Convention*').

This was the first treaty to apply the compatibility rule to stocks of highly migratory species, with the aim of ensuring their long-term conservation and sustainable use.[34] However, the matter of achieving compatible CMMs presented significant complexities. Given the region's geographical characteristics—being predominantly sea-locked, with a high density of abutting EEZs and extensive archipelagic waters—coastal States advocated for WCPFC's mandate to be confined to the high seas. DWFS, however, opposed any attempt to constrain the Commission's authority to promote conservation and management throughout the regulatory area. As the WCPFC *Convention* was expected to set a new high-level international standard for regional cooperation in accordance with the Agreement, it was required to fulfil two overarching objectives: first, to establish an effective system for the sustainable regulation of stocks throughout their migratory range; and second, to equitably balance the interests of DWFS and coastal States.[35] Disagreements over these issues ultimately led to the WCPFC *Convention* reflecting an indeterminate stance in its treaty text with respect to key functions, such as the allocation of fishing opportunities and the setting of TACs.[36] Nonetheless, in practice, the Commission adopts CMMs with the intention that they apply across the full biological range of the fishery resources.[37]

This underscores the critical recognition that effective CMMs can only be achieved through cooperation.[38] In this respect, the international measures adopted by the Commission are explicitly applicable 'throughout the range of

Oude-Elferink–Rothwell (Eds) *Ocean Management in the 21st Century* (2004) 196–200.

34 WCPFC *Convention* 2.

35 United States Department of State, 'Press Release on Successful Conclusion of Western Pacific Fisheries Negotiations' (6/9/2000); and *see* Murphy, 'Conservation of Fish in the Western and Central Pacific Ocean', AJIL (2001) 154–155.

36 Tarte, 'A Duty to Cooperate: Building a regional regime for the conservation and management of highly migratory fish stocks in the western and central Pacific', *Ocean Yearbook* (2002) 269*ff.*

37 *e.g.* WCPFC:CMM/1-2023 (¶42) & CMM/1-2018 (¶¶39–44).

38 WCPFC *Convention*, eighth preambular recital; and *see* the common understanding on the need to develop sustainably the tuna resources of the region previously reached in WCPFC MHLC, 'Majuro Declaration' (13/6/1997).

the stocks', with the objective of conserving and managing highly migratory species 'in their entirety'.[39] Accordingly, the conservation and management principles, along with the resulting measures, 'shall be applied by coastal States within areas under national jurisdiction' within the regulatory area, 'in the exercise of their sovereign rights for the purpose of exploring and exploiting, conserving and managing highly migratory fish stocks'.[40] The aim of managing stocks as a unified whole is evident in the explication of the compatibility rule, which provides that '[CMMs] established for the high seas and those adopted for areas under national jurisdiction shall be compatible in order to ensure conservation and management of highly migratory fish stocks in their entirety'.[41] In adopting such measures, States are required to take into account the 'biological unity' of the stocks, their other biological characteristics, the relationship between the distribution of the stocks, and the geographical particularities of the region, including the extent to which the stocks occur and are exploited in areas under national jurisdiction.[42] It must also be ensured that such measures do not adversely affect living marine resources 'as a whole'.[43]

The balance of interests between DWFS and coastal States is further safeguarded through provisions requiring the Commission, in pursuing compatible CMMs, to apply international measures respecting the same stocks in high seas areas in accordance with both the Convention and the Agreement, while also ensuring that national measures adopted under article 61 LOSC are not undermined.[44] Entering into negotiations with a view to managing stocks in accordance with their biological unity gives rise to a duty to cooperate in achieving compatible CMMs, which may require mutual accommodations within the respective jurisdictional sea areas to account for the interests of all parties involved. On the one hand, DWFS need to recognize that international CMMs should reflect the interests of coastal States. On the other hand, coastal States are required to exercise restraint in adopting CMMs within their national jurisdiction where such measures would undermine the effectiveness of international measures applied across the regulatory area.[45] These provisions thereby provide a basis for negotiations aimed at balancing the respective fishing interests of States in a fair and transparent manner.[46] While WCPFC

---

39 WCPFC *Convention* 3(3) & 5.
40 *ibid.*, 7(1).
41 *ibid.*, 8(1).
42 *ibid.*, 8(2)(a).
43 *ibid.*, 8(2)(e).
44 *ibid.*, 8(2)(b).
45 *ibid.*, 8(3).
46 Botet, 'Filling in One of the Last Pieces of the Ocean: Regulating tuna in the western and central Pacific Ocean', VJIL (2001) 801–802.

practice reflects the understanding that the obligation to achieve compatible CMMs applies within EEZs, a continuing legal challenge concerns the applicability of such measures to territorial seas and archipelagic waters. It has been recommended that this issue, which is critical to the implementation of effective CMMs spanning the migratory range of the species, be addressed through the adoption of an interpretive declaration under the WCPFC *Convention*.[47]

6.3.2 South East Atlantic Fisheries Organisation

The ocean region of the southeast Atlantic was previously regulated by the International Commission on South East Atlantic Fisheries, established under the *Convention on the Conservation of the Living Resources of the Southeast Atlantic* (1969). However, over time, the Commission's operations declined, and it became inactive by the early 1990s.[48] Prompted by the successful conclusion of the *Fish Stocks Conference*, the coastal States of the region convened exploratory meetings later that same year, which eventually led to formal negotiations with DWFS in 1997 and culminated in the adoption of the *Convention on the Conservation and Management of Fishery Resources in the South East Atlantic Ocean* (2001) ('SEAFO *Convention*'). This treaty established the South East Atlantic Fisheries Organisation ('SEAFO'), with the objective of ensuring the long-term conservation and sustainable use of straddling fish stocks in the region.[49] Stocks of interest primarily include species of cephalopods, cartilaginous fishes, and bony fishes, such as *Dissostichus eleginoides* (Patagonian toothfish) and deep-water *Hoplostethus atlanticus* (Orange roughy).

Although SEAFO is the first regional regime for the management of straddling stocks established following the adoption of the Agreement, it has given only limited effect to the conservation and management principles set out therein. This reflects how unresolved differences that had emerged during the *Fish Stocks Conference* between DWFS and coastal States endured and subsequently spilled over into the regional negotiations leading to the establishment of SEAFO. Aspects of the Agreement that had generated considerable debate during the Conference—such as the general conservation and management principles, including the compatibility rule, as well as the compliance and enforcement arrangements—were deliberately toned down to facilitate

[47] WCPFC, 'Performance Review Panel Report: 2012', ¶3(4.1).

[48] *see* Gertenbach, 'Fishery Convention Areas in the Southeast Atlantic and Adjacent Seas: Overlapping issues', *Acta Juridica* (1986) 51 and Miller–Molenaar, 'The SEAFC Convention: A comparative analysis in a developing coastal State perspective', *Ocean Yearbook* (2006) 305.

[49] SEAFO *Convention* 1(1) & 2.

compromise. This approach was adopted in the expectation that consensus on these matters could more readily be achieved within the organization at a later stage, when the political stakes were lower and the risk of setting a broader precedent less acute.[50]

With the above in mind—and despite the fact that the regulatory area comprises only high seas[51]—the SEAFO *Convention* nonetheless affirms the intention to cooperate with coastal States to ensure the adoption of compatible CMMs.[52] The requirement of compatibility is articulated in deliberately neutral terms, evidently crafted so as not to disturb the carefully balanced formulation of article 7 FSA. It provides that 'compatibility shall be ensured in such a way which does not undermine measures established in accordance with articles 61 and 119 [LOSC]', thereby reinforcing the shared duty of DWFS and coastal States to cooperate in achieving compatible CMMs both in high seas areas and within EEZs.[53] Although SEAFO's performance was reviewed in 2010 and again in 2016, the review panels did not record any specific observations with respect to the criterion of adopting compatible CMMs as set out in article 7 FSA.[54] This may be attributed to the low level of fishing effort in the region and the limited number of straddling stocks targeted at the time.[55] Nonetheless, it has been pointed out that, given the SEAFO *Convention* was adopted prior to the entry into force of the Agreement—and in the absence of certainty that all future SEAFO members would become parties to the latter—numerous latent inconsistencies between the two instruments remain and are likely to surface as regulatory practice continues to evolve.[56]

6.3.3 Southern Indian Ocean Fisheries Arrangement

A similar approach was employed a few years later through the *Southern Indian Ocean Fisheries Agreement* (2006) ('SIOFA *Agreement*'), which established the Southern Indian Ocean Fisheries Arrangement ('SIOFA') with the objective of ensuring the long-term conservation and sustainable use of

50 Sydnes, 'New Regional Fisheries Management Regimes: Establishing the south east Atlantic fisheries organisation', *Marine Policy* (2001) 361–362.

51 SEAFO *Convention* 4.

52 *ibid.*, fourth and ninth preambular recitals.

53 *ibid.*, 19(1).

54 SEAFO, 'Performance Review Panel Report: 2016', Annex 1:70.

55 *ibid.*, ¶4(1.5).

56 Jackson, 'The Convention on the Conservation and Management of Fishery Resources in the South East Atlantic Ocean, 2001: An introduction', IJMCL (2002) 46–47.

primarily demersal fish stocks.[57] Although the Agreement excludes waters under national jurisdiction,[58] it invokes the compatibility rule by assigning to the Meeting of Parties—the principal operational body of the Arrangement—the responsibility of promoting cooperation and coordination among member States. Importantly, it requires that CMMs for straddling stocks occurring in areas under national jurisdiction adjacent to the regulatory area be compatible with international measures adopted for the high seas.[59] However, issues of compatibility have arisen in relation to measures adopted by adjoining regional regimes. For example, CMMs adopted by SIOFA for the Patagonian toothfish have been considered incompatible with those established by the neighbouring Commission on the Conservation of Antarctic Marine Living Resources, which has demonstrated greater success in regulating the stock within its area of competence. Against this backdrop, SIOFA members have stressed the 'legal and moral obligation' of the organization to manage the stock in its entirety and to fully account for obligations under international law, including the requirement under the Agreement to adopt compatible measures and to take into consideration existing measures already established in high seas areas with respect to the same stock.[60] Another example concerns the incompatibility of management measures with those adopted by the Indian Ocean Tuna Commission. This has, at times, resulted in specific instances of inadequate compliance with CMMs for fish stocks of mutual interest and, more broadly, has raised concerns about the confusion arising from overlapping governance in sea areas where both legal frameworks apply.[61] However, the issue of compatibility has not, in its own right, been substantively addressed, nor was it included as a benchmark criterion for assessing the adoption of CMMs in the terms of reference of the first Performance Review Panel, conducted during 2022–2023.[62] Nevertheless, the Panel underscored the importance of ensuring compatible CMMs across the regulatory areas of adjacent regional regimes, to be achieved either through co-management of the stocks or by strengthening inter-regime cooperation, as appropriate.[63]

---

57 SIOFA *Agreement* 1(f) & 2.

58 *ibid.*, 3.

59 *ibid.*, sixth preambular recital & 6(1)(g).

60 *see* the position recorded by Australia in SIOFA, 'Report of the Sixth Meeting of Parties' (5/7/2019), ¶12(7).

61 *e.g.* the comments by the European Union, Japan, and the Chinese Taipei as a participating Fishing Entity, minutes in the SIOFA, 'Report of the Eighth Meeting of Parties' (9/7/2021), ¶¶55–56 & 207–208.

62 *see* Agenda item 10 in SIOFA, 'Report of the Ninth Meeting of Parties' (8/7/2022), ¶¶203–204 and Annex S:163–164.

63 SIOFA, 'Performance Review Panel Report: 2023', ¶¶11 & 42.

6.3.4 South Pacific Regional Fisheries Management Organisation

One of the most detailed and comprehensive applications of the compatibility rule is found in the *Convention on the Conservation and Management of High Seas Fishery Resources in the South Pacific Ocean* (2009) ('SPRFMO *Convention*'), which was adopted by consensus after three years of international consultations spanning nine negotiation rounds, from 6/11/2006 to 14/11/2009. The purpose of these consultations was to enable interested States to cooperate—consistent with international law and best practices—in addressing the regulatory gap concerning the conservation and management of non-highly migratory fish stocks, as well as the protection of marine biodiversity in areas beyond national jurisdiction in the south Pacific Ocean. The South Pacific Regional Fisheries Management Organisation ('SPRFMO') is widely regarded as an enhanced institutional model, particularly in terms of decision-making and organizational effectiveness, with the capacity to address key challenges in the sustainable regulation of fisheries.[64] It is responsible for managing a range of straddling fish stocks, including both demersal and pelagic species with extensive distribution across the region. Of particular commercial importance are deep-water orange roughy and *Trachurus murphyi* (Chilean jack mackerel), which occur within EEZs and adjacent high seas areas. The recovery of these stocks is a strategic priority for the organization.

The formulation of the compatibility rule in article 4 SPRFMO *Convention* exemplifies the normative neutrality of its content, thereby maintaining the balance of fishing interests enshrined in article 7 FSA. Its objective is to ensure the long-term conservation and sustainable use of fishery resources through the application of the principles of ecosystem and precautionary approach, and, in so doing, to protect the marine ecosystems in which these resources occur.[65] In giving effect to this objective—and recognizing its application to high seas areas adjacent to and beyond national jurisdiction[66]—members are obliged to ensure that CMMs adopted by the SPRFMO Commission for the high seas are compatible with those applied to the same stocks in areas under national jurisdiction.[67] The neutrality of the compatibility rule is further reinforced by the expression of a reciprocal obligation to establish compatible CMMs for both areas under national jurisdiction and adjacent high seas, with the

64 Schiffman, 'The South Pacific Regional Fisheries Management Organization: An improved model of decision-making for fisheries conservation?', *Journal of Environmental Studies and Sciences* (2013) 209.

65 SPRFMO *Convention* 2.

66 *ibid.*, 5.

67 *ibid.*, 3(1)(a) *lit.*(vi).

objective of conserving and managing straddling stocks 'in their entirety'.[68] Accordingly, the convention affirms the corresponding duty of cooperation, assigning responsibility to DWFS for adopting measures applicable to their vessels operating on the high seas, and to coastal States for adopting measures in areas under their national jurisdiction.[69]

The normative condition of neutrality in the compatibility rule, as set out in article 4 SPRFMO *Convention*, is unequivocal. Nevertheless, the positions taken by DWFS and coastal States during the consultations reveal important nuances regarding the requirement of compatible CMMs. Chile and Peru voiced concerns about the inclusion of the compatibility rule, indicating that they did not view its adoption as entirely appropriate. Their position maintained that any regulatory regime should give due priority to existing national measures—particularly those adopted by coastal States within their EEZs—before the introduction of new measures by the organization. Nonetheless, Chile ultimately agreed to the inclusion of the compatibility rule 'in a spirit of compromise', on the condition that the national measures already in place would not be undermined.[70] A more moderate stance was taken by Russia, which argued that while any future CMMs should be limited to high seas areas beyond national jurisdiction, the coordination of national and international measures should nonetheless be undertaken on the basis of the compatibility rule.[71] The consultations ultimately yielded a balanced outcome by rejecting any formulation of the rule that might suggest an extension of coastal State interests into high seas areas. For instance, Chile had proposed amendments stipulating that CMMs adopted for the regulatory area should not undermine the effectiveness of coastal State measures, without including a corresponding safeguard to prevent the latter from undermining international ones.[72] Instead, a neutral formulation of the rule was adopted, with support from the European

---

68 *ibid.*, 4(2).

69 *ibid.*, 4(1).

70 *see* Consultations on the Proposed Pacific Regional Fisheries Management Organisation documents 'SP/06/INF/04 "Suggestions and Amendments by Chile"' (2008); 'SP/06/INF/06 "Letter of Peru to the Chair"' (2008); and 'SP/06/ INF/08 "Peru Proposals"' (2008).

71 *see* Consultations on the Proposed Pacific Regional Fisheries Management Organisation 'Statement by the Russian Delegation at the Plenary Meeting of 22/5/2009', and the accompanying proposed amendments contained in 'SP/07/ INF/09 "Russian Federation Proposal to the Revision of the Convention"' (2009).

72 *compare* Consultations on the Proposed Pacific Regional Fisheries Management Organisation documents 'SP/06/KInf/05 "Chilean Proposed Amendments"' (2008), draft article 3(h), *with* the initial version of the compatibility rule in 'SP/06/WP/1 "Fourth Revision of the Draft Agreement"' (1/9/2008), draft article 4.

Union and other delegations, who emphasized that it should reaffirm the balance of interests between DWFS and coastal States.[73]

In addition to the clearly neutral formulation of the compatibility rule, the SPRFMO *Convention* also establishes institutional mechanisms to operationalize the rule, including provisions that allow the SPRFMO Commission to be vested with the authority to adopt compatible CMMs covering the entire biological range of fishery resources, both within and beyond EEZs. The Commission is tasked with promoting the compatibility of CMMs in the regulatory area, including 'adjacent areas under national jurisdiction and adjacent areas of [the] high seas'.[74] The primary focus of these measures is to ensure the long-term sustainability of fishery resources and to advance their responsible utilization.[75] With respect to resources that straddle the regulatory area and sea areas under the national jurisdiction of a coastal member State, the Commission is mandated to establish appropriate CMMs for the regulatory area, giving due regard to—and ensuring they do not undermine the effectiveness of—existing measures adopted by the relevant coastal State.[76] In these circumstances, the coastal State in question is required to cooperate in coordinating the respective measures.[77] Moreover, with the express consent of that State, the Commission may set TAC limits and TAFE levels applicable across the entire distribution range of the stocks.[78] A clear illustration of the rule's application in areas under national jurisdiction is found in the repeated instances where Chile and Ecuador have agreed to the establishment of TACs for jack mackerel stocks.[79] Beyond its standard decision-making procedures, the SPRFMO Commission holds an important emergency function: it may adopt CMMs on an urgent basis, relying on the best available scientific evidence, when fishing activities present a serious threat to the sustainability of fishery resources or are likely to cause significant adverse impacts on

73 *see* Consultations on the Proposed Pacific Regional Fisheries Management Organisation 'SP/06/INF/02 "EU, 6th Meeting for the Establishment of a New South Pacific RFMO"' (2008) and 'SP/07/INF/04 "EU, 7th Meeting for the Establishment of a New South Pacific RFMO"' (2009); and also, the likeminded proposal by the Cook Islands in 'SP/06/INF18 "Proposal Article 4 Compatibility of Conservation and Management Measures"' (2008).

74 SPRFMO *Convention* 8(f).

75 *ibid.*, 20(1).

76 *ibid.*, 4(3).

77 *ibid.*, 20(4)(a) *lit.*(i).

78 *ibid.*, 20(4)(a) *litt.*(ii)–(iii).

79 *e.g.* SPRFMO:CMM/1-2024; CMM/1-2023; CMM/1-2022; CMM/1-2021; CMM/1-2020; CMM/1-2019; CMM/1-2018; CMM/1-2017; CMM/4.1-2016; CMM/3.1-2015; & CMM/2.1-2014.

their condition.[80] Such emergency measures are not subject to the objection procedure but may be challenged through the dispute settlement procedures under the SPRFMO *Convention*.[81] The establishment of SPRFMO has been hailed as a landmark development, offering a compelling example of how the Agreement can be effectively applied in a regional setting. Its provisions on compatible CMMs are consistent with, and reflective of, ecosystem-based fisheries management.[82]

6.3.5 North Pacific Fisheries Commission

The north Pacific Ocean encompasses some of the most biologically productive marine areas in the world. Historically, the richness of its ecosystem allowed the States of the Pacific Rim to exercise freedom of access to high seas fisheries without a formalized regional management regime. This changed with the establishment of the North Pacific Fisheries Commission ('NPFC') in 2015, following nearly a decade of international negotiations among States with fishing interests in the region. The *Convention on the Conservation and Management of High Seas Fisheries Resources in the North Pacific Ocean* (2012) ('NPFC *Convention*') sets out the objective of ensuring the long-term conservation and sustainable use of fishery resources within the regulatory area, while also protecting and preserving the vulnerable marine ecosystems in which these resources occur.[83] Although NPFC was initially formed in response to growing global concerns about the ecosystem impacts of high seas fishing activities, the Commission's focus has increasingly shifted to managing the rapid expansion of pelagic fish stocks into the region, many of which were not covered by pre-existing RFMO/As.

The NPFC *Convention* incorporates the compatibility rule among its general principles. It requires members, either individually or collectively as appropriate, to ensure, in accordance with article 7 FSA, that CMMs established for straddling stocks on the high seas and those adopted in areas under national jurisdiction are 'compatible in order to ensure conservation and management of these fisheries resources in their entirety'.[84] As the NPFC regulatory area

---

80 SPRFMO *Convention* 20(5)(a).

81 *ibid.*, 20(5)(b).

82 *see* SPRFMO, 'Performance Review Panel Report: 2018', ¶¶55–56; *nb.*, at its 12th Commission Meeting (29/1 to 2/2/2024), the SPRFMO Commission adopted Decision 6-2024, which launched the Second Performance Review. The agreed terms of reference include, as one of the key criteria, the review of CMMs in light of their compatibility with measures adopted in adjacent EEZs.

83 NPFC *Convention* 2.

84 NPFC *Convention* 3(i).

is defined to include 'only waters of the high seas',[85] coastal member States are encouraged to adopt compatible CMMs in sea areas under their national jurisdiction. In practice, however, coastal members have, on several occasions, been explicitly called upon to adopt such measures, especially with respect to priority fishery resources identified through ongoing stock assessments.[86] To that effect, NPFC has reaffirmed its members' commitment to establishing conservation measures for straddling stocks in high seas areas consistent with the Agreement, while also acknowledging the application of the compatibility rule to measures adopted both on the high seas and in areas under national jurisdiction.[87] It should be noted that the application of the compatibility rule has, in certain instances, given rise to concerns, particularly where verification has not been possible regarding how coastal State measures within national jurisdiction were assessed for compatibility with those adopted by the Commission for the high seas. This issue underscores the need for coastal member States to monitor and report on their compliance with the implementation of relevant measures within their EEZs.[88]

There appears to be a certain degree of interpretive latitude within the NPFC's functions, which include the authority 'to adopt, where necessary', and based on the best available scientific information and advice, 'CMMs for species belonging to the same ecosystem or dependent upon or associated with the target stocks', as well as management strategies for any fishery resources and for such associated or dependent species 'as may be necessary' to achieve the convention's objective.[89] The open-ended nature of this functional mandate—especially when contrasted with other functions expressly limited to the regulatory area—allows for the inclusion of EEZs within the spatial scope of NPFC measures where considered 'necessary'. In practice, this has enabled NPFC to request that coastal member States implement measures within their areas of national jurisdiction that align with those adopted for the high seas.[90]

6.3.6 Agreement to Prevent Unregulated High Seas Fisheries in the Central Arctic Ocean

The central Arctic Ocean stands out as one of the most foreseeable arenas for compatibility disputes, although substantial commercial fishing in its high

---

85 *ibid.*, 1(f) in conjunction with 4(1).
86 *e.g.* NPFC:CMM/8-2017 & CMM/7-2016.
87 *e.g.* NPFC:CMM/7-2019.
88 *see* NPFC, 'Performance Review Panel Report: 2022', ¶¶108–109 & 214.
89 NPFC *Convention* 7(1) *litt.*(c)–(d).
90 *e.g.* NPFC:CMM/7-2023 & CMM/11-2023.

seas areas is unlikely so long as they remain ice-covered.[91] The *Agreement to Prevent Unregulated High Seas Fisheries in the Central Arctic Ocean* (2018) ('CAOFA') constitutes an ambitious framework for taking precautionary action to ensure that any future fishing in the region is conducted sustainably.[92] Its principal objective is to prevent unregulated fishing in high seas areas through the application of the precautionary approach, as part of a long-term strategy aimed at protecting and preserving healthy marine ecosystems and ensuring the conservation and sustainable use of fish stocks. Member States are obliged to cooperate in achieving compatible CMMs for fish stocks occurring in sea areas both within and beyond national jurisdiction, consistent with article 7 FSA.[93] However, concerns have already been raised regarding the effective implementation of such measures, particularly in view of potential spatial overlaps with the NEAFC regulatory area. Such overlaps may pose challenges to coherent management, particularly given the ecosystem-related interactions and interdependencies between target stocks and associated or dependent marine species.[94] Scholarly analysis has noted that the inclusion of the compatibility rule under CAOFA is intended to apply equally to DWFS and coastal States, emphasizing that restrictions imposed solely on fishing in high seas areas—without corresponding measures in sea areas under national jurisdiction—may fail to satisfy the compatibility requirements of the Agreement.[95]

6.3.7 Permanent Commission for the South Pacific

The only notable departure from the otherwise balanced and neutral formulation of the compatibility rule at the regional level is found in the formulation adopted under the *Acuerdo Marco para la Conservación de los Recursos vivos marinos en la alta Mar del Pacífico Sudeste* (2000) ('*Galápagos Agreement*'), which explicitly favours the fishing interests of the coastal States in the south-east Pacific Ocean. This development stems from the conventional regime established by the *Santiago Declaration* (1952) and the subsequent creation of

---

91 Østhagen, 'Fish, Not Oil, at the Heart of (Future) Arctic Resource Conflicts', *Arctic Yearbook* (2020) 43.

92 Balton, 'Implementing the New Arctic Fisheries Agreement', in Heidar (Ed) *New Knowledge and Changing Circumstances in the Law of the Sea* (2020) 429.

93 CAOFA, 2 & 3(6).

94 Schatz–Proelss–Liu, 'The 2018 Agreement to Prevent Unregulated High Seas Fisheries in the Central Arctic Ocean: A critical analysis', IJMCL (2019) 213–215.

95 Heidar, 'The Legal Framework for High Seas Fisheries in the Central Arctic Ocean', in Nordquist–Moore–Long (Eds) *International Marine Economy* (2017) 185.

CPPS among Chile, Colombia, Ecuador and Peru.[96] Although CPPS functions primarily as an advisory body—alongside OLDEPESCA, whose competence encompasses sea areas under national jurisdiction—to coordinate a common maritime policy and strengthen its members' negotiating capacity against third States, the *Galápagos Agreement* contains provisions with clear regulatory implications. Its stated objective is the conservation of living marine resources in the high seas of the southeast Pacific, with particular reference to straddling and highly migratory fish stocks, as well as associated or dependent living marine species.[97]

In line with this objective, the compatibility rule is framed in an outward-facing manner that accords 'preferential status' to CMMs adopted by coastal States. This approach is portrayed as justified by the ecological interconnectedness of fish stocks within coastal marine ecosystems and the impacts of fishing on associated or dependent species. The agreement stipulates that it applies exclusively to the high seas,[98] while expressly subordinating international measures adopted for those areas to the national measures implemented by coastal States within their EEZs. Specifically, it requires that CMMs applicable to the high seas must not be less stringent than the corresponding in-zone national measures for the same stock, so as to avoid undermining their effectiveness, and must, in all cases, be fully compatible with the relevant domestic regime.[99] Furthermore, it provides that, in applying conservation principles, due regard must be given to the fact that, in accordance with international law, the freedom to fish on the high seas is subject, inter alia, to the rights, duties, and interests of coastal States, including their conservation and administrative regulations applicable to living marine resources in adjacent high seas areas.[100]

A unilateral approach of this kind, however, is 'not beyond doubt'[101] and cannot be regarded as setting, or even influencing, the normative content of the compatibility rule. As far as the compatibility rule is concerned, the regional regime established under the *Galápagos Agreement* is inconsistent not only with the Agreement but also with the Convention itself. In attributing preferential rights to coastal States in high seas areas, the agreement exceeds the

---

[96] *cit.* Chapter 2 (*fn*79); the Commission's authority was further strengthened by the *Convención sobre Personalidad Jurídica Internacional de la Comisión Permanente del Pacífico Sur* (1966).

[97] *Galápagos Agreement* 4, in conjunction with 1(13).

[98] *ibid.*, 3.

[99] *ibid.*, 5(1)(e).

[100] *ibid.*, 5(2).

[101] Molenaar, 'Addressing Regulatory Gaps in High Seas Fisheries', IJMCL (2005) 546.

scope of what is allowed under the Convention's fisheries provisions.[102] Had the Part XV LOSC proceedings not been discontinued in the *Swordfish stocks* dispute, the European Union was prepared to challenge the legality of the regime. Notably, it intended to contest whether the regime had been negotiated in accordance with the Convention and whether its substantive provisions were consistent with, inter alia, articles 64 and 116–119 LOSC.[103] Admittedly, there have been several indications that legal claims arising from the interpretation of these provisions by the four coastal States—particularly in relation to assertions of exclusive fishing rights and *de facto* extensions of national jurisdiction—were not initially intended to be enforceable against third States.[104] Moreover, internal divergences have occasionally surfaced between the CPPS collective agenda and the individual positions of its member States, calling into question the overall consistency of these claims.

Without question, the most compelling illustration of the *Galápagos Agreement*'s ineffectiveness is its continued difficulty in entering into force, despite having a narrow base of signatories. Only Chile on 12/11/2001 and Ecuador on 11/6/2002 have deposited instruments of ratification. Even a subsequent amending protocol aimed at facilitating its entry into force by reducing the requisite number of ratifications from four to three resulted in the same standstill. In contrast, Chile ratified the SPRFMO *Convention* on 25/7/2012, thereby providing the requisite number of contracting parties for that treaty to enter into force a month later. Ecuador acceded on 10/6/2015, followed by Peru's ratification of its signature on 21/1/2016. Although Colombia has signed but not ratified the SPRFMO Convention, it remains a participating State in SPRFMO meetings. It is worth noting that the SPRFMO *Convention* not only overlaps with the regulatory area of the *Galápagos Agreement* but also formulates the compatibility rule in distinctly different—and significantly more neutral—terms. These regional dynamics have rendered the *Galápagos Agreement* effectively a dead letter, unless CPPS member States reassess their fisheries policy in light of contemporary approaches to the sustainable regulation of fish stocks. Their joint proclamation of a new international strategy at the UN Conference on Sustainable Development *Rio+20* in June 2012 signalled the beginning of a cautious re-engagement with DWFS, as reflected in the commitment of CPPS to consider:

---

102 Oude-Elferink, 'What if the United Nations Convention on the Law of the Sea had Entered into Force Unamended: Business as usual or dystopia?', in Venzke–Heller (Eds) *Contingency in International Law* (2021) 226.

103 *Conservation and Sustainable Exploitation of Swordfish Stocks* (Constitution of Chamber) 2000, 2§3 *litt.*(a)–(h).

104 *Maritime Dispute* (Judgment) 2014, §116.

> el enfoque ecosistémico, el principio de precaución y los instrumentos internacionales destinados a la protección de los mares y océanos, respetando las políticas nacionales y mecanismos internos de cada país. Esta orientación será aplicada en la zona de soberanía y jurisdicción de 200 millas de los países miembros de la CPPS, los que están comprometidos a promover esos principios en el marco de los acuerdos internacionales que se apliquen a la zona de alta mar adyacente.[105]

Although there have been more recent calls for CPPS to play a constructive role at the regional level—particularly in achieving SDG 14[106]—these have not yet produced a concrete approach to interpreting the compatibility rule in a manner consistent with its normative content under article 7 FSA.

6.4 APPROACHES IN REGIONAL REGIMES PREDATING THE 1995 UNITED NATIONS FISH STOCKS AGREEMENT

The neutral normativity of the compatibility rule has also been reaffirmed in the context of several reform processes undertaken by RFMO/As that predate the Agreement. In some cases, performance reviews—although not resulting in formal legal revisions—have exerted political pressure for change within regional regimes by publicly disseminating findings in line with best practices for the sustainable regulation of fish stocks.[107] Other regimes, while having amended their constitutive treaties to update their conservation and management principles, do not include explicit provisions on the adoption of compatible CMMs. Nonetheless, as these organizations remain competent in terms of both spatial scope and species coverage for the conservation and management of straddling and highly migratory fish stocks, they will be briefly introduced

---

[105] CPPS, 'Declaración de los Países Miembros de la Comisión Permanente del Pacífico Sur en el Marco de la Reunión de Río+20' (13/6/2012). Author's translation: 'the ecosystem approach, the precautionary principle, and the international instruments aimed at the protection of the seas and oceans, respecting the national policies and internal mechanisms of each State. This orientation shall be applied within the 200-mile zone of sovereignty and jurisdiction of the member States of the CPPS, which are committed to promoting those principles within the framework of the international agreements applicable to the adjacent high seas area.'

[106] *e.g.* CPPS, 'Conmemoración 70 Aniversario de la Comisión Permanente del Pacifico Sur' (9/9/2022).

[107] Hoel, 'Performance Reviews of Regional Fisheries Management Organizations', in Russell–VanderZwaag (Eds) *Recasting Transboundary Fisheries Management Arrangements in Light of Sustainability Principles* (2011) 471.

here and examined in more detail in the next part of the chapter with particular emphasis on their dispute settlement procedures.

6.4.1 General Fisheries Commission for the Mediterranean

The General Fisheries Commission for the Mediterranean ('GFCM'), established under the *Agreement for the Establishment of a General Fisheries Council for the Mediterranean* (1949) ('GFCM *Agreement*'), is mandated to promote the development, conservation, rational management, and optimal utilization of fishery resources in the Mediterranean and the Black Sea. As the regulatory area encompasses all marine waters in the region—and given the absence of formally declared EEZs or broadly equivalent special-purpose protection zones—the effective regulation of common interest marine living resources, particularly small pelagic and demersal stocks distributed across national and high seas areas, is addressed through the adoption of CMMs grounded in the principles of ecosystem and precautionary approach.[108] This regulatory approach has rendered the issue of jurisdictional compatibility of lesser or no immediate relevance.[109]

6.4.2 Inter-American Tropical Tuna Commission

IATTC, which is responsible for regulating tuna and tuna-like species in the eastern Pacific Ocean, was the first RFMO to revise its conventional regime in light of the general conservation and management principles provided in the Agreement. Originally established as a bilateral arrangement to coordinate marine scientific research under the *Convention between the United States of America and the Republic of Costa Rica for the Establishment of an Inter-American Tropical Tuna Commission* (1949), its mandate gradually expanded and was eventually formalized with regulatory authority through the *Convention for the Strengthening of the Inter-American Tropical Tuna Commission Established by the 1949 Convention between the United States of America and the Republic of Costa Rica* (2003) ('IATTC *Antigua Convention*'). Under the revised regime, IATTC aims to ensure the long-term conservation and sustainable use of fish stocks within its regulatory area, in accordance with international law.[110] While the treaty explicitly incorporates the precautionary approach, it does not refer to the ecosystem approach in its text.

---

108 GFCM *Agreement* 5(c) & 8(b), and 'GFCM 2030 Strategy for Sustainable Fisheries and Aquaculture in the Mediterranean and the Black Sea' (9/7/2021), Target 1.

109 GFCM, 'Performance Review Panel Report: 2011', ¶4(1.6).

110 IATTC *Antigua Convention* II.

However, the IATTC *Antigua Convention* explicitly provides for the application of the compatibility rule, requiring that '[CMMs] established for the high seas and those adopted for areas under national jurisdiction shall be compatible'.[111] In addition, it mandates periodic review of the status and effectiveness of CMMs not only for targeted stocks but also for non-targeted, associated, or dependent species. To this end, a dedicated committee is tasked with reviewing and monitoring compliance with measures adopted by the IATTC Commission, including cooperative arrangements, and making recommendations to promote the compatibility of measures among members.[112] Species of particular concern include bigeye, bluefin, skipjack, and yellowfin tuna, which are Pacific-wide stocks that are predominantly harvested in purse seine fisheries. As a result, efforts to harmonize the respective CMMs of IATTC and WCPFC have continued to face difficulties. In its 2016 performance review, the Panel observed that, given the variation in stock outcomes for managed species, deeper inter-regional cooperation is needed. This should involve harmonizing stock assessment methodologies and undertaking regular joint assessments to improve understanding of stock status across their full biological range. Such coordinated efforts would facilitate the identification and resolution of potential compatibility issues arising from divergent fishing practices or the differing implementation of CMMs across the respective regulatory areas of IATTC and WCPFC.[113]

6.4.3 International Commission for the Conservation of Atlantic Tunas

The International Commission for the Conservation of Atlantic Tunas ('ICCAT') was established under the *International Convention for the Conservation of Atlantic Tunas* (1966) ('ICCAT *Convention*') and is mandated to regulate tuna and tuna-like species in the Atlantic Ocean. Since its establishment, the conventional regime has undergone substantial amendments to broaden its scope and update its conservation and management objectives. The revision—introduced by the *Protocol to amend the International Convention for the Conservation of Atlantic Tunas* (2019) ('ICCAT *Mallorca Convention*')—marked a significant step towards harmonizing the regime with contemporary developments in international fisheries law. Although the revised treaty text does not explicitly mention the compatibility rule, the regulatory area covers both the high seas and sea areas under national jurisdiction,

[111] *ibid.*, V(2).
[112] *ibid.*, VII, XVIII & Annex III.
[113] IATTC, 'Performance Review Panel Report: 2016', ¶22.

including the territorial seas and EEZs of its coastal State members.[114] This jurisdictional breadth enables ICCAT to give effect to the ecosystem approach in accordance with relevant internationally agreed standards, best practices, and procedures, and to adopt science-based CMMs applicable across the entire biological range of the fishery resources.[115]

The previous conventional regime had expressly allowed each contracting party to determine its own 'rights, claims or views' regarding the extent of its fisheries jurisdiction under international law.[116] This might have implied that coastal member States retained discretion in interpreting the extent to which ICCAT measures applied within their EEZs. As a result, depending on a coastal State's position, such a provision could potentially conflict with the obligation under the Agreement to ensure the compatibility of CMMs throughout the migratory range of the stocks.[117] The amended regime introduced significant regulatory clarity on this important issue, bringing it in line with the requirements of article 7 FSA. It is now beyond question that ICCAT's regulatory area encompasses all sea areas adjoining the Atlantic Ocean, with international measures applying uniformly across both EEZs and the high seas. By defining its mandate in this way, ICCAT enjoys a distinct institutional advantage over other tuna RFMOs. This framework enables the consistent implementation of CMMs throughout the regulatory area and facilitates the integrated management of each stock as a biological unit, notwithstanding the wide migratory range of the species concerned.[118]

6.4.4 Commission on the Conservation of Antarctic Marine Living Resources

Special mention should be made of the Commission on the Conservation of Antarctic Marine Living Resources ('CCAMLR'), established under the *Convention on the Conservation of Antarctic Marine Living Resources* (1980) ('CCAMLR *Convention*'), which regulates all marine living resources within the Antarctic marine ecosystem. While CCAMLR has set global benchmarks for the conduct and management of fisheries, it differs markedly from other regulatory schemes in that its principal objective is conservation—including rational use—rather than the sustainable harvesting of fishery resources.[119]

---

114 ICCAT *Mallorca Convention* I.

115 *ibid.*, IV & IX.

116 ICCAT *Convention* II.

117 ICCAT, 'Performance Review Panel Report: 2009', ¶I(4).

118 ICCAT, 'Performance Review Panel Report: 2016', ¶2(5).

119 CCAMLR *Convention* II; and *see* Press–Hodgson-Johnston–Constable, 'The principles of the Convention on the Conservation of Antarctic Marine Living

Nevertheless, the CCAMLR *Convention* contains several explicit provisions that correspond to the functions typically associated with RFMO/As. Under its mandate, CCAMLR has adopted CMMs for various Antarctic finfish species, as well as for straddling stocks such as *Champsocephalus gunnari* (Icefish mackerel), *Euphausia superba* (Antarctic krill) and Patagonian toothfish, as discussed earlier. Member States are obligated to take appropriate measures within their jurisdiction or control to ensure compliance with the convention and with CMMs adopted under its framework.[120]

Markedly, this regional regime—although predating both the Agreement and the Convention—embodies the concepts of the ecosystem and precautionary approaches,[121] and adopts measures aimed at implementing them.[122] It also establishes a requirement to harmonize CMMs, providing that CCAMLR shall seek cooperation with members exercising jurisdiction in sea areas adjacent to its regulatory area for the conservation of stocks, or associated species that occur both in areas under national jurisdiction and on the high seas.[123] While the ecosystem approach clearly entails ensuring the compatibility of measures across the full biological range of fishery resources, practical challenges frequently arise in determining appropriate measures throughout the regulatory area and, where necessary, across both high seas and sea areas under the national jurisdiction of coastal member States. This issue is especially significant with respect to aligning monitoring and stock assessment measures across the jurisdictional regimes. Many CCAMLR measures are inapplicable within sea areas under the national jurisdiction of coastal member States owing to their reservations, and no consistently applied mechanisms exist to verify whether national measures adopted in those areas are compatible with, or equivalent to, the Commission's international measures.[124] In this regard, fulfilling CCAMLR's conservation objectives requires a comprehensive suite of MCS measures, capable of governing the entire 'chain of custody of catch all the way from the hook to the plate'.[125]

Resources: Why its commission is not a regional fisheries management organisation', in Liu–Brooks–Qin (Eds) *Governing Marine Living Resources in the Polar Regions* (2019) 9.

120 CCAMLR *Convention* IX & XXI.

121 *ibid.*, II(c) & IX(i).

122 CCAMLR, 'Performance Review Panel Report: 2017', ¶¶4(34) & 5(40).

123 CCAMLR *Convention* XI.

124 CCAMLR, 'Performance Review Panel Report: 2008', ¶¶4(1)–4(3).

125 CCAMLR, 'Performance Review Panel Report: 2017', ¶5(51) *et seq.*

6.4.5 North-East Atlantic Fisheries Commission

As one of the world's most resource-rich marine regions, the northeast Atlantic hosts major straddling stocks, such as *Clupea harengus* (Herring), *Micromesistius poutassou* (Blue whiting), and *Sebastes mentella* (Pelagic redfish), as well as various groups of *Scomber scombrus* (Mackerel), including *Trachurus trachurus* (Atlantic horse mackerel), all of which are exploited both in EEZs and in adjacent high seas areas. Concentrated efforts towards cooperative CMMs in the northeast Atlantic region trace back to the *North-East Atlantic Fisheries Convention* (1959), which was superseded by the current regime under the *Convention on Future Multilateral Cooperation in North-East Atlantic Fisheries* (1980) ('NEAFC *Convention*').[126] The convention was amended in 2004 following a proposal by the European Union to introduce recommendations for dispute settlement procedures, and again in 2006 pursuant to a proposal by Iceland aimed at updating, among other elements, the regime's objectives and general conservation and management principles. The 2006 amendments entered into force for all contracting parties on 29/10/2013.

The revised objective mandates NEAFC, in exercising its functions, to ensure the long-term conservation and optimum utilization of fishery resources in the regulatory area.[127] While NEAFC may adopt CMMs throughout the entire regulatory area, such measures are issued as recommendations based on scientific advice from the International Council for the Exploration of the Sea. However, its regulatory authority in sea areas under national jurisdiction is conditional upon the request and support of the relevant coastal State.[128] For high seas areas adjacent to and beyond EEZ, NEAFC measures are adopted by a qualified majority.[129] Although the amended treaty text does not explicitly incorporate the compatibility rule, it retains the requirement of consistency, which was originally set out in the 1980 constitutive convention. This refers chiefly to the consistency of adopted measures with the scientific advice provided to the organization. In exercising its functions, NEAFC is required to ensure consistency between:

(a) any recommendation that applies to stocks occurring both within an area under the jurisdiction of a contracting party and beyond, or any recommendation that would have an effect through species inter-relationships on a stock or group of

---

[126] On the transition to the new regional regime *see* Sevaly, 'The Evolution of High Seas Fisheries Management in the North-East Atlantic', OCM (1997) 85.
[127] NEAFC *Convention* 2.
[128] *ibid.*, 6.
[129] *ibid.*, 5(1).

stocks occurring in whole or in part within an area under the jurisdiction of that party; and

(b) any measures and decisions taken by such party for the management and conservation of that stock or group of stocks with respect to fisheries conducted within the area under its jurisdiction.[130]

In the context of performance reviews, the application of the above requirement has been found to fall short of fulfilling the aim of achieving compatible CMMs as envisaged under the Agreement. While NEAFC is effectively obligated to promote coordination through cooperation to ensure consistency between in-zone national measures adopted by coastal States and international measures adopted for the high seas, this requirement does not, in practice, keep pace with the normative neutrality of the compatibility rule as set out in article 7 FSA.[131] In cases where coastal States fail to agree on CMMs applicable throughout the distribution range of a stock, they remain free to prescribe unilaterally in-zone national measures, even if these are not necessarily compatible with the international measures adopted for adjacent high seas areas. These situations have raised concern, particularly with respect to the management of several straddling stocks, where consistency could only be ensured through prior agreement on conservation measures, including TAC limits and TAFE levels. Persistent failures to reach agreement on compatible CMMs—especially for the portions of stocks occurring within EEZs—have been clearly identified as a matter requiring urgent solution.[132]

It follows that the requirement of consistency, arguably retained to reinforce regime linkages in favour of coastal State fishing interests,[133] is increasingly at odds with the normative content of the compatibility rule under the Agreement. The latter aims, in principle, to establish a balance of interests, thereby allowing for the adjustment of either national or international measures. This tension is likely to feature prominently in a future performance review, especially as calls intensify for the terms of reference to incorporate criteria that address the need for fundamental governance reform aligned with the principle of ecosystem approach. Such reform is particularly urgent in light of the long-standing mismanagement of pelagic stocks and the persistent challenges in NEAFC's relationship with coastal member States.[134] In evaluating

130 *ibid.*, 5(2).

131 NEAFC, 'Performance Review Panel Report: 2006', ¶3(3.7).

132 NEAFC, 'Performance Review Panel Report: 2014', ¶3(5.3).

133 *see* Churchill, 'Managing Straddling Fish Stocks in the North-East Atlantic: A multiplicity of instruments and regime linkages – but how effective a management?', in Stokke (Ed.) *Governing High Seas Fisheries* (2001) 238.

134 NEAFC, 'Report of the 42nd Annual Meeting of the North-East Atlantic Fisheries Commission' (17/11/2023), ¶13; and *see* the comments recorded by the

whether the consistency requirement genuinely upholds the balance envisaged by the compatibility rule, it is instructive to consider, as set out immediately below, a contrasting approach reflected in the practice of another RFMO operating in the north Atlantic Ocean.

6.4.6 Northwest Atlantic Fisheries Organization

As discussed earlier, NAFO holds the mandate to conserve and manage straddling fish stocks in the northwest Atlantic Ocean. The extensive amendments to the NAFO *Convention* marked the first formal step in reforming the regional regime following the entry into force of the Agreement. These amendments, which came into effect on 18/5/2017, modernized NAFO by incorporating the principle of an ecosystem approach, streamlining its decision-making processes, and introducing dispute settlement procedures. Yet, with respect to achieving compatible CMMs, the NAFO *Convention* retains a provision akin to that found in the NEAFC *Convention*, establishing a consistency requirement. Having first appeared in article XI of the 1978 constitutive convention, the provision was carried over unchanged into the amended treaty text. Article VI(11)(a) NAFO *Convention* continues to stipulate that the NAFO Commission, in exercising its functions, shall seek to ensure consistency between:

> (i) any measure that applies to a stock or group of stocks found both within the Regulatory Area and within an area under national jurisdiction of a coastal State, or any measure that would have an effect through species interrelationships on a stock or group of stocks found in whole or in part within an area under national jurisdiction of a coastal State; and
> (ii) any actions taken by a coastal State for the management and conservation of that stock or group of stocks with respect to fishing activities conducted within the area under its national jurisdiction.

Both the NAFO Commission and the coastal member States are required to promote the coordination of their respective CMMs, with each State additionally obliged to inform the Commission of its relevant actions.[135] The consistency requirement has been evaluated in the course of NAFO's performance reviews. In 2009, NAFO established a Performance Assessment Working Group comprising both internal and external experts, which has since conducted two reviews—completed in 2011 and 2018, respectively—each contributing key findings relevant to the compatibility rule. The 2011 Panel noted that, although the amended treaty text includes provisions aimed at achieving

European Union and Norway, in Annex C, and by the non-governmental organizations Marine Stewardship Council and Pew Charitable Trusts, in Annex F.

135 NAFO *Convention*, as amended, VI(11)(b).

consistency of CMMs, these are neither as mandatory nor as precise as the compatibility requirements in article 7 FSA. More importantly, it observed that the responsibility lies primarily with the NAFO Commission to ensure consistency between the international measures adopted for the high seas and the actions taken by coastal States in sea areas under their national jurisdictions. However, both the Commission and coastal States are expected to coordinate their respective measures. The language of the provision does not clearly define the respective scope of the obligation placed on the NAFO Commission and on coastal States to ensure consistency in their measures. Therefore, the Panel recommended that NAFO adopt policy guidance to reinforce its commitment to achieving compatibility and consider clarifying the respective responsibilities of coastal member States and the NAFO Commission in coordinating their measures and actions accordingly.[136] The 2018 Panel assessed the organization's performance over the period 2011–2017, with particular attention to actions taken in response to the recommendations of the first review. Concerning the consistency of CMMs, the Panel emphasized—as a matter of high priority—the importance of aligning adopted measures with scientific advice and implementing these through multiannual management strategies and delivery plans. Regarding the adoption of compatible CMMs, the Panel recommended developing mechanisms to operationalize article VI(11) NAFO *Convention*. Although some progress had been made to ensure internal consistency of CMMs measures, this remains an area requiring further attention. It was also noted with regret that a 2016 proposal for coastal member States to notify the NAFO Executive Secretary of their measures for stocks in sea areas under their national jurisdiction failed to reach consensus.[137]

6.4.7 Indian Ocean Tuna Commission

The Indian Ocean Tuna Commission ('IOTC') was established under the *Agreement for the Establishment of the Indian Ocean Tuna Commission* (1993) ('IOTC *Agreement*') to promote cooperation among its members in regulating stocks of highly migratory species—particularly tuna and tuna-like species—in the Indian Ocean. Its primary objective is to ensure the conservation and optimum utilization of these stocks through science-based management and to encourage their sustainable development across the regulatory area.[138] The IOTC *Agreement* includes a clause stipulating that CMMs 'shall not prejudice the exercise of sovereign rights of a coastal State in accordance

136 NAFO, 'Performance Review Panel Report: 2011', ¶3(2.5).

137 NAFO, 'Performance Review Panel Report: 2018', ¶4(b).

138 IOTC *Agreement* V & IX.

with the international law of the sea' with respect to the exploration, exploitation, conservation, and management of stocks within areas up to 200 nm under its jurisdiction.[139] This clause diverges from the species-based approach adopted in both the Convention and the Agreement, as well as in most regional regimes governing highly migratory stocks. As the IOTC *Agreement* predates the Agreement, it does not reference the compatibility rule, nor does the clause in question impose clear obligations to ensure that CMMs adopted in sea areas under national jurisdiction are compatible with those applied on the high seas, in line with article 7 FSA.

In its 2016 performance review, the Panel provided detailed commentary on this issue, stressing that CMMs must encompass the full biological range of fishery resources, irrespective of the jurisdictional regimes of the sea areas through which the stocks migrate. Accordingly, measures should be implemented in a harmonized manner by all members concerned. While the IOTC *Agreement* does not explicitly state whether international measures apply in sea areas under national jurisdiction, coastal member States remain obligated to adopt and implement compatible measures in conformity with article 7 FSA. In practice, IOTC adopts CMMs intended to apply across the migratory range, with coastal member States prescribing corresponding measures in sea areas under their national jurisdiction.[140]

6.4.8 Commission for the Conservation of the Southern Bluefin Tuna

CCSBT, responsible for the single species of southern bluefin tuna, does not define a regulatory area in geographical terms. Consequently, CMMs apply directly to the stock regardless of whether it is located in sea areas under national jurisdiction or on the high seas.[141] Although the CCSBT *Convention* does not expressly provide for the ecosystem and precautionary approaches, both principles have been integrated into the Commission's management procedures, which are pre-agreed rules that enable the adjustment of measures based on current monitoring data and scientific assessments.[142] Since CCSBT measures are not spatially differentiated, the issue of compatibility does not formally arise. Once CMMs are adopted through the decision-making process, they are intended to apply throughout the species' biological range.

---

139 *ibid.*, XVI.

140 *see* IOTC, 'Performance Review Panel Report: 2009', ¶¶4(1.6)–4(1.7) and 'Performance Review Panel Report: 2016', ¶¶3(13) & 4(9).

141 CCSBT *Convention* 1.

142 *e.g.* the 'Cape Town Procedure' pursuant to CCSBT resolutions on the 'Adoption of a Management Procedure' (17/10/2019) and 'Allocation of the Global Total Allowable Catch' (12/10/2023).

Nonetheless, CCSBT has previously faced glaring problems due to uncertainty in the scientific data underpinning its stock assessments, to the extent that reliable determinations of stock trends were impeded.[143] In practice, implementation remains the principal concern regarding the effective application of compatible measures. Although CMMs have been adopted, their implementation in sea areas under national jurisdiction has not always been sufficiently robust to inspire confidence. Furthermore, CCSBT has been urged to address compatibility challenges in aligning its MCS measures for ecologically related species with those of other RFMO/As in areas where their regulatory competence overlaps.[144]

6.5 COMPULSORY SETTLEMENT PROCEDURES REGARDING COMPATIBILITY DISPUTES IN REGIONAL REGIMES

As intergovernmental bodies, RFMO/As derive their performance chiefly from the extent to which member States are willing to cooperate and reach agreement on the sustainable regulation of fish stocks. However, cooperation by agreement has often proven difficult in the context of achieving compatible CMMs, particularly where consensus is required, meaning typically the absence of any formal objection. This challenge is compounded in regimes where consensus is the sole decision-making rule or where members are permitted to 'opt out'. Commonly referred to as 'objection procedures' or 'non-acceptance procedures', such mechanisms allow dissenting members to exempt themselves from the application of a measure in exchange for not blocking its adoption. Given that members may have competing, yet legitimate, interests—which often cannot be simultaneously accommodated in an economically viable manner—decisions concerning the allocation of fishing opportunities are among the most contentious. These tensions can stall progress on conservation strategies.[145] When this practice is used solely to entrench rigid policy positions in defiance of sound scientific advice, it risks undermining the entire regulatory scheme by frustrating the principles of ecosystem and precautionary approach. Thus, one of the most vexing issues in the operation of many regional regimes arises from the uneasy relationship

143 CCSBT, 'Performance Review Panel Report: 2008', ¶3.1.

144 CCSBT, 'Performance Review Panel Report: 2021', ¶¶28, 55–56 & Attachment C(G21).

145 Engler-Palma, 'Allocation of Fishing Opportunities in Regional Fisheries Management Organizations: From power to law?', in Russell–VanderZwaag (Eds) *Recasting Transboundary Fisheries Management Arrangements in Light of Sustainability Principles* (2011) 476.

between decision-making and dispute settlement, even though these processes are fundamentally linked and jointly crucial to the effective governance of fisheries.[146] This was vividly illustrated by the sharp escalation of the unresolved TAC quota dispute between Canada and the European Union, which reached its apex when Canada seized the Spanish trawler *Estai* in the NAFO regulatory area during the *Fish Stocks Conference* negotiations. Under similar circumstances, opting out of CMMs has triggered other long-standing fisheries disputes around the world.[147] This problem is particularly acute in regional regimes where decision-making procedures are not reinforced by mechanisms for compulsory and binding dispute settlement of disagreements.

The *Fish Stocks Conference* paid close attention to this issue in seeking appropriate remedies.[148] The Agreement addresses the interrelationship between decision-making and dispute settlement procedures through a sequence of three interlinked steps. First, the Agreement recognizes that dispute prevention falls within the substantive scope of the duty to cooperate, requiring States to 'agree on efficient and expeditious decision-making procedures' within RFMO/As and, where necessary, to strengthen those already in place.[149] Second, in fulfilling their duty to cooperate through RFMO/As, States 'shall promote the peaceful settlement of disputes in accordance with Part VIII [FSA]'.[150] This obligation encompasses both binding and non-binding procedures and may include voluntary or compulsory mechanisms. As a key feature, the use of *ad hoc* expert panels to resolve in-house disputes of a technical nature regarding CMMs may yield more expeditious outcomes, while offering the parties a constructive opportunity to reassess their positions during the resolution process.[151] Third, the Agreement envisages the use of compulsory and binding procedures established under general, regional, or bilateral treaties, which may apply in lieu of Part VIII FSA. Failing that, recourse may be had to the procedures under Part VIII FSA, which apply

146 Swan, *Regional Fishery Bodies and Governance* (2000) 36.

147 *see* Churchill, 'Legal Uncertainties in International High Seas Fisheries Management', *Fisheries Research* (1998) 227–229.

148 A/Conf.164/INF/2 (1993), *cit.* Chapter 2 (*fn*120), ¶88.

149 FSA 28.

150 FSA 10(k).

151 FSA 29; and *see* Pulvenis de Séligny-Maurel, 'Regional Fisheries Bodies and Regional Fisheries Management Organizations and the Settlement of Disputes Concerning Marine Living Resources', in Del Castillo (Ed.) *Law of the Sea, From Grotius to the International Tribunal for the Law of the Sea* (2015) 698 and Rayfuse, 'Settling Disputes in Regional Fisheries Management Organisations: Dealing with objections', in Fabri *et al.* (Eds) *A Bridge Over Troubled Waters* (2020) 249.

mutatis mutandis the provisions of section 2 Part XV LOSC, thereby enabling binding and final decisions on disputes concerning the interpretation or application of RFMO/As treaties.[152]

Achieving compatible CMMs across jurisdictional regimes depends primarily on the sincere commitment of members to cooperate through the decision-making processes of RFMO/As. Nonetheless, principled disagreements may arise in the interpretation or application of the compatibility rule. In this event, the *FSA Review Conference* has called attention to the need to avoid unilateral CMMs, particularly where the States concerned are parties to the same RFMO/A.[153] It has further been recognized by the parties to the Agreement that the interplay between decision-making and dispute settlement procedures is vital not only for securing compatible CMMs but also for safeguarding the rights and interests of both DWFS and coastal States.[154]

6.5.1 Review Panels

In recent years, many RFMO/As have increasingly adopted a combination of flexible decision-making procedures. These permit voting once consensus prospects have been exhausted and/or allow members to lodge objections to decisions initially taken by the main organ of the regime (i.e. the commission), thereby triggering review by an *ad hoc* panel of experts. Objections may be based on claims that a measure is inconsistent with the constitutive treaty, the Convention, or the Agreement, or that it unjustifiably discriminates in form or effect against the objecting member.[155] To ensure transparency and impartiality, some regimes form these panels externally from the RFMO/A framework. SPRFMO has already applied this approach, with Russia[156] and Ecuador[157] referring their objections to CMMs—regarding TAC and quota allocations for jack mackerel stocks—to adjudication, as disputes over the interpretation or

[152] FSA 30(2).

[153] *FSA Review Conference* 2016, ¶¶44–46 and Outcome A5.

[154] *see* the comments recorded in the *Informal Consultations FSA* ICSP8/UNFSA/REP/INF.6 (16/4/2009), ¶23.

[155] *e.g.* WCPFC *Convention* 20(6) & Annex II.

[156] *see* the 2013 and 2023 *Review Panel established under the Convention on the Conservation and Management of High Seas Fishery Resources in the South Pacific Ocean* (Findings/Recommendations).

[157] *see* the 2018 *Review Panel established under the Convention on the Conservation and Management of High Seas Fishery Resources in the South Pacific Ocean* (Findings/Recommendations).

application of the SPRFMO *Convention*.[158] The prevailing trend in regional regimes suggests that submitting a disagreement over the compatibility of CMMs to panel review proceedings does not preclude recourse to the compulsory procedures entailing binding decisions under Part VIII FSA. However, the precise relationship between panel review proceedings and the dispute settlement mechanisms of Part VIII FSA is to be determined by the *lex specialis* procedural rules of the relevant RFMO/A's constitutive treaty, which may, for instance, require the prior exhaustion of internal review mechanisms.[159] While review panels may form part of a wide array of dispute resolution options depending on the specific treaty framework, and may often be preferred by members seeking durable solutions, this procedural avenue generally operates without prejudice to the initiation of adjudication proceedings under Part VIII FSA. Accordingly, a compatibility dispute remains referable to the binding procedures established therein. Should this occur, where RFMO/As do not have independent legal standing before the settlement forum, the dispute would proceed between the member State(s) supporting the contested measure and the objecting member State(s). As this procedural configuration has not yet been tested, several questions remain uncertain, particularly regarding the direct legal effect of a judgment rendered by the dispute settlement forum and the nature or scope of the remedies that may be granted.[160] Alternatively, to avoid contentious litigation between RFMO/A members, parties could consider invoking the advisory jurisdiction of ITLOS.[161] A compelling prospect—subject to necessary amendments of the conventional regime—would be to confer upon the Tribunal a quasi-appellate function, modelled on the jurisdiction exercised by ICJ through special references.[162]

[158] SPRFMO *Convention* 17(6) and Annex II; *qv.* Urrutia, 'The Objection Procedure in the South Pacific Regional Fisheries Management Organisation', in Kunoy–Heidar–Yiallourides (Eds) *International Fisheries Law* (2024) 106.

[159] *e.g.* IATTC *Antigua Convention* XXV(3) provides, with respect to technical disputes, that the panel shall confer with the members concerned and 'shall endeavor to resolve the dispute expeditiously without recourse to binding procedures for the settlement of disputes'.

[160] Schatz, '*Ad Hoc* Expert Panels: RFMOs', *Max Planck Encyclopedia of International Procedural Law* (2022) 3806.

[161] *see* the formal statement by ITLOS Deputy Registrar Doo-Young Kim in FAO 'First Meeting of Regional Fishery Body Secretariats in the Network' (13/3/2007), 37–38.

[162] ICJ Statute 30, in conjunction with the ICJ *Rules of the Tribunal* (as adopted on 14/4/1978 and subsequently amended), Rule 87.

6.5.2 Regimes Established after the 1995 United Nations Fish Stocks Agreement

The intention to establish an inseparable legal link between the normative neutrality of the compatibility rule and the principle of compulsory adjudication has been evident in the practice of most regional regimes formed after the adoption of the Agreement. Preserving the carefully negotiated language of the substantive requirements in article 7 FSA—reflecting a balanced approach to the interests of DWFS and coastal States—has underscored the importance of ensuring access to dispute settlement mechanisms when agreement on compatible CMMs cannot be reached through existing decision-making processes.[163] To this end, the majority of regional regimes established post-Agreement have incorporated, *in concreto*, the procedures of Part VIII FSA through a general dispute settlement clause. With only minor variations in the conventional language across fisheries-related constitutive treaties, these clauses typically stipulate that 'the provisions relating to the settlement of disputes set out in Part VIII of the 1995 Agreement apply, *mutatis mutandis*, to any dispute between Contracting Parties, whether or not they are also Parties to the 1995 Agreement'.[164]

Crucially, these procedures also cover disputes concerning the adoption of compatible CMMs, including the establishment of TAC limits and TAFE levels, following a review of such decisions. For example, in the context of NPFC, where an objection has been raised and the decision in question is found not to discriminate in form or in fact against the objecting member, nor to be inconsistent with the NPFC *Convention*, the Convention, or the Agreement—and where the objecting member's alternative measures are not deemed equivalent in effect—that member may initiate dispute settlement proceedings under Part VIII FSA.[165] Likewise, SEAFO provides for compulsory settlement procedures in cases where members disagree on issues related to the interpretation or application of the SEAFO *Convention*, including disputes concerning the adoption of compatible CMMs.[166] In the latter case, the implementation of the implicated measures does not prejudice the unilateral right of any member

---

[163] McDorman, 'Implementing Existing Tools: Turning words into actions – decision-making processes of RFMOs', IJMCL (2005) 436 & 442.

[164] *e.g.* NPFC *Convention* 19, with a more recent incorporation by reference appearing in CAOFA 7; *qv.* Schatz, 'The Settlement of Disputes concerning Conservation of Fish Stocks in the Arctic and Antarctic High Seas: Towards comprehensive compulsory jurisdiction?', in Liu–Brooks–Qin (Eds) *Governing Marine Living Resources in the Polar Regions* (2019) 196.

[165] NPFC *Convention* 9(1) *litt.*(j)–(iii).

[166] SEAFO *Convention* 24(2).

to invoke the dispute settlement procedures once all other means for resolving the dispute have been exhausted.[167] In particular, if a dispute is not referred or resolved within a reasonable period of time through such other means, either party may submit the matter unilaterally:

> for binding decision in accordance with procedures for the settlement of disputes provided in Part XV of the 1982 Convention or, where the dispute concerns one or more straddling stocks, by provisions set out in Part VIII of the 1995 Agreement. The relevant part of the 1982 Convention and the 1995 Agreement shall apply whether or not the parties to the dispute are also Parties to these instruments.[168]

In the same manner, SIOFA incorporates compulsory procedures entailing binding decisions by directing its members to the dispute settlement procedures of Part VIII FSA.[169] These procedures are also applicable to disputes involving a fishing entity that has expressed its commitment to be bound by the terms of the Agreement. Where a dispute of this kind cannot be resolved by other 'amicable' means, it shall, at the request of any party, be submitted to 'final and binding arbitration' following the relevant rules of PCA.[170] In a similar way, both the WCPFC *Convention* and SPRFMO *Convention* incorporate the dispute settlement procedures of Part VIII FSA through an identical clause. This provision, while succinct, unequivocally affirms that where a dispute remains unresolved through other means—which may include recourse to an *ad hoc* expert panel in the case of technical disagreements—the procedures of Part VIII FSA shall apply *mutatis mutandis* to any dispute between members.[171]

Given these incorporations by reference, it is instructive to examine how the principle of compulsory adjudication has been received in regional regimes. A notable example arises from Japan's opposition to the inclusion of compulsory settlement procedures for compatibility disputes at WCPFC MHLC. In reviving a dormant disagreement with FFA over the interpretation of article 64 LOSC,[172] Japan issued a statement expressing its position on the proposed dispute settlement procedures. This statement provides a useful vantage point

---

167 *ibid.*, 23(3).

168 *ibid.*, 24(4).

169 SIOFA *Agreement* 20(1).

170 *ibid.*, 20(2); *qv.* Scanlon, 'Incorporating Taiwan in International Fisheries Management: The Southern Indian Ocean Fisheries Agreement experience', ODIL (2017) 35; the same approach is followed among other regional regimes in SPRFMO *Convention* 4, in conjunction with Annex IV, and NPFC *Convention* 15.

171 WCPFC *Convention* 31 and SPRFMO *Convention* 34.

172 *see* Haward, 'Management of Marine Living Resources: International and regional perspectives on transboundary issues', in Blake *et al.* (Eds) *International*

for exploring the nuanced interplay between the compulsory dispute settlement provisions under the Convention, the Agreement, and—by extension—the *lex specialis* of regional regimes established pursuant to the Agreement. At the final stage of the negotiations, the Japanese delegation asserted that disputes concerning the interpretation or application of the WCPFC *Convention* should first be settled diplomatically within the WCPFC framework and only be referred to compulsory procedures with the consent of the parties in dispute.[173] This approach, however, was not accepted. In response, the former Chairman of the *Fish Stocks Conference* Satya Nandan, who was also the Chairman of WCPFC MHLC in the period 1997–2000, playing a central role in drafting the WCPFC *Convention*,[174] issued a counterstatement. Regretting Japan's position as 'an unfortunate aspect', Chairman Nandan affirmed that the WCPFC *Convention* had been drafted in line with the spirit and letter of the Agreement, as to uphold the carefully negotiated balance of interests between DWFS and coastal States, and he further opined that:

> A fundamental precept in the regime is that there is provision for compulsory and binding dispute settlement arising from disputes over the interpretation and application of the convention. These procedures are based on the widely accepted norms contained in the [Convention], which is the paramount instrument governing all ocean-related activities. The wide acceptance of those procedures was recognized by their inclusion in the [Agreement] and they are now incorporated in this convention.[175]

The above clear statement provides yet another critical insight into the drafting intent behind the Agreement, which is the indubitable application of compulsory procedures entailing binding decisions in the context of compatibility disputes.

This leads directly to the question of how far the remit of these procedures extends. Do they apply solely to high seas components of CMMs, or—considering the progressive development of the law on the sustainable regulation of fish stocks—are they also applicable to the EEZ components? On this exact point, the informed reflections offered by Michael Lodge, who served as the Secretary of WCPFC MHLC in the period 1997–2000, are particularly significant with regard to the scope of the fisheries limitation. Essentially arguing

---

Boundaries and Environmental Security, Frameworks for Regional Cooperation (1997) 53.

173 WCPFC MHLC, 'Report of the Seventh Session' (30/8 to 5/9/2000), Annex 3:20.

174 Hu, 'Father of the WCPFC Convention: Ambassador Satya N. Nandan', in Lodge–Nordquist (Eds) *Peaceful Order in the World's Oceans* (2014) 382.

175 WCPFC MHLC (2000), *supra cit.* (*fn*173), Annex 8:74.

for a restrictive construction of the limitation's subject matter, he perceptively noted that:

> [T]he traditional view is that, under article 297(3) LOSC, disputes concerning the exercise of sovereign rights within the EEZ are excluded from compulsory binding settlement procedures. It is arguable, however, that the effect of the Agreement, which places clear obligations on coastal States with regard to the management of EEZ stocks, is to make coastal States accountable for ineffective management of such stocks, where this result from a failure to observe the basic principles for conservation and management [...] This would require a narrower interpretation of article 297(3) LOSC to cover only the exercise of coastal State discretion on matters which are purely of EEZ concern.[176]

The proposed narrow interpretation fully aligns with the normative drafting of the compatibility rule in both the Agreement and the WCPFC *Convention*, the latter of which explicitly incorporates the dispute settlement procedures of Part VIII FSA. Limiting the remit of compulsory and binding procedures to high seas components alone would contradict the overarching objective of the WCPFC *Convention*, which is to ensure the long-term conservation and sustainable use of highly migratory fish stocks across their entire migratory range. In this respect, the Agreement qualifies the rights of all States by requiring that, through cooperation, CMMs adopted for the high seas and those applied within EEZs be compatible, with those rights being further circumscribed under the WCPFC *Convention*.[177]

Opposition to the compulsory settlement of disputes within regional regimes was maintained by coastal States that advocated for extending their fishing interests into high seas areas. During the *Fish Stocks Conference*, these States sought unsuccessfully to undermine the textual pattern of clausal embeddedness in the drafting of the compatibility rule. After the adoption of the Agreement, Chile—followed by Peru—attempted to amend the dispute settlement clause of the SPRFMO *Convention* by proposing the removal of any reference to the compulsory procedures under Part VIII FSA.[178] In general, it can be observed that States dissatisfied with the inclusion of the principle of compulsory adjudication in the Agreement later sought to dilute its effect in

---

[176] Lodge, 'The Draft Convention for the Conservation and Management of Highly Migratory Fish Stocks in the Western and Central Pacific Ocean', in Nordquist–Moore (Eds) *Current Fisheries Issues and the Food and Agriculture Organization of the United Nations* (2000) 27.

[177] Larocque, 'The Convention on the Conservation and Management of Highly Migratory Fish Stocks in the Western and Central Pacific Ocean: Can tuna promote development of Pacific island nations?', *Asia Pacific* LPJ (2003) 106–108.

[178] SP/06/INF/04 (2008) *supra cit.* (*fn*70), draft article 34(1).

the constitutive treaties of regional regimes where coastal interests are more prominently represented. A clear manifestation of this practice is found in the *Galápagos Agreement*, which—as might be expected given the territorialist posture of its signatories—adopts a light-touch approach by merely referring to article 33 UNC and other applicable international instruments *inter partes*. Disputes are to be referred by agreement to conciliation or a technical body, depending on their nature, unless both parties agree on a different procedure. If the available non-binding means are exhausted, or if the parties fail to agree on recourse to other means—such as ICJ or ITLOS—either party may seek to submit the dispute to arbitration.[179] The elusiveness of the provisions concerning the remit of binding procedures becomes even more pronounced in the context of compatibility disputes, owing to a clear-cut limitation which stipulates that:

> Sin perjuicio de las normas aplicables de conformidad con el derecho internacional, en ningún caso podrán someterse a los procedimientos previstos en el párrafo 3, las controversias relativas al ejercicio de los derechos soberanos de los Estados ribereños dentro de sus respectivas zonas bajo jurisdicción nacional.[180]

The above limitation—explicitly stating that in no case shall disputes concerning the exercise of sovereign rights by the coastal State in sea areas under its national jurisdiction be submitted to any of the procedures under ICJ, ITLOS, or arbitration—is further reinforced by a clause that preserves each party's position regarding the nature and scope of rights in its maritime zones, as well as its stance on any treaties addressing such matters.[181]

6.5.3 Regimes Predating the 1995 United Nations Fish Stocks Agreement

Regional regimes predating the Agreement contain dispute settlement clauses that do not clearly mandate the use of compulsory and binding procedures. A representative example is IATTC, whose constitutive treaty was revised to include a provision stating that, 'if a dispute is not settled through consultation within a reasonable period', the parties concerned 'shall consult among themselves as soon as possible in order to settle the dispute through

---

179 *Galápagos Agreement* 14(1)–(3).
180 *ibid.*, 14(4). Author's translation: 'Without prejudice to the applicable rules in accordance with international law, in no case may be submitted to the procedures set out in paragraph 3, the disputes relating to the exercise of the sovereign rights of the coastal States within their respective zones under national jurisdiction.'
181 *ibid.*, 15.

any peaceful means they may agree upon, in accordance with international law'.[182] Similarly, the clauses employed by GFCM and IOTC require that, failing resolution through non-binding recommendations of an *ad hoc* committee, fisheries disputes be referred to ICJ or arbitration, unless the parties mutually agree on an alternative method of settlement.[183] In other regimes, members are obliged to 'consult among themselves with a view to having the dispute resolved by negotiation, inquiry, mediation, conciliation, arbitration, judicial settlement or other peaceful means of their own choice',[184] or specify further that a dispute may only be referred to binding procedures at 'the joint request of the parties'[185] or 'with the consent in each case of all parties to the dispute'.[186]

As it will be recalled from Chapter 3, this latter type of wording was interpreted by the arbitral tribunal in the *Southern Bluefin Tuna Cases* as precluding recourse to any dispute settlement procedure not accepted by all parties, including those under section 2 Part XV LOSC. While the tribunal ultimately considered that Part XV LOSC may fall short of establishing a truly comprehensive dispute settlement regime, it nevertheless alluded, in *obiter dictum*, to the importance of effectively implementing the Convention's relevant provisions under the Agreement, and to reinforcing the compulsory procedures applicable to this category of fisheries disputes. Notably, this understanding is reflected in academic commentary by those with first-hand involvement in the proceedings. Bill Mansfield, who acted as counsel and advocate for the applicants, interprets the tribunal's final comment as an indication that the procedural aspects of the award are unlikely to carry lasting significance following the entry into force of the Agreement.[187] In a similar vein—and particularly pertinent to this discussion—are the observations of Chusei Yamada, the Japanese arbitrator in the case. He argued that the award did not compromise the Agreement's principle of compulsory adjudication, cautioning that Japan might need to 'think twice before deciding to ratify it' if the result conflicted with its interests.[188] Of the same mind, Tullio Treves—then serving as a Judge

[182] IATTC *Antigua Convention* XXV(2).

[183] GFCM *Agreement* 19 and IOTC *Agreement* XXIII.

[184] *Convention on the Conservation and Management of Pollock in the Central Bering Sea* (1994) XIII.

[185] ICCAT *Mallorca Convention* X.

[186] CCAMLR *Convention* XXV and CCSBT *Convention* 16.

[187] Mansfield, 'Compulsory Dispute Settlement After the Southern Bluefin Tuna Award', in Oude-Elferink–Rothwell (Eds) *Ocean Management in the 21st Century* (2004) 270.

[188] Yamada, 'Priority Application of Successive Treaties Relating to the Same Subject Matter: The southern bluefin tuna', in Ando–McWhinney–Wolfrum (Eds)

of ITLOS during the provisional measures phase—reached a comparable conclusion, suggesting that had the Agreement been in force for all parties, the outcome of the case would likely have been different.[189]

These views concur with the broader understanding that the procedures established under the Agreement prevail over those in the constitutive treaties of regional regimes that do not confer compulsory jurisdiction on a forum capable of issuing binding and final decisions.[190] By making the procedures of Part XV LOSC applicable *mutatis mutandis* to regional regimes, the Agreement does not derogate from the general principle of free choice of means but rather supplements it by introducing a binding procedural framework. The award in the *Southern Bluefin Tuna Cases* highlighted the importance of this development, describing it as marking a notable shift towards compulsory dispute settlement. This trend has been further amplified by the explicit incorporation of *mutatis mutandis* clauses in post-Agreement regimes, such as the WCPFC *Convention*. These developments exemplify regional regimes that serve the purpose of the Agreement in giving effect to the Convention and, in accordance with article 288(2) LOSC and article 30(2) FSA, confer compulsory jurisdiction on a court or tribunal under section 2 Part XV LOSC and Part VIII FSA, respectively.[191] In practice, this *pari passu* procedural evolution is further evidenced in the revisions undertaken by NEAFC and NAFO to align their dispute settlement procedures with the requirements of the Agreement.

As previously discussed, the NEAFC *Convention* was amended in 2004 following a proposal by the European Union to include a new article 18*bis*, empowering the NEAFC Commission to recommend the establishment of dispute settlement procedures. A procedure was accordingly recommended for disputes concerning straddling stocks between parties. Where such a dispute is not resolved by recourse to peaceful means of their own choice, it may be referred 'to compulsory procedures entailing binding decisions',

Liber Amicorum Judge Shigeru Oda (2002) 769.

189 Treves, 'A System for Law of the Sea Dispute Settlement', in Freestone–Barnes–Ong (Eds) *The Law of the Sea, Progress and Prospects* (2006) 422.

190 *see* Oxman, 'Complementary Agreements and Compulsory Jurisdiction', AJIL (2001) 306 and Röben, 'The Southern Bluefin Tuna Cases: Re-regionalization of the settlement of law of the sea disputes?', ZaöRV (2002) 61*ff.*

191 On this particular aspect *see* the analysis by Barbara Kwiatkowska in 'International Decisions – Southern Bluefin Tuna', AJIL (2001) 170 and 'The Australia and New Zealand *v.* Japan – Southern Bluefin Tuna (Jurisdiction and Admissibility) Award of the First Law of the Sea Convention Annex VII Arbitral Tribunal', IJMCL (2001) 289–290, wherein it is emphasized that the statement by Japan at the WCPFC MHLC negotiations essentially refuted its own argument previously made in the context of the *Southern Bluefin Tuna Cases*.

to be governed *mutatis mutandis* by Part VIII FSA. Specifically, parties are required to cooperate to prevent disputes. However, once a dispute arises, they must seek to resolve it expeditiously—whether through voluntary or compulsory means—where it concerns the interpretation or application of the NEAFC *Convention*. These procedures extend to disputes arising from disagreements over compatible CMMs, including those linked to objections or terminations of acceptance under the decision-making process. If a dispute concerns the interpretation or application of a recommendation adopted by the NEAFC Commission, the parties may refer the matter to an *ad hoc* panel constituted under procedural rules to be adopted by the Commission. If not resolved by that means, either party may submit the dispute to the compulsory and binding procedures under Part VIII FSA. The forum seised shall apply, as appropriate, the relevant provisions of the NEAFC *Convention*, the Convention, and the Agreement, along with other applicable rules of international law consistent with these treaties and the relevant NEAFC recommendations, with a view to ensuring the conservation and optimum utilization of the fish stocks concerned. Importantly, the recommended procedure stipulates that the relevant provisions of the Convention and the Agreement shall apply *mutatis mutandis*, regardless of whether the parties to the dispute are formally contracted under those respective treaties.[192] While this amendment secured the required three-fourths majority of all contracting parties in 2013, it did not enter into force due to an objection lodged by Russia on 31/7/2013.[193]

The availability of dispute settlement procedures has been acknowledged positively during NEAFC's performance review process. However, the specific issue concerning the remit of compulsory and binding procedures—particularly in relation to the substantive requirement of consistency—was not directly addressed.[194] The 2014 Panel concluded that the procedures were fully consistent with Part VIII FSA, albeit in somewhat indistinct terms, noting that '[t]he mechanisms for settlement of disputes entailing binding decisions apply to disputes concerning CMMs adopted for the regulatory area, as disputes relating to CMMs in areas under national jurisdiction of coastal States are governed by article 297(3)(a) LOSC'.[195] This comment appears to refer to article 29 FSA, implying that it would be important to clarify that the *ad hoc* expert panel for technical disputes is not analogous to the panel of experts under

---

192 NEAFC, 'Report of the 23rd Annual Meeting of the North-East Atlantic Fisheries Commission' (12/11/2004), Volume I:(27–29) & Volume II:(Annex K).

193 NEAFC *Convention* 19; and *see* further NEAFC Report (2023), *supra cit.* (*fn*134), ¶5.

194 *see* NEAFC, 'Performance Review Panel Report: 2006', ¶3(4.4) and 'Performance Review Panel Report: 2014', ¶5(2.1).

195 NEAFC, 'Performance Review Panel Report: 2014', ¶5(2.1:109).

the special arbitration procedure in Annex VIII LOSC, which renders binding decisions. It was therefore recommended that NEAFC consider specifying what constitutes a 'technical' dispute appropriate for submission to an *ad hoc* panel, and which other types of disputes might be better suited to different dispute settlement procedures.[196]

The NAFO *Convention*, having undergone an extended negotiation process culminating in its 2007 amendment, now features a robust mechanism for the settlement of disputes. This applies to general disagreements concerning the interpretation or application of the treaty and to disputes, in particular, arising from the use of the objection procedure to NAFO Commission decisions on CMMs and their implementation.[197] In this category markedly fall disputes regarding the consistency of CMMs. The relevant amendments were carefully developed over more than six years by the NAFO Working Group on Dispute Settlement Procedures.[198] Article XV of the amended NAFO *Convention* affirms the principle of free choice of means for dispute resolution, while also providing for the option of referring a dispute to an *ad hoc* panel. The panel's report may contain recommendations deemed appropriate for resolving the dispute. However, these are not binding, and the objecting party may maintain its objection at the conclusion of the process.[199] Where no settlement is reached by these means, or if the parties fail to agree on another means of resolution, any party may submit the dispute to 'compulsory proceedings entailing binding decisions pursuant to section 2 Part XV [LOSC] or Part VIII [FSA]'.[200]

The principle of compulsory adjudication is placed beyond doubt by the closing paragraph of article XV NAFO *Convention*, which unequivocally attests to its applicability to compatibility disputes. It states that '[n]othing in this convention shall be argued or construed to prevent a Contracting Party to a dispute … from submitting the dispute to compulsory procedures entailing binding decisions against another State Party pursuant to section 2 Part XV [LOSC] … or article 30 [FSA]'.[201] Intriguingly, the provision refers directly

---

196 *ibid.*, ¶5(2.1:110).

197 NAFO *Convention*, as amended, XV and Annex II 'Rules concerning the *ad hoc* panel procedure pursuant to Article XV'.

198 *see* the drafts containing the dispute settlement provisions in the reports issued by the Working Group on Dispute Settlement Procedures during the Copenhagen (29–31/5/2000); Dartmouth (12–14/6/2001) and Lunenburg (12–17/9/2006) sessions, as well as the minutes of the Lisbon annual meeting (24–28/9/2007) when the final version of the article containing the dispute settlement procedures was formally adopted.

199 NAFO *Convention*, as amended, XV(3)–(7).

200 *ibid.*, XV (6) & (8).

201 *ibid.*, XV(12).

to 'section 2', rather than to Part XV LOSC as a whole, thereby—arguably—omitting the limitations outlined in section 3 on the applicability of compulsory procedures. As noted by the 2011 Panel, 'this provision is useful in keeping all potentially available means of dispute settlement open to the parties'.[202] Of particular significance is the interpretation offered regarding the fisheries limitation under article 297(3) LOSC. In assessing the coherence of the NAFO *Convention* with other relevant international legal regimes on fisheries, the Panel examined the applicable provisions of the Convention and concluded that:

> [L]OSC Part XV requires fishery disputes concerning the interpretation or application of the Convention, with the exception of *domestic* fisheries, to be settled in accordance with the compulsory procedures leading to binding decisions when no settlement has been reached by recourse to other alternative mechanisms of dispute resolution such as negotiation or conciliation. The exception provided for in article 297(3)(a) [LOSC] relates to any dispute relating to the coastal State's sovereign rights with respect to the living resources *in* the exclusive economic zone or their exercise. This effectively excludes *domestic* fisheries from the compulsory provisions but would not preclude a coastal State from relying on these procedures against, for example, a flag State whose vessel was *fishing in its EEZ* in breach of the fisheries provisions of the Convention.[203]

This finding, issued through a formally institutionalized process, ought to be carefully weighed in the context of any future NAFO dispute involving disagreements over compatible CMMs, with due regard to balancing the interests of both DWFS and coastal States.

6.6 CONCLUSION

RFMO/As play a pivotal role in regulating fish stocks by providing institutional frameworks and facilitating cooperation to achieve long-term conservation and sustainable use in their respective ocean regions. The preponderant multilateral treaty practice in implementing the compatibility rule within regional regimes has upheld the normative neutrality of the substantive obligations in article 7 FSA, striking a careful balance between the interests of DWFS and coastal States. The establishment of an inseparable legal relationship between the normative neutrality of the compatibility rule and the principle of compulsory adjudication has been evident in the practices of most regimes following the adoption of the Agreement. Unresolved disagreements over compatible CMMs are subject to compulsory and binding procedures in accordance with

---

[202] NAFO, 'Performance Review Panel Report: 2011', ¶3(2.11).

[203] *ibid.*, Appendix VII (178); emphasis added.

Part VIII FSA, ensuring that each fish stock is managed as a unified whole and throughout its biological range. Still, in some cases, effective cooperation between RFMO/As with neighbouring or overlapping regulatory areas will become increasingly critical. Ensuring compatibility of regional CMMs for both targeted and associated or dependent stocks in the context of large marine ecosystems remains a pressing challenge for the future.

7. Final remarks: taking fish stocks to the room of proceedings—when to agree to disagree on compatible conservation and management measures

The compatibility rule possesses a *sui generis* legal character derived from its normativity of having progressively developed the rights and obligations of States to ensure the long-term conservation and sustainable use of straddling and highly migratory stocks through the effective implementation of the Convention. Since these stocks cannot be conserved at levels capable of producing MSY and soundly managed either on the high seas or in sea areas under national jurisdiction alone, the duty of DWFS and coastal States to cooperate in achieving compatible CMMs across the full extent of their distribution and migratory range becomes paramount.

The substantive formulation of the rule in article 7 FSA, informed by the long treaty pedigree of provisions on the conservation and management of fish stocks, deliberately employs vague legislative language to address its spatial scope. Drafting ambiguity is a common practice at international conferences, enhancing flexibility and efficiency of the negotiations; albeit at the expense of 'textual determinacy'.[1] The technique often produces legal rules that encapsulate carefully crafted and artfully formulated compromises.[2] These rules are

[1] The term is used here to refer to the quality of the text in clearly conveying the permissive boundaries of the rule through its wording; *qv*. Gottlieb, *The Logic of Choice* (1968) 38–42 and Franck, *The Power of Legitimacy Among Nations* (1990) 52.

[2] For the allowance of logical vagueness in law-making processes *see* Fawcett, 'The Legal Character of International Agreements', BYBIL (1953) 381; Elias–Lim, *The Paradox of Consensualism in International Law* (1998) 258–260; Scott, 'Ambiguity *versus* Precision: The changing role of terminology in conference diplomacy', in Kurbalija–Slavik (Eds) *Language and Diplomacy* (2001) 153; and Schane, *Language and the Law* (2006) 12–53.

thus left in a state of relative indeterminacy, with their content to be refined, by default, through third-party judicial processes in the event of unresolved disagreement.[3] In line with this rulemaking logic, neutral interpretations—corroborated by the preponderant practice of RFMO/As—regard the interplay between article 3(1) FSA and paragraphs (1) and (2) of article 7 FSA as deliberately creating a constructive ambiguity regarding the spatial scope of the rule. The main *desideratum* of this ambiguity is to allow sufficient latitude for States to agree on compatible CMMs that assure the biological integrity of stocks across the full extent of their distribution and migratory range, while remaining responsive to the science–policy interface and strengthening cooperation at all levels to improve global fisheries outcomes. It neither rules out the extension of coastal measures into high seas areas nor precludes the application of international measures in sea areas under the national jurisdiction of coastal States, where necessary for the long-term conservation and sustainable use of stocks. So long as the States involved in the negotiations maintain a shared understanding and legitimate expectation, the ambiguity surrounding the rule's spatial scope remains dormant. Should the parties, despite conducting negotiations *bona fide*, fail to agree on compatible CMMs, the procedural provision in paragraph (4) of article 7 FSA prescribes the course of action to be followed.

While the precise modalities for referring a dispute to international adjudication must be carefully assessed against the applicable adjectival *lex specialis* on a case-by-case basis, this book concludes that article 7(4) FSA operates as a built-in adjudicative jurisdiction clause. Subject only to the *ratione temporis* requirement of having negotiated for a reasonable period of time, the applicant party can unilaterally refer the compatibility dispute to the compulsory procedures entailing binding decisions under Part VIII FSA. Where, nevertheless, in a contested jurisdiction case, article 32 FSA is invoked against the principle of compulsory adjudication, it is for the court or tribunal seised of the case to determine *ad casum* whether the subject matter of the dispute falls within the exclusionary scope of the fisheries limitation. It is realistic to expect

[3] On the legislative economy of implicitly delegating the task of interpretation to third-party adjudication *see* Lauterpacht, 'Codification and Development of International Law', AJIL (1955) 19–20; Simma, 'Consent: Strains in the treaty system', in Macdonald–Johnston (Eds) *The Structure and Process of International Law* (1983) 491–492; Brus, *Third Party Dispute Settlement in an Interdependent World* (1995) 96–127 & 178–190; Ginsburg, 'Bounded Discretion in International Judicial Lawmaking', VJIL (2005) 641*ff*; Koskenniemi, *From Apology to Utopia* (2006) 39–41, 489 *et seq*; and Klein–Kroon, 'Settlement of International Environmental Law Disputes', in Fitzmaurice–Brus–Merkouris (Eds) *Research Handbook on International Environmental Law* (2021) 249.

the respondent party to raise such a jurisdictional objection and, most likely, request that it be addressed as a preliminary matter. According to the settled *jurisprudence constante* on this issue, the party raising jurisdictional objections is entitled to have these answered at the preliminary stage of the proceedings, unless the forum does not have full knowledge of the facts necessary to determine its precise nature or the objection in question is so interwoven with the merits of the case that it can only be dealt with in conjunction with the merits. Otherwise, there is a risk of adjudicating on questions that pertain to the merits or prejudging issues closely connected thereto. In such cases, it should join these issues to the merits if it determines that the facts and arguments supporting the preliminary objections are, to a significant extent, the same as those on which the merits depend, or if the elucidation of the relevant facts and their legal consequences requires it.[4] A dispute arising from a disagreement over the spatial scope of compatible CMMs will plausibly implicate jurisdictional questions that cannot be effectively resolved without delving into substantive issues closely tied to the merits. A jurisdictional objection of this kind is therefore unlikely to be of an exclusively preliminary character.

The above conclusion fully aligns with the general rule that a treaty must be interpreted in good faith, in accordance with the ordinary meaning of its terms within their textual context, and in light of the treaty's object and purpose.[5] The entry point for interpreting article 7 FSA lies in the principle of textuality, namely, that the meaning of a treaty is grounded in its actual text, which, by being reduced to writing, captures the authentic common expression of the

4 *see* in *Pajzs, Czáky, Esterházy* (Preliminary Objections) 1936, p.9; *Panevezys–Saldutiskis Railway Case* (Preliminary Objections) 1938, p.7; *Right of Passage over Indian Territory* (Preliminary Objections) 1957, p.150; *Barcelona Traction, Light and Power Co. Ltd* (Preliminary Objections) 1964, p.39*ff*; *Military and Paramilitary Activities in and against Nicaragua* (Jurisdiction) 1984, §76 and in (Merits) 1986, §§37–41; *Guyana v. Suriname Arbitration* (Preliminary Objections) 2005, *dispositif* (2); *Territorial and Maritime Dispute* (Preliminary Objections) 2007, §§45–52; and *Chagos MPA* (Application to Bifurcate Proceedings) 2013, *dispositif* (1)–(2).

5 VCLT 31, reflecting on this point also customary international law as per the guiding dictums inter alia in *Arbitral Award of 31 July 1989* (Judgment) 1991, §48; *Oil Platforms* (Preliminary Objections) 1996, §23; *Kasikili/Sedudu Island* (Judgment) 1999, §18; *Sovereignty over Pulau Ligitan and Pulau Sipadan* (Judgment) 2002, §37; *Responsibilities and Obligations of States Sponsoring Persons and Entities with Respect to Activities in the Area* (Advisory Opinion) 2011, §57; *Alleged Violations of Sovereign Rights and Maritime Spaces in the Caribbean Sea* (Preliminary Objections) 2016, §35.

parties.[6] Accordingly, the process of interpretation, i.e. the argumentation to discern and assign legal meaning, as well as to determine the legal effect and purpose of a written rule, begins with elucidating the treaty text itself, rather than inquiring *ab initio* into the individual intentions of the parties.[7] The text-centred method pursued in this book was not confined just to a plain reading of the compatibility rule. In addition to parsing its linguistic statement in terms of grammatical composition and syntactic logic, the rule's morphology was closely scrutinized to examine the linguistic structure of article 7 FSA in the context of the Agreement's treaty text as a whole.[8]

The 'structure of article', as an interpretative concept, has been acknowledged on several occasions in the juridical logic of courts and tribunals, both in substantive and procedural contexts. It has been practically applied to construe the exact legal meaning and assess the imposition of corresponding obligations,[9] as well as to establish the degree and quality of their ensuing bindingness.[10] Additionally, it has been used to examine the drafting intention evident in both the structure of an article and the wording of its provisions,[11] and to confirm that such interpretation neither contradicts nor is in any other way discordant with the structure built into the article itself.[12] Furthermore, considerable emphasis has been placed on the structure of articles for justifying the construction of procedural preconditions contained in jurisdictional

[6] *see* in *Territorial Dispute* (Judgment) 1994, §41; *Boundary Dispute between Argentina and Chile concerning the Delimitation of the Frontier Line between Boundary Post 62 and Mount Fitzroy* (Decision) 1994, p.104; and *Taxes on Alcoholic Beverages–Japan* (Appellate) 1996, p.17.

[7] A/CN.4/191 'ILC Draft Articles on the Law of Treaties with Commentaries' (4/5 to 19/7/1966) [1966*YbILC*(II) 220–223].

[8] The morphology of the text, which comprises also the structural formation upon which a written rule rests, can furnish crucial material for interpretative exercises; *qv.* Focsaneanu, 'Les Langues Comme Moyen d'Expression du Droit International', *Annuaire Français* (1970) 256 followed by de Casadevante-Romani, *Sovereignty and Interpretation of International Norms* (2007) 40–41.

[9] *see* in *Asylum Case* (Judgment) 1950, p.279; *Application of the Convention on the Prevention and Punishment of the Crime of Genocide* (Merits) 2007, §174; and *Application of the Interim Accord of 13 September 1995* (Judgment) 2011, §97.

[10] *Oil Platforms* (Merits) 2003, §81.

[11] *see* in *Anglo-French Continental Shelf Arbitration* (Decision) 1977, §64 and *Request for Interpretation of the Judgment of 11 June 1998 in the Case concerning the Land and Maritime Boundary between Cameroon and Nigeria* (Judgment) 1999, §12.

[12] *see* in *Constitution of the Maritime Safety Committee of the Inter-Governmental Maritime Consultative Organization* (Advisory Opinion) 1960, pp.150, & 160–166 and *OSPAR Arbitration* (Decision) 2003, §134.

clauses—most notably, the expiration of time limits before unilaterally referring a dispute to settlement procedures[13]—or for assigning specific meaning to terms employed in the wording of its provisions, aiding in the determination of either *prima facie* or general compulsory jurisdiction.[14]

As a significant step in the progressive development of conservation and management law, the substantive formulation of the compatibility rule was made contingent upon the strict requirement that any disagreement over CMMs be submitted to compulsory third-party procedures entailing binding decisions. The principle of compulsory adjudication is intrinsically enshrined as a *conditio sine qua non*, shaping the structure of article 7 FSA through a textual pattern of clausal embeddedness, which renders the procedural stipulations inseparable from its substance. This textual morphology, in turn, reflects the declarative construction of the rule, conveying to prospective interpreters the *totum pro parte* impression of its *sui generis* legal character. It is thus the substantive idiosyncrasy of the compatibility rule itself that originally necessitated this legal drafting, and not the other way around; as if the embedded clause in paragraph (4), peculiar as it may seem, were a textual defect impinging on the clarity or determinacy of article 7 FSA. On the contrary, the compulsoriness principle—serving as an indispensable condition for the application of a progressively developed rule—has been acknowledged in the legislative history of articles 6(5), 7(4) and 8(2) FCC, as well as articles 74/83(2) LOSC. Moreover, the available jurisprudential pronouncements on the procedural stipulations in the latter set of provisions have consistently upheld this correlation. All these provisions have been purposefully constructed as the articles to which form part, respectively, were adopted under the compulsoriness principle. The perception of inseparability between the substantive and procedural constituent elements was reinforced in the structural formation of these articles through the use of embedded clauses.

Paragraph (4) is the cornerstone provision that underpins the formulation of the compatibility rule in article 7 FSA, rather than a cosmetic provision destined to suffer the interpretative fate of a *clausula vel dispositio inutilis per præsumptionem remotam, vel causam ex post facto non fulcitur*.[15] Contrary to

13 *Application of the International Convention on the Elimination of all Forms of Racial Discrimination* (Provisional Measures) 2008, §114.

14 *see* in *Nottebohm Case* (Preliminary Objections) 1953, pp.121–122; *Aegean Sea Continental Shelf* (Jurisdiction) 1978, §56; and *Maritime Delimitation in the Indian Ocean* (Preliminary Objections) 2017, §121.

15 According to Bacon's traditional Regula XXI, this construction canon posits that a useless clause or disposition is not supported by a remote presumption or by a cause arising afterwards, which may induce an operation of idle words; *qv.* Montagu, *The Works of Francis Bacon* (1826) 66–70.

such suspect reading that only reduces a provision to surplusage, a meaningful and certain cause that fulfils the object and purpose of the treaty shall be given in accordance with the principle of effectiveness. As expressed through the maxim *ut res magis valeat quam pereat*, which has been invariably affirmed in international jurisprudence, the forum is not at liberty to adopt an interpretation that would render any part of the treaty text redundant or ineffective.[16] The text is presumed to have been intended to carry the fullest legal weight and practical effect that is consistent with its wording, insofar as the treaty language is not strained and the interpretative function remains within its proper bounds. Accordingly, where the rule is deemed ambiguous, a textual construction should be sought that affords a practical effect to its provisions, whereas otherwise there would be none, based on the legitimate assumption of a shared expectation and common intention among the parties,[17] which is contemporaneous with the treaty's adoption.[18]

This approach likewise applies to jurisdictional clauses in instances of uncertainty regarding their appropriate procedural effect. Where a grammatical reading proves inconclusive, such clauses should be interpreted in a natural and reasonable manner that is faithful to the letter and spirit of the treaty. Due consideration should be given to the original intention of the parties, as that is expressed in the actual wording, to ensure that the forum neither exceeds its conferred jurisdiction nor fails to exercise it fully.[19] ITLOS may prove to

---

[16] *see* in *Cayuga Indians* (Award) 1926, p.184; *Competence of Assembly regarding Admission to the United Nations* (Advisory Opinion) 1950, p.8; *Territorial Dispute* (Judgment) 1994, §41; *Standards for Reformulated and Conventional Gasoline–United States* (Appellate) 1996, p.23; *Arbitration regarding the Iron Rhine Railway* (Award) 2005, §49 and for an expanded analysis together with a broad array of other cases addressing the principle *see* Fitzmaurice, 'The Law and Procedure of the International Court of Justice: Treaty interpretation and certain other treaty points', BYBIL (1951) 8*ff*; Jennings–Watts, *Oppenheim's International Law* (1992) 1280; Fitzmaurice, 'Dynamic (Evolutive) Interpretation of Treaties', *Hague Yearbook of International Law* (2008) 113–120; and Lekkas–Merkouris–Peat, 'The Interpretative Practice of the International Court of Justice', *Yearbook UN Law* (2023) 352–356.

[17] *see* Merrills, 'Two Approaches to Treaty Interpretation', *Aust.* YbIL (1968) 56; and *compare* McDougal, 'The International Law Commission's Draft Articles Upon Interpretation: Textuality *redivivus*', AJIL (1967) 992 *with* Fitzmaurice, '*Vae Victis* or Woe to the Negotiators! Your Treaty or Our Interpretation of It', AJIL (1971) 358.

[18] *Dispute Regarding Navigational and Related Rights* (Judgment) 2009, §63.

[19] *see* in *Free Zones of Upper Savoy and the District of Gex* (Order) 1929, p.13 followed by *Corfu Channel* (Merits) 1949, pp.23–24; *Interpretation of Peace Treaties (Second phase)* (Advisory Opinion) 1950, p.229; *Anglo-Iranian*

be a forum inclined to assert comprehensive jurisdiction, as it has commendably done in the past, by reasoning that failing to resolve an impasse between States would be contrary to the object and purpose of the dispute settlement procedures in Part XV LOSC,[20] or—in particular circumstances—it has found that declining to address the merits of a dispute would be at odds with justice.[21]

It is worth emphasizing, however, that the principle of effectiveness, on its own, does not furnish a ready-made solution in resolving interpretive tensions between a primary rule and its exclusions.[22] Assuming, *arguendo* if need be, that such a tension is latently present in the interplay between articles 7(4) and 32 FSA, owing to the constructive ambiguity regarding the rule's spatial scope, the ultimate solution will depend not only on the wording of the specific provisions but also on the object and purpose of the treaty. But what is the object and purpose in light of which articles 7 and 32 FSA should be interpreted and applied, bearing in mind that article 4 FSA requires the Agreement to 'be interpreted and applied in the context of and in a manner consistent with the Convention'? As has been pointedly argued elsewhere, the Convention does not clearly define its object and purpose, leaving it open to purposive interpretations that are both reflective of its underlying principles and values—even where these may be only imperfectly captured in its provisions—and responsive to developments in general international law.[23] In marked contrast, the Agreement utterly expresses in article 2 FSA the objective of ensuring 'the long-term conservation and sustainable use of straddling fish stocks and highly migratory fish stocks through effective implementation of the relevant provisions of the Convention'. In the words of Chairman Nandan 'the long-term

Oil Co. Case (Preliminary Objections) 1952, p.104; *Fisheries Jurisdiction Cases* (Jurisdiction) 1973, §§17–23; *United States Diplomatic and Consular Staff in Tehran* (Jurisdiction and Admissibility) 1980, §§45–54; and *Continental Shelf* (Judgment) 1985, §19; and *see* further in Lauterpacht, 'Restrictive Interpretation and the Principle of Effectiveness in the Interpretation of Treaties', BYBIL (1949) 71–75; McNair, *The Law of Treaties* (1961) 365; and Shearer, *Starke's International Law* (1994) 480–481.

20 *Delimitation of the Maritime Boundary in the Bay of Bengal* (Judgment) 2012, §§391–394.

21 *The "M/V Saiga" Case (No.2)* (Judgment) 1999, §73(d).

22 *see* Lauterpacht, *The Development of International Law by the International Court* (1958) 229 and Orakhelashvili, *The Interpretation of Acts and Rules in Public International Law* (2008) 424–425.

23 *see* Lijnzaad, 'As Time Goes By: A preliminary inquiry into the "object and purpose" of the Law of the Sea Convention', in Siig–Feldtmann–Billing (Eds) *The United Nations Convention on the Law of the Sea* (2023) 264 and Lewis, 'The "Constitution for the Oceans"? The Law of the Sea Convention as a Living Treaty', ICLQ (2025) 26*ff.*

sustainability of fishery resources reflects a strong common purpose [among the negotiating parties], and the end result [that they] achieved in [adopting] the Agreement'.[24] This is the animating legal *telos* that gives meaning to the treaty text, as a whole, and which in the specific context of the compatibility rule manifests itself in the obligation of States to cooperate in order to achieve compatible CMMs and preserve the biological unity of the stocks across the full extent of their distribution and migratory range. This is 'precisely the purpose' of article 7 FSA.[25]

This book therefore has advanced an interpretation that aligns with both the language and structure of article 7 FSA. A purely grammatical reading of the fisheries limitation precludes an *a priori* presumption that compatibility disputes are excluded from the compulsory procedures entailing binding decisions of section 2 Part XV LOSC, which are to apply *mutatis mutandis* under Part VIII FSA. A reference made under paragraph (4) of article 7 FSA is not caught *ipso facto* by article 32 FSA. To determine the applicable scope of article 32 FSA, the forum seised of the case must acknowledge the interpretative weight carried by the embedded clause and then proceed to examine the legal relationship between the rule and the limitation. By intendment of the progressively developed law to 'ensure the long-term conservation and sustainable use' of fish stocks and, in furtherance of that objective, to also ensure cooperation between States even amidst disagreement, article 32 FSA should be given a construction that restricts its scope solely to domestic stocks or to EEZ components of CMMs that have been categorically and explicitly reserved in their entirety to the coastal State's jurisdiction. Any broader interpretation of these exclusions may undermine the objective of achieving compatible CMMs across jurisdictional regimes, and will ultimately defeat the very object and purpose of the Agreement.

In pursuing the legal *telos* based on the object and purpose of the Agreement, recourse to the general principle of equity may play a critical role in interpreting the compatibility rule *cas d'espèce*. While often described as largely 'subjective or intuitive' and typically defined in similarly elusive terms—such as fairness, justice, and reasonableness, which lack objectively identifiable normative content or concrete meaning when considered in isolation from their specific factual context—equity has long been recognized as an influential legal notion permeating judicial reasoning.[26] When applied *infra legem*, it

---

[24] Nandan–Dalaker, *Reflections on the Making of the Modern Law of the Sea* (2021) 213.

[25] Edeson–Freestone–Gudmundsdottir, *Legislating for Sustainable Fisheries* (2001) 35.

[26] *see* Schachter, 'The Nature and Process of Legal Development in International Society', in Macdonald–Johnston (Eds) *The Structure and Process*

allows the forum to choose from among several possible interpretations of the law the one that, in light of the circumstances of the case, produces an equitable result. As one commentator aptly observed, equity *infra legem* serves as the link between a specific rule, or provision of law, and the realities to which it applies.[27] This should not be mistaken for a legalistic subterfuge to arbitrarily reason *ex aequo et bono*, as its invocation is grounded in considerations of equity inherent to the normative character of the rule itself; that is, arising from within its own structure.[28] On this crucial methodological distinction, ICJ has clarified that, 'it is not a matter of finding simply an equitable solution, but an equitable solution derived from the applicable law … it is not a question of applying equity simply as a matter of abstract justice, but of applying a rule of law which itself requires the application of equitable principles'.[29]

Equity *infra legem* can be pertinently applied to rules of a *jus aequum* legal nature. In contrast to rules of a *jus strictum* nature, which must be interpreted strictly as they establish absolute rights, *jus aequum* rules provide for relative rights, requiring their exercise to be fair, just, and reasonable. As such, they are amenable to equitable interpretations.[30] The claim that the law governing the conservation of natural resources is remote from *jus aequum* because 'the pledges of good faith [therein] are not pledges of "good faith" in its true connotative sense'[31] is not a viable argument against applying equity *infra legem*. The law of the sea has gradually evolved into a public order that moderates State authority by balancing national and international interests. Its purpose is to optimize ocean utilization to the greatest extent possible, while maintaining a secure, adaptable, and equitable balance between exclusive and inclusive uses of the marine environment, without undermining sustainable development as

---

of International Law (1983) 797; Lowe, 'The Role of Equity in International Law', *Aust.* YbIL (1988) 54; and Titi, *The Function of Equity in International Law* (2021) 76*ff.*

27 Lachs, 'Equity in Arbitration and in Judicial Settlement of Disputes', LJIL (1993) 326.

28 *see* in *Free Zones of Upper Savoy and District of Gex (Second Phase)* (Order) 1930, pp.10–11; *Continental Shelf* (Judgment) 1982, §71; and *Continental Shelf* (Judgment) 1985, §45.

29 *Fisheries Jurisdiction Cases* (Merits) 1974, §78 in affirming its dictum previously stated in *North Sea Continental Shelf Cases* (Judgment) 1969, §88; and followed in *Frontier Dispute* (Judgment) 1986, §28.

30 Schwarzenberger, *The Dynamics of International Law* (1976) 32–55 & 56 *et seq.*

31 Chattopadhyay, 'Equity in International Law: Its growth and development', *Georgia Journal of International and Comparative Law* (1975) 401.

a communal aspiration.[32] In this ongoing process, equity has become particularly relevant in regulating common-pool marine living resources, especially considering the challenges that the Convention's zoning governance model poses to their cooperative conservation and management.[33]

While good faith does not, in and of itself, create rights—and corresponding *erga omnes partes* obligations—where none otherwise exist, it remains a fundamental equitable principle that governs their performance and guards against abusive practices.[34] In this regard, a fair, just, and reasonable exercise of a right is one that is appropriate and necessary to fulfil the purpose for which it is conferred by the treaty. Exercising such a right in a manner that prejudices the interests of another contracting party should be regarded as inconsistent with the execution of treaty obligations and may amount to a *mala fides* breach of the treaty.[35] In the fisheries context, the doctrine of *abus des droits* was explicitly invoked by both sides in the *Swordfish stocks* dispute.[36] Notably, the Agreement unequivocally calls upon its parties, and imposes a

---

32 *see* McDougal–Burke, *The Public Order of the Oceans* (1962) 1–14; Pardo, 'The Law of the Sea: Its past and its future', OLR (1984) 16–17; Payoyo, *Cries of the Sea* (1997) 71*ff*; Edeson, 'Towards Long-term Sustainable Use: Some recent developments in the legal regime of fisheries', in Boyle–Freestone (Eds) *International Law and Sustainable Development* (1999) 172–173; Kreća, 'A Few Remarks on the Theoretical Basis of the New Law of the Sea', in Nisuke–McWhinney–Wolfrum (Eds) *Liber Amicorum Judge Shigeru Oda* (2002) 1209–1214; and Mendenhall, 'Drivers and Mechanisms of Change in UNCLOS: An international relations research agenda', in Alger–Sumaila (Eds) *A Research Agenda for Sustainable Ocean Governance* (2025) 135.

33 Franck–Sughrue, 'The International Role of Equity-as-Fairness', *Georgetown Law Journal* (1993) 572 & 582*ff*; and Schäli, 'Intergenerational Justice and the Concept of Common Concern in Marine Resource Allocation and Ocean Governance', in Cottier–Lalani–Siziba (Eds) *Intergenerational Equity* (2019) 74*ff*.

34 *see* in *Nuclear Tests* (Judgment) 1974, §46; *Border and Transborder Armed Actions* (Jurisdiction and Admissibility) 1988, §94; *Import Prohibition of Certain Shrimp and Shrimp Products–United States* (Appellate) 1998, §158; *The "M/V Louisa" Case* (Judgment) 2013, §137; *The "M/V Virginia G" Case* (Judgment) 2014, §§396–399; and *The "M/V Norstar" Case* (Merits) 2019, §241.

35 Lauterpacht, *The Function of Law in the International Community* (1933) 294–314; Cheng, *General Principles of Law as applied by International Courts and Tribunals* (1953) 121–136; Kiss, *L'Abus de Droit en Droit International* (1953) 184*ff*; Taylor, 'The Content of the Rule against Abuse of Rights in International Law', BYBIL (1972) 323; Kolb, 'Principles as Sources of International Law (with Special Reference to Good Faith)', *Netherlands International Law Review* (2006) 24 *et seq*; and Paulsson, *The Unruly Notion of Abuse of Rights* (2020) 70–73.

36 *Conservation and Sustainable Exploitation of Swordfish Stocks* (Constitution of Chamber) 2000, §2(3) *litt.*(d) & (g).

positive obligation on them, to fulfil 'in good faith the obligations assumed' and 'exercise [their] rights recognized ... in a manner which would not constitute an abuse of right'.[37] The procedural stipulation embedded in paragraph (4) of article 7 FSA safeguards the balance of interests between DWFS and coastal States, particularly by forestalling potential abuses of rights within the respective jurisdictional regimes. By serving as a linchpin for the compulsory settlement of disputes arising from disagreements over the spatial scope of the compatibility rule, the provision encourages greater restraint in auto-interpretation and helps deter unilateral actions by States that might otherwise too lightly discharge their duty of cooperation in achieving effective CMMs spanning the full distribution and migratory range of stocks.

In the years ahead, the dispute settlement process—particularly by means of third-party adjudication—is poised to play an increasingly constructive role in resolving compatibility disputes. Although disagreement may be inevitable owing to the complexity of conserving and managing these resources, it should not bring an end to efforts to achieve compatible CMMs. In the era of elusive sustainability, it will be imperative that such disagreements be addressed promptly and objectively through recourse to compulsory procedures entailing binding decisions that promote a comprehensive solution, help to mitigate friction, and foster mutual trust among the parties involved. Ultimately, this will ensure effective and equitable cooperation between DWFS and coastal States, advancing their common interest in the sustainable regulation of straddling and highly migratory fish stocks.

[37] FSA 34.

Selected bibliography

Adede, Andronico, 'Settlement of Disputes Arising under the Law of the Sea Convention', AJIL 69(4) 798–818 (1975)

Adede, Andronico, 'Prolegomena to the Disputes Settlement Part of the Law of the Sea Convention', NYUJ *Int'l LP* 10(2) 253–394 (1977)

Adede, Andronico, 'Toward the Formulation of the Rule of Delimitation of Sea Boundaries Between States with Adjacent or Opposite Coasts', VJIL 19(2) 207–256 (1978)

Adede, Andronico, 'Law of the Sea – The Integration of the System of Settlement of Disputes under the Draft Convention as a Whole', AJIL 72(1) 84–112 (1978)

Adede, Andronico, 'Streamlining the System for Settlement of Disputes under the Law of the Sea Convention', *Pace Law Review* 1(1) 15–58 (1980)

Adede, Andronico, 'The Basic Structure of the Disputes Settlement Part of the Law of the Sea Convention', ODIL 11(1/2) 125–148 (1982)

Adede, Andronico, *The System for Settlement of Disputes under the United Nations Convention on the Law of the Sea: A drafting history and a commentary* (Nijhoff, 1987)

Aguilar, Andrés, 'The Patrimonial Sea or Economic Zone Concept', SDLR 11(3) 579–602 (1974)

Ahwoi, Kwamena, 'The Exclusive Economic Zone: An analysis of the concept as contained in the Revised Single Negotiations Text', *University of Ghana Law Journal* 14(2) 137–164 (1977)

Aikman, Colin, 'Island Nations of the South Pacific and Jurisdiction over Highly Migratory Species', VUWLR 17(1) 101–124 (1987)

Akaha, Tsuneo, 'From Conflict to Cooperation: Fishery relations in the Sea of Japan', PRL&PJ 1(2) 225–280 (1992)

Akaha, Tsuneo, 'Japan–South Korea Fishery Agreement of 1998: Pursuing pragmatic interests without compromising sovereignty', 249–285 in Harry Scheiber (Ed.), *Law of the Sea: The common heritage and emerging challenges* (Nijhoff, 2000)

Alexander, Lewis, 'Regionalism and the Law of the Sea: The case of semi-enclosed seas', ODIL 2(2) 151–186 (1974)

Alexander, Lewis – Hodgson, Robert, 'The Impact of the 200-Mile Economic Zone on the Law of the Sea', SDLR 12(3) 569–599 (1975)

Allen, Edward, 'Control of Fisheries Beyond Three Miles', *Washington* LRSBJ 14(2) 91–98 (1939)

Allen, Edward, 'Legal Limits of Coastal Fishery Protection', *Washington* LRSBJ 21(1) 1–4 (1946)

Allen, Stephen, 'Article 297 of the United Nations Convention on the Law of the Sea and the Scope of Mandatory Jurisdiction', ODIL 48(3/4) 313–330 (2017)

Allott, Phillip, 'Power Sharing in the Law of the Sea', AJIL 77(1) 1–30 (1983)

Allott, Phillip, 'Interpretation – An Exact Art', 373–392 in Andrea Bianchi – Daniel Peat – Matthew Windsor (Eds), *Interpretation in International Law* (OUP, 2015)
Anand, Ram, 'Tyranny of the Freedom of the Seas Doctrine', *International Studies* 12(3) 416–429 (1973)
Anderson, Chandler, 'The Final Outcome of the Fisheries Arbitration', AJIL 7(1) 1–16 (1913)
Anderson, David, 'The Straddling Stocks Agreement of 1995: An initial assessment', ICLQ 45(2) 463–475 (1996)
Anderson, David, *Modern Law of the Sea: Selected essays* (Nijhoff, 2008)
Anderson, David, 'Peaceful Settlement of Disputes under UNCLOS', 385–416 in Jill Barrett – Richard Barnes (Eds), *Law of the Sea: UNCLOS as a living treaty* (British Institute of International and Comparative Law, 2016)
Anderson, Winston, *The Law of the Sea in the Caribbean* (Brill–Nijhoff, 2022)
Applebaum, Bob, 'The UN Conference on Straddling Fish Stocks and Highly Migratory Fish Stocks: The current Canadian perspective', 299–303 in Myron Nordquist – John Moore (Eds), *Entry into Force of the Law of the Sea Convention* (Nijhoff, 1995)
Aqorau Transform, 'The Draft Convention for the Conservation and Management of Highly Migratory Fish Stocks in the Western and Central Pacific Ocean', IJMCL 15(1) 111–150 (2000)
Attard, David, *The Exclusive Economic Zone in International Law* (Clarendon Press, 1987)
Azmi, Kamal – Hanich, Quentin, 'Mapping Interests in the Tuna Fisheries of the Western and Central Pacific Ocean', OCM 212 (105779) 1–16 (2021)
Ball, Milner, 'Law of the Sea: Expression of solidarity', SDLR 19(3) 461–473 (1982)
Balton, David, 'Strengthening the Law of the Sea: The new agreement on straddling fish stocks and highly migratory fish stocks', ODIL 27(2) 125–151 (1996)
Balton, David, 'Implementing the New Arctic Fisheries Agreement', 429–445 in Tomas Heidar (Ed.), *New Knowledge and Changing Circumstances in the Law of the Sea* (Brill–Nijhoff, 2020)
Bankes, Nigel, 'Legislative and Enforcement Jurisdiction of the Coastal State with respect to Fisheries in the Exclusive Economic Zone', 73–103 in Jensen Øystein (Ed.), *The Development of the Law of the Sea Convention: The role of international courts and tribunals* (EEP, 2021)
Bankes, Nigel, 'The Jurisdiction of the Dispute Settlement Bodies of the Law of the Sea Convention with Respect to Other Treaties', ODIL 52(4) 346–380 (2021)
Bardonnet, Daniel, 'Frontières Terrestres et Frontières Maritimes', *Annuaire Français* [35] 1–64 (1989)
Barnes, Richard, 'Fisheries and Marine Biodiversity', 542–563 in Malgosia Fitzmaurice – David Ong – Panos Merkouris (Eds), *Research Handbook on International Environmental Law* (EEP, 2010)
Barnes, Richard – Rosello, Mercedes, 'Fisheries and Maritime Security: Understanding and enhancing the connection', 48–82 in Malcolm Evans – Sofia Galani (Eds), *Maritime Security and the Law of the Sea: Help or hindrance?* (EEP, 2020)
Barston, Ronald, 'United Nations Conference on Straddling and Highly Migratory Fish Stocks', *Marine Policy* 19(2) 159–166 (1995)
Barston, Ronald, 'The Law of the Sea and Regional Fisheries Organisations', IJMCL 14(3) 333–352 (1999)
Beckman, Robert – Sim, Christine, 'Maritime Boundary Disputes and Compulsory Dispute Settlement: Recent developments and unresolved issues', 228–250 in

Myron Nordquist – John Moore – Ronán Long (Eds), *Legal Order in the World's Oceans* (Brill–Nijhoff, 2017)
Beringen, Ethan, 'International Fisheries as the "Whale in the Room" at the Negotiations for a New Instrument for Biodiversity Beyond National Jurisdiction', 108–141 in Liu Nengye – Scott Shirley (Eds), *The Law of the Sea and the Planetary Crisis* (Routledge, 2025)
Bernhardt, Peter, 'Compulsory Dispute Settlement in the Law of the Sea Negotiations: A reassessment', VJIL 19(1) 69–105 (1978)
Bianchi, Andrea – Zarbiyev, Fuad, *Demystifying Treaty Interpretation* (CUP, 2024)
Bishop, William, 'International Law Commission Draft Articles on Fisheries', AJIL 50(3) 627–636 (1956)
Bishop, William, 'The 1958 Geneva Convention on Fishing and Conservation of the Living Resources of the High Seas', *Columbia Law Review* 62(7) 1206–1229 (1962)
Bjørndal, Trond – Kaitala, Veijo – Munro, Gordon – Lindroos, Marko, 'The Management of High Seas Fisheries', *Annals of Operational Research* 94(1) 183–196 (2000)
Bjørndal, Trond – Munro, Gordon, *Brexit and the Consequences for Fisheries Management in the North Sea* (Centre for Applied Research at Norwegian School of Economics, 2020)
Borg, Simone, *Conservation on the High Seas: Harmonizing international regimes for the sustainable use of living resources* (EEP, 2012)
Borg, Simone, 'The Influence of International Case Law on Aspects of International Law Relating to the Conservation of Living Marine Resources Beyond National Jurisdiction', YbIEL [23] 44–79 (2012)
Botet, Violanda, 'Filling in One of the Last Pieces of the Ocean: Regulating tuna in the western and central Pacific Ocean', VJIL 41(4) 787–814 (2001)
Bowett, Derek, 'The Second United Nations Conference on the Law of the Sea', ICLQ 9(3) 415–435 (1960)
Bowett, Derek, *The Law of the Sea* (MUP, 1967)
Boyle, Alan, 'The Principle of Co-operation: The environment', 120–136 in Vaughan Lowe – Colin Warbrick (Eds), *The United Nations and the Principles of International Law: Essays in memory of Michael Akehurst* (Routledge, 1994)
Boyle, Alan, 'Dispute Settlement and the Law of the Sea Convention: Problems of fragmentation and jurisdiction', ICLQ 46(1) 37–54 (1997)
Boyle, Alan, 'Problems of Compulsory Jurisdiction and the Settlement of Disputes Relating to Straddling Fish Stocks', IJMCL 14(1) 1–25 (1999)
Boyle, Alan, 'Problems of Compulsory Jurisdiction and the Settlement of Disputes Relating to Straddling Fish Stocks', 91–120 in Olav Stokke (Ed.), *Governing High Seas Fisheries: The Interplay of global and regional regimes* (OUP, 2001)
Boyle, Alan, 'The *Southern Bluefin Tuna* Arbitration', ICLQ 50(2) 447–452 (2001)
Boyle, Alan, 'Further Development of the Law of the Sea Convention: Mechanisms for change', ICLQ 54(3) 563–584 (2005)
Boyle, Alan, 'UNCLOS Dispute Settlement and the Uses and Abuses of Part XV', *Revue Belge De Droit International* 47(1) 182–204 (2014)
Boyle, Alan – Redgwell, Catherine, *Birnie, Boyle, and Redgwell's International Law and the Environment* (OUP, 2021 4th edition)
Brodd, Mary, 'A Common Heritage Approach to Fisheries through Regional Controls', NYUJ *Int'l LP* 10(1) 171–202 (1977)
Brown, Chester, *A Common Law of International Adjudication* (OUP, 2007)

Brown, Edward, 'The Continental Shelf and the Exclusive Economic Zone: The problem of delimitation at UNCLOS/III', *Maritime Policy and Management* 4(6) 377–408 (1977)

Brown, Edward, 'Delimitation of Offshore Areas: Hard labour and bitter fruits at UNCLOS/III', *Marine Policy* 5(3) 172–184 (1981)

Brown, Edward, *The International Law of the Sea*, Volume I (Dartmouth Publishing, 1994)

Brown, Edward, 'Dispute Settlement and the Law of the Sea: The UN convention regime', *Marine Policy* 21(1) 17–43 (1997)

Brown, Howard, 'United Nations Conference on Straddling Fish Stocks and Highly Migratory Fish Stocks: An analysis of international environmental law and the conference's final agreement', *Vermont Law Review* 21(2) 547–590 (1996)

Brus, Marcel, *Third Party Dispute Settlement in an Interdependent World: Developing a theoretical framework* (Nijhoff, 1995)

Buga, Irina, 'Between Stability and Change in the Law of the Sea Convention: Subsequent practice, treaty modification, and regime interaction', 46–68 in Donald Rothwell – Alex Oude-Elferink – Karen Scott – Tim Stephens (Eds), *Oxford Handbook of the Law of the Sea* (OUP, 2015)

Burke, William, 'Some Comments on the 1958 Conventions', *American Society of International Law Proceedings* [53] 197–206 (1959)

Burke, William, *The Soviet Union and the Law of the Sea* (Johns Hopkins Press, 1971)

Burke, William, 'Highly Migratory Species in the New Law of the Sea', ODIL 14(3) 273–314 (1984)

Burke, William, 'The Law of the Sea Convention Provisions on Conditions of Access to Fisheries Subject to National Jurisdiction', OLR 63(1) 73–120 (1984)

Burke, William, 'Coastal State Fishery Regulation under International Law: A comment on the *La Bretagne* Award of July 17, 1986 (The Arbitration between Canada and France)', SDLR 25(3) 495–534 (1988)

Burke, William, 'Fishing in the Bering Sea Donut: Straddling stocks and the new international law of fisheries', ELQ 16(1) 285–310 (1989)

Burke, William, *Fisheries Regulations under Extended Jurisdiction and International Law* (FAO, 1992)

Burke, William, *The New International Law of Fisheries, UNCLOS 1982 and Beyond* (Clarendon Press, 1994)

Burke, William, 'Importance of the 1982 UN Convention on the Law of the Sea and its Future Development', ODIL 27(1) 1–4 (1996)

Burke, William, 'Compatibility and Precaution in the 1995 Straddling Stock Agreement', 105–126 in Harry Scheiber (Ed.), *Law of the Sea: The common heritage and emerging challenges* (Nijhoff, 2000)

Burke, William, 'Evolution in the Fisheries Provisions of UNCLOS', 1355–1362 in Nisuke Ando – Edward McWhinney – Rüdiger Wolfrum (Eds), *Liber Amicorum Judge Shigeru Oda* (Kluwer, 2002)

Buzan, Barry, '"*United We Stand* …": Informal negotiating groups at UNCLOS/III', *Marine Policy* 4(3) 183–204 (1980)

Caddy, John – Griffiths, Ray, *Living Marine Resources and their Sustainable Development: Some environmental and institutional perspectives* (FAO, 1995)

Caminos, Hugo, 'The Jurisdiction and Procedure of the International Tribunal for the Law of the Sea: An overview', 257–273 in Jon Van Dyke – Sherry Broder – Seokwoo Lee – Jin-Hyun Paik (Eds), *Governing Ocean Resources: New challenges and emerging regimes* (Brill, 2013)

Cardoso, Fernando, 'Considerations on Some Global Institutional Challenges within the Context of the Conservation and Management of Marine Living Resources', 111–120 in Marta Ribeiro – Fernando Bastos – Tore Henriksen (Eds), *Global Challenges and the Law of the Sea* (Springer, 2020)
Carr, Christopher, 'Transformations in the Law Governing Highly Migratory Species: 1970 to the present', 55–94 in David Caron – Harry Scheiber (Eds), *Bringing New Law to Ocean Waters* (Brill, 2004)
Carstensen, Nils, 'A Re-Internationalisation of Dispute Settlement in Law of the Sea', ZaöRV [62] 73–76 (2002)
Casado-Raigón, Rafael, *La Pesca en Alta Mar* (Junta de Andalucía, Consejería de Agricultura y Pesca, 1994)
Cassese, Antonio, *International Law* (OUP, 2005 2nd edition)
Castilla, Juan-Carlos – Orrego-Vicuña, Francisco, 'Highly Migratory Species and the Coordination of Fishery Policies within Certain Exclusive Economic Zones: The South Pacific', *Ocean Management* 9(1) 21–33 (1984)
Chandra, Satish, *Law of the Sea* (Mittal Publications, 1985)
Charney, Jonathan, 'Entry into Force of the 1982 Convention on the Law of the Sea', VJIL 35(2) 381–404 (1995)
Charney, Jonathan, 'The Implications of Expanding International Dispute Settlement Systems: The 1982 Convention on the Law of the Sea', AJIL 90(1) 69–75 (1996)
Chattopadhyay, Subir, 'Equity in International Law: Its growth and development', *Georgia Journal of International and Comparative Law* 5(2) 381–406 (1975)
Cheng, Bin, *General Principles of Law as applied by International Courts and Tribunals* (Stevens & Sons, 1953)
Cheng, Bin, 'On the Nature and Sources of International Law', 203–233 in Bin Cheng (Ed.), *International Law: Teaching and practice* (Stevens & Sons, 1982)
Cheung, William – Pauly, Daniel – Sumaila, Rashid, 'Hope or Despair Revisited: Assessing progress and new challenges in global fisheries', *Fish and Fisheries* 26(2) 257–269 (2025)
Cho, Hoon, 'Is Compulsory Jurisdiction Under UNCLOS Part XV Expanding? A Critical Review of the Claims', ODIL 55 (1/2) 185–205 (2024)
Christie, Donna, 'It Don't Come EEZ: The failure and future of coastal state fisheries management', *Journal of Transnational Law and Policy* 14(1) 1–36 (2004)
Christy, Francis, 'Fisheries and the New Conventions on the Law of the Sea', SDLR 7(3) 455–468 (1970)
Christy, Francis, 'Transitions in the Management and Distribution of International Fisheries', *International Organization* 31(2) 235–265 (1977)
Churchill, Robin, 'The Fisheries Jurisdiction Cases: The contribution of the International Court of Justice to the debate on coastal States' fisheries rights', ICLQ 24(1) 82–105 (1975)
Churchill, Robin, 'Legal Uncertainties in International High Seas Fisheries Management', *Fisheries Research* 37(1/3) 225–237 (1998)
Churchill, Robin, 'The Barents Sea Loophole Agreement: A "coastal State" solution to a straddling stock problem', IJMCL 14(4) 467–490 (1999)
Churchill, Robin, 'International Tribunal for the Law of the Sea the Southern Bluefin Tuna Cases (*New Zealand v. Japan; Australia v. Japan*): Order for Provisional Measures of 27 August 1999', ICLQ 49(4) 979–990 (2000)
Churchill, Robin, 'Managing Straddling Fish Stocks in the North-East Atlantic: A multiplicity of instruments and regime linkages – but how effective a management?',

235–272 in Olav Stokke (Ed.), *Governing High Seas Fisheries: The Interplay of global and regional regimes* (OUP, 2001)
Churchill, Robin, 'The Role of the International Court of Justice in Maritime Boundary Delimitation', 125–141 in Alex Oude-Elferink – Donald Rothwell (Eds), *Ocean Management in the 21st Century: Institutional frameworks and responses* (Nijhoff, 2004)
Churchill, Robin, 'Some Reflections on the Operation of the Dispute Settlement System of the UN Convention on the Law of the Sea During its First Decade', 388–416 in David Freestone – Richard Barnes – David Ong (Eds), *The Law of the Sea, Progress and Prospects* (OUP, 2006)
Churchill, Robin, 'The Jurisprudence of the International Tribunal for the Law of the Sea Relating to Fisheries: Is there much in the net?', IJMCL 22(3) 383–424 (2007)
Churchill, Robin, 'Trends in Dispute Settlement in the Law of the Sea: Towards the increasing availability of compulsory means', 143–172 in Duncan French – Matthew Saul – Nigel White (Eds), *International Law and Dispute Settlement: New problems and techniques* (Hart, 2012)
Churchill, Robin, 'The General Dispute Settlement System of the UN Convention on the Law of the Sea: Overview, context and use', ODIL 48(3/4) 217–239 (2017)
Churchill, Robin, 'Fisheries Management in European Union and United Kingdom Waters after Brexit: A change for the better?', *Ocean Yearbook* [36] 287–313 (2022)
Churchill, Robin – Lowe, Vaughan, *The Law of the Sea* (MUP, 1999 3rd edition)
Churchill, Robin – Lowe, Vaughan – Sander, Amy, *The Law of the Sea* (MUP, 2022 4th edition)
Churchill, Robin – Ulfstein, Geir, *Marine Management in Disputed Areas: The case of the Barents Sea* (Routledge, 1992)
Clingan, Thomas, 'Emerging Law of the Sea: The economic zone dilemma', SDLR 14(3) 530–547 (1977)
Clingan, Thomas, 'The Changing Global Pattern of Fisheries Management', *Lawyer of the Americas* 10(3) 658–685 (1978)
Clingan, Thomas, 'Mar Presencial: Deja vu all over again? A response to Francisco Orrego-Vicuña', ODIL 24(1) 93–97 (1993)
Cochrane, Kevern, 'Reconciling Sustainability, Economic Efficiency and Equity in Marine Fisheries: Has there been progress in the last 20 years?', *Fish and Fisheries* 22(2) 298–323 (2021)
Cochrane, Kevern – Garcia, Serge (Eds), *A Fishery Manager's Guidebook: Management measures and their application* (Wiley-Blackwell, 2009 2nd edition)
Colburn, Jamison, 'Turbot Wars: Straddling stocks, regime theory, and a new UN Agreement', *Journal of Transnational Law and Policy* 6(2) 323–366 (1997)
Collier, John – Lowe, Vaughan, *The Settlement of Disputes in International Law: Institutions and procedures* (OUP, 1999)
Collins, Richard, 'Navigating Choppy Waters: UNCLOS dispute settlement coming of age?', 137–160 in Russell Buchan – Daniel Franchini – Nicholas Tsagourias (Eds), *The Changing Character of International Dispute Settlement: Challenges and prospects* (CUP, 2023)
Colson, David – Hoyle, Peggy, 'Satisfying the Procedural Prerequisites to the Compulsory Dispute Settlement Mechanisms of the 1982 Law of the Sea Convention: Did the Southern Bluefin Tuna tribunal get it right?', ODIL 34(1) 59–82 (2003)
Copes, Parzival, 'The Impact of UNCLOS/III on Management of the World's Fisheries', *Marine Policy* 5(3) 217–228 (1981)

Craik, Neil, 'The Duty to Cooperate in International Environmental Law: Constraining State discretion through due respect', YbIEL [30] 22–44 (2019)
Crawford, James, *Brownlie's Principles of Public International Law* (OUP, 2019 9th edition)
Dahmani, Mohamed, *The Fisheries Regime of the Exclusive Economic Zone* (Nijhoff, 1987)
Dalby, Simon, *Rethinking Environmental Security* (EEP, 2022)
Dalton, Jane, 'The Chilean Mar Presencial: A harmless concept or a dangerous precedent?', IJMCL 8(3) 397–418 (1993)
Darwin, Henry, 'General Introduction', 57–75 in Humphrey Waldock (Ed.), *International Disputes: The legal aspects* (Europa Publications, 1972)
d'Aspremont, Jean, 'Towards an International Law of Brigandage: Interpretative engineering for the regulation of natural resources exploitation', *Asian* JIL 3(1) 1–24 (2013)
Davies, Paul, 'The EC/Canadian Fisheries Dispute in the Northwest Atlantic', ICLQ 44(4) 927–939 (1995)
Davies, Peter – Redgwell, Catherine, 'The International Legal Regulation of Straddling Fish Stocks', BYBIL [67] 199–274 (1997)
Davis, Ruth – Hanich, Quentin, 'Developing an Equitable and Ecosystem-Based Approach to Fisheries Management', 124–148 in Harry Scheiber – James Kraska – Moon-Sang Kwon (Eds), *Science, Technology, and New Challenges to Ocean Law* (Brill–Nijhoff, 2015)
Davis, Ruth – Hanich, Quentin, 'Transparency in Fisheries Conservation and Management Measures', *Marine Policy* [136] 104088–104104 (2022)
Day, Douglas, 'Tending the Achilles' Heel of NAFO: Canada acts to protect the nose and tail of the Grand Banks', *Marine Policy* 19(4) 257–270 (1995)
Day, Douglas, 'Addressing the Weakness of High Seas Fisheries Management in the Northwest Atlantic', OCM 35(2) 69–84 (1997)
de Brabandere, Eric, 'The Principles and Methods of International Dispute Settlement', 528–554 in Malcolm Evans (Ed.), *International Law* (OUP, 2024 6th edition)
de Casadevante-Romani, Carlos, *Sovereignty and Interpretation of International Norms* (Springer, 2007)
de Fontaubert, Charlotte, 'The Politics of Negotiation at the United Nations Conference on Straddling Fish Stocks and Highly Migratory Fish Stocks', OCM 29(2) 79–91 (1995)
de Fontaubert, Charlotte, 'The United Nations Conference on Straddling Fish Stocks and Highly Migratory Fish Stocks: Another step in the implementation of the Law of the Sea Convention', *Ocean Yearbook* [12] 82–91 (1996)
de Klemm, Cyril, 'Migratory Species in International Law', *Natural Resources Journal* 29(4) 935–978 (1989)
de la Fayette, Louise, 'The Fisheries Jurisdiction Case (Spain *v.* Canada), Judgment on Jurisdiction of 4 December 1998', ICLQ 48(3) 664–672 (1999)
de Mestral, Armand, 'Deux Récents Accords Bilatéraux en Matière de Pêche en 1977', *Canad. YbIL* [15] 287–300 (1977)
de Mestral, Armand, 'Compulsory Dispute Settlement in the Third United Nations Convention on the Law of the Sea: A Canadian perspective', 169–188 in Thomas Buergenthal (Ed.), *Contemporary Issues in International Law: Essays in honor of Louis B. Sohn* (Engel, 1984)
de Yturriaga, José-Antonio, *The International Regime of Fisheries: From UNCLOS 1982 to the presential sea* (Nijhoff, 1997)

Dean, Arthur, 'The Geneva Conference on the Law of the Sea: What was accomplished', AJIL 52(4) 607–628 (1958)
Dean, Arthur, 'The Second Geneva Conference on the Law of the Sea: The Fight for Freedom of the Seas', AJIL 54(4) 751–789 (1960)
Dhokalia, Ramaa, *The Codification of Public International Law* (MUP, 1970)
Diz, Daniela, 'The Interface Between the BBNJ Agreement and RFMOs', 309–325 in Bjørn Kunoy – Tomas Heidar – Constantinos Yiallourides (Eds), *International Fisheries Law: Persistent and emerging challenges* (Routledge, 2024)
Doulman, David, *Structure and Process of the 1993–1995 United Nations Conference on Straddling Fish Stocks and Highly Migratory Fish Stocks* (FAO, 1995)
Dunlap, William, 'Bering Sea', IJMCL 10(1) 114–135 (1995)
Dupont, Alan – Baker, Christopher, 'East Asia's Maritime Disputes: Fishing in troubled waters', *Washington Quarterly* 37(1) 79–98 (2014)
Dupuy, Pierre-Marie, 'Evolutionary Interpretation of Treaties: Between memory and prophecy', 123–137 in Enzo Cannizzaro (Ed.), *The Law of Treaties Beyond the Vienna Convention* (OUP, 2011)
Dupuy, René-Jean, *The Law of the Sea: Current problems* (Sythoff, 1974)
Edeson, William, 'Towards Long-term Sustainable Use: Some recent developments in the legal regime of fisheries', 165–204 in Alan Boyle – David Freestone (Eds), *International Law and Sustainable Development: Past achievements and future challenges* (OUP, 1999)
Edeson, William, 'Sustainable Use of Marine Living Resources', ZaöRV [63] 355–375 (2003)
Edeson, William, 'A Brief Introduction to the Principal Provisions of the International Legal Regime Governing Fisheries in the EEZ', 17–30 in Syma Ebbin – Håkon Hoel – Are Sydnes (Eds), *A Sea Change: The Exclusive Economic Zone and governance institutions for living marine resources* (Springer, 2005)
Edeson, William – Freestone, David – Gudmundsdottir, Elly, *Legislating for Sustainable Fisheries: A guide to implementing the 1993 FAO Compliance Agreement and 1995 UN Fish Stocks Agreement* (World Bank Publications, 2001)
Eiriksson, Gudmundur, *The International Tribunal for the Law of the Sea* (Nijhoff, 2001)
Eiriksson, Gudmundur, 'Satya N. Nandan's Role in Drafting the Informal Single Negotiating Text: Aspects of the preparatory work for UNCLOS', 37–50 in Michael Lodge – Myron Nordquist (Eds), *Peaceful Order in the World's Oceans: Essays in honor of Satya N. Nandan* (Brill–Nijhoff, 2014)
Elias, Olufemi – Lim, Chin, *The Paradox of Consensualism in International Law* (Kluwer, 1998)
Engler-Palma, Maria, 'Allocation of Fishing Opportunities in Regional Fisheries Management Organizations: From power to law?', 473–518 in Dawn Russell – David VanderZwaag (Eds), *Recasting Transboundary Fisheries Management Arrangements in Light of Sustainability Principles* (Brill, 2011)
Erasmus, Gerhard, 'Dispute Settlement in the Law of the Sea', *Acta Juridica* [25] 15–27 (1986)
Evans, Malcolm, 'The Southern Bluefin Tuna Dispute: Provisional thinking on provisional measures?', YbIEL [10] 7–14 (2000)
Extavour, Winston, *The Exclusive Economic Zone: A study of the evolution and progressive development of the international law of the sea* (Institut Universitaire de Hautes Etudes Internationales, 1979)

Fahner, Johannes, 'In Dubio Mitius: Advancing clarity and modesty in treaty interpretation', EJIL 32(3) 835–861 (2021)

Fauteux, Paul, 'L'Initiative Juridique Canadienne sur la Pêche en Haute Mer', *Canad. YbIL* [31] 33–87 (1994)

Fawcett, James, 'The Legal Character of International Agreements', BYBIL [30] 381–400 (1953)

Fietta, Stephen – Cleverly, Robin, *A Practitioner's Guide to Maritime Boundary Delimitation* (OUP, 2016)

Finlay, Luke, 'The Proposed New Convention on the Law of the Sea: A candid appraisal', *Syracuse Journal of International Law and Commerce* 7(2) 135–152 (1980)

Finus, Michael – Schneider, Raoul, 'Scope and Compatibility of Measures in International Fisheries Agreements', *Oxford Economic Papers* 67(4) 865–888 (2015)

Fitzmaurice, Gerald, 'The Law and Procedure of the International Court of Justice: Treaty interpretation and certain other treaty points', BYBIL [28] 1–28 (1951)

Fitzmaurice, Gerald, 'The Law and Procedure of the International Court of Justice, 1951–4: Treaty interpretation and certain other treaty points', BYBIL [33] 203–293 (1957)

Fitzmaurice, Gerald, '*Vae Victis* or Woe to the Negotiators! Your Treaty or Our Interpretation of It', AJIL 65(2) 358–373 (1971)

Fitzmaurice, Malgosia, 'Dynamic (Evolutive) Interpretation of Treaties', *Hague Yearbook of International Law* [21] 101–153 (2008)

Fitzmaurice, Malgosia, 'The Practical Working of the Law of Treaties', 144–178 in Malcolm Evans (Ed.), *International Law* (OUP, 2024 6th edition)

Fleischer, Carl, 'The Right to a 200-Mile Exclusive Economic Zone or a Special Fishery Zone', SDLR 14(3) 548–583 (1977)

Focsaneanu, Lazar, 'Les Langues Comme Moyen d'Expression du Droit International', *Annuaire Français* [16] 256–274 (1970)

Fortier, Yves, 'From Confrontation to Cooperation on the High Seas: Recent developments in international law concerning the conservation of marine resources', 1377–1390 in Nisuke Ando – Edward McWhinney – Rüdiger Wolfrum (Eds), *Liber Amicorum Judge Shigeru Oda* (Kluwer, 2002)

Foster, Caroline, *Science and the Precautionary Principle in International Courts and Tribunals: Expert evidence, burden of proof and finality* (CUP, 2011)

Franck, Thomas, *The Power of Legitimacy Among Nations* (OUP, 1990)

Franck, Thomas – Sughrue, Dennis, 'The International Role of Equity-as-Fairness', *Georgetown Law Journal* 81(3) 563–596 (1993)

Franckx, Erik, '*Pacta Tertiis* and the Agreement for the Implementation of the Straddling and Highly Migratory Fish Stocks Provisions of the United Nations Convention on the Law of the Sea', *Tulane Journal of International and Comparative Law* 8(1) 49–81 (2000)

Franckx, Erik, '200-Mile Limit: Between creeping jurisdiction and creeping common heritage?', *George Washington International Law Review* 39(3) 467–498 (2007)

Franckx, Erik – Benatar, Marco, 'The Duty to Co-Operate for States Bordering Enclosed or Semi-Enclosed Seas', *Chinese (Taiwan) Yearbook of International Law and Affairs* [31] 66–81 (2013)

Freestone, David – Makuch, Zen, 'The New International Environmental Law of Fisheries: The 1995 United Nations Straddling Stocks Agreement', YbIEL [7] 3–51 (1998)

Freestone, David – Oude-Elferink, Alex, 'Flexibility and Innovation in the Law of the Sea – Will the LOS Convention Amendment Procedures Ever be Used?', 169–221 in Alex Oude-Elferink (Ed.), *Stability and Change in the Law of the Sea: The role of LOS Convention* (Nijhoff, 2005)

Fuller, Susanna – Schleit, Kathryn, 'The Future of Managing Fisheries and the Global Commons through Regional Fisheries Management Organizations: Steps toward global stewardship', 361–366 in Werle Dirk – Paul Boudreau – Mary Brooks – Michael Butler – Anthony Charles – Scott Coffen-Smout – David Griffiths – Ian McAllister – Moira McConnell – Ian Porter – Susan Rolston – Peter Wells (Eds), *The Future of Ocean Governance and Capacity Development: Essays in honor of Elisabeth Mann Borgese* (Brill–Nijhoff, 2018)

Gamble, John, 'The Law of the Sea Conference: Dispute settlement in perspective', *Vanderbilt Journal of Transnational Law* 9(2) 323–342 (1976)

Gamble, John, 'The 1982 UN Convention on the Law of the Sea: Binding dispute settlement?', *Boston University International Law Journal* 9(2) 39–58 (1991)

Ganz, David, 'The United Nations and the Law of the Sea', ICLQ 26(1) 1–53 (1977)

Garcia, Serge, 'The Precautionary Approach to Fisheries: Progressive review and main issues (1995–2000)', 479–557 in John Moore – Myron Nordquist (Eds), *Current Fisheries Issues and the Food and Agriculture Organization of the United Nations* (Nijhoff, 2000)

García-Amador, Francisco, *The Exploitation and Conservation of the Resources of the Sea* (Sijthoff, 1959)

García-Amador, Francisco, 'The Latin American Contribution to the Development of the Law of the Sea', AJIL 68(1) 33–50 (1974)

García-Revillo, Miguel, *The Contentious and Advisory Jurisdiction of the International Tribunal for the Law of the Sea* (Brill–Nijhoff, 2015)

Gardiner, Richard, *Treaty Interpretation* (OUP, 2015 2nd edition)

Gautier, Philippe, 'The International Tribunal for the Law of the Sea: Activities in 2006', CJIL 6(2) 389–402 (2007)

Gavouneli, Maria, *Functional Jurisdiction in the Law of the Sea* (Nijhoff, 2007)

Gertenbach, Dormehl, 'Fishery Convention Areas in the Southeast Atlantic and Adjacent Seas: Overlapping issues', *Acta Juridica* [25] 51–60 (1986)

Ginsburg, Tom, 'Bounded Discretion in International Judicial Lawmaking', VJIL 45(3) 631–674 (2005)

Gjerde, Kristina – Kaoru, Yoshiaki – McGillivary, Philip – Mirovitskaya, Natalia – Pontecorvo, Giulio, 'Living Resource Problems: The North Pacific', 50–85 in James Broadus – Raphael Vartanov (Eds), *The Oceans and Environmental Security: Shared US and Russian perspectives* (Island Press, 1994)

Goldie, Louis, 'The Oceans' Resources and International Law – Possible Developments in Regional Fisheries Management', CJTL 8(1) 1–53 (1969)

Goldie, Louis, 'The Management of Ocean Resources: Regimes for structuring the maritime environment', 155–247 in Cyril Black – Richard Falk (Eds), *The Future of the International Legal Order: The structure of the international environment* (Princeton University Press, 1972)

Goldie, Louis, 'Equity and the International Management of Transboundary Resources', *Journal of Natural Resources* 25(3) 665–669 (1985)

Goltz, Jon, 'The Sea of Okhotsk Peanut Hole: How the United Nations draft agreement on straddling stocks might preserve the pollack fishery', PRL&PJ 4(2) 443–478 (1995)

Goodman, Camille, *Coastal State Jurisdiction over Living Resources in the Exclusive Economic Zone* (OUP, 2022)
Goodman, Camille, 'Compulsory Settlement of EEZ Fisheries Enforcement Disputes under UNCLOS: *Swallowing the rule* or *balancing the equation*?', GöJIL 13(1) 27–80 (2023)
Gotlieb, Allan, 'The Canadian Contribution to the Concept of a Fishing Zone in International Law', *Canad. YbIL* [2] 55–76 (1964)
Gottlieb, Gidon, *The Logic of Choice: An investigation of the concepts of the rule and rationality* (Allen & Unwin, 1968)
Gross, Leo, 'States as Organs of International Law and the Problem of Autointerpretation', 59–88 in George Lipsky (Ed.), *Law and Politics in the World Community: Essays on Hans Kelsen's pure theory and related problems in international law* (University of California Press, 1953)
Gudmundsson, Thorir, 'Cod War on the High Seas: Norwegian–Icelandic dispute over "Loophole" fishing in the Barents Sea', NJIL 64(4) 557–573 (1995)
Guilfoyle, Douglas, 'The *South China Sea Award*: How should we read the UN Convention on the Law of the Sea?', *Asian* JIL 8(1) 51–63 (2018)
Guilfoyle, Douglas – Mossop, Joanna, 'The Extent and Legitimacy of the Judicial Function in UNCLOS Dispute Settlement', ICLQ 73(3) 641–667 (2024)
Haas, Bianca – Haward, Marcus – McGee, Jeffrey – Fleming, Aysha, 'The Influence of Performance Reviews on Regional Fisheries Management Organizations', *International Council for the Exploration of the Sea Journal of Marine Science* 76(6) 2082–2089 (2019)
Haas, Peter, 'Constructing Environmental Conflicts from Resource Scarcity', *Global Environmental Politics* 2(1) 1–11 (2002)
Hagiwara, Kazuki, 'Sustainable Development before International Courts and Tribunals: Duty to cooperate and States' good faith', 167–187 in Christina Voigt (Ed.), *International Judicial Practice on the Environment: Questions of legitimacy* (CUP, 2019)
Hanich, Quentin – Ota, Yoshitaka, 'Moving Beyond Rights-Based Management: A transparent approach to distributing the conservation burden and benefit in tuna fisheries', IJMCL 28(1) 135–170 (2013)
Hanich, Quentin – Campbell, Brooke – Bailey, Megan – Molenaar, Erik, 'Research into Fisheries Equity and Fairness – Addressing Conservation Burden Concerns in Transboundary Fisheries', *Marine Policy* [51] 302–304 (2015)
Hardin, Garrett, 'The Tragedy of the Commons', *Science* [162] 12–43 (1968)
Harrison, James, *Making the Law of the Sea: A study in the development of international law* (CUP, 2011)
Harrison, James, 'Defining Disputes and Characterizing Claims: Subject-matter jurisdiction in Law of the Sea Convention litigation', ODIL 48(3–4) 270–284 (2017)
Harrison, James, *Saving the Oceans through Law: The international legal framework for the protection of the marine environment* (OUP, 2017)
Harrison, James, 'Key Challenges Relating to the Governance of Regional Fisheries', 79–102 in Richard Caddell – Erik Molenaar (Eds), *Strengthening International Fisheries Law in an Era of Changing Oceans* (Hart, 2019)
Harrison, James, 'Emerging Dispute Settlement Principles and Practices in RFMOs', 88–105 in Bjørn Kunoy – Tomas Heidar – Constantinos Yiallourides (Eds), *International Fisheries Law: Persistent and emerging challenges* (Routledge, 2024)
Haward, Marcus, 'Management of Marine Living Resources: International and regional perspectives on transboundary issues', 41–55 in Gerald Blake – Chia Sien – Carl

Grundy-Warr – Martin Pratt – Clive Schofield (Eds), *International Boundaries and Environmental Security, Frameworks for Regional Cooperation* (Kluwer, 1997)
Hayashi, Moritaka, 'Soviet Policy on International Regulation of High Seas Fisheries', *Cornell International Law Journal* 5(1) 131–160 (1972)
Hayashi, Moritaka, 'United Nations Conference on Straddling Fish Stocks and Highly Migratory Fish Stocks: An analysis of the 1993 sessions', *Ocean Yearbook* [11] 20–45 (1994)
Hayashi, Moritaka, 'The 1995 Agreement on the Conservation and Management of Straddling and Highly Migratory Fish Stocks: Significance for the Law of the Sea Convention', OCM 29(1) 51–69 (1995)
Hayashi, Moritaka, 'The 1995 UN Fish Stocks Agreement and the Law of the Sea', 37–53 in Davor Vidas – Willy Østreng (Eds), *Order for the Oceans at the Turn of the Century* (Nijhoff, 1999)
Hayashi, Moritaka, 'The Straddling and Highly Migratory Fish Stocks Agreement', 55–83 in Ellen Hey (Ed.), *Developments in International Fisheries Law* (Kluwer, 1999)
Hayashi, Moritaka, 'Three Decades' Progress in High Seas Fisheries Governance: Towards a common heritage regime?', 375–397 in John Moore – Myron Nordquist – Said Mahmoudi (Eds), *The Stockholm Declaration and Law of the Marine Environment* (Nijhoff, 2003)
Heidar, Tomas, 'The Legal Framework for High Seas Fisheries in the Central Arctic Ocean', 179–203 in Myron Nordquist – John Moore – Ronán Long (Eds), *International Marine Economy: Law and policy* (Brill–Nijhoff, 2017)
Henkin, Louis – Pugh, Richard – Schachter, Oscar – Smit, Hans, *International Law: Cases and materials* (West Publishing, 1980)
Henriksen, Tore, 'Revisiting the Freedom of Fishing and Legal Obligations on States Not Party to Regional Fisheries Management Organizations', ODIL 40(1) 80–96 (2009)
Henriksen, Tore, 'Allocation of Fishing Rights: Principles and alternative procedures', 522–558 in Myron Nordquist – John Moore – Ronán Long (Eds), *Challenges of the Changing Arctic: Continental shelf, navigation, and fisheries* (Brill–Nijhoff, 2016)
Henriksen, Tore, 'The FAO and Ocean Governance', 3–27 in David Attard – Malgosia Fitzmaurice – Alexandros Ntovas (Eds), *UN Specialized Agencies and Global Ocean Governance: The IMLI treatise on global ocean governance* (OUP, 2018)
Henriksen, Tore – Hoel, Håkon, 'Determining Allocation: From paper to practice in the distribution of fishing rights between countries', ODIL 42(1) 66–93 (2011)
Henriksen, Tore – Hønneland, Geir – Sydnes, Are, *Law and Politics in Ocean Governance: The UN Fish Stocks Agreement and regional fisheries management regimes* (Nijhoff, 2006)
Hey, Ellen, *The Regime for the Exploitation of Transboundary Marine Fisheries Resources* (Nijhoff, 1989)
Hey, Ellen, 'The Persistence of a Concept: Maximum sustainable yield', 89–97 in David Freestone (Ed.), *The 1982 Law of the Sea Convention at 30: Successes, challenges and new agendas* (Brill, 2013)
Higgenson, Charles, 'The Law of the Sea Convention and the Protection of Fisheries', GIELR 7(4) 771–774 (1995)
Highet, Keith, 'Maritime Boundary Disputes and Settlement', 745–777 in Mochtar Kusuma-Atmadja – Thomas Mensah – Bernard Oxman (Eds), *Sustainable Development and Preservation of the Oceans: The challenges of UNCLOS and Agenda 21* (LOSI UH, 1997)

Hoekman, Bernard – Mavroidis, Petros – Sasmal, Sunayana, 'Managing Externalities in the WTO: The Agreement on Fisheries Subsidies', *Journal of International Economic Law* 26(2) 266–284 (2023)

Hoel, Håkon, 'Performance Reviews of Regional Fisheries Management Organizations', 449–472 in Dawn Russell – David VanderZwaag (Eds), *Recasting Transboundary Fisheries Management Arrangements in Light of Sustainability Principles* (Brill, 2011)

Hoel, Håkon, 'The Importance of Marine Science in Sustainable Fisheries: The role of the 1995 UN Fish Stocks Agreement', 379–395 in Myron Nordquist – John Moore – Ronán Long (Eds), *Legal Order in the World's Oceans* (Brill–Nijhoff, 2017)

Hoel, Håkon – Kvalvik, Ingrid, 'The Allocation of Scarce Natural Resources: The case of fisheries', *Marine Policy* 30(4) 347–356 (2006)

Holst, Rozemarijn, *Change in the Law of the Sea: Context, mechanisms and practice* (Brill–Nijhoff, 2022)

Hu, Nien-Tsu, 'Father of the WCPFC Convention: Ambassador Satya N. Nandan', 381–403 in Michael Lodge – Myron Nordquist (Eds), *Peaceful Order in the World's Oceans: Essays in honor of Satya N. Nandan* (Brill–Nijhoff, 2014)

Huck, Winfried, *Sustainable Development Goals: Article-by-article commentary* (Nomos, 2022)

Hudson, Carolyn, 'Fishery and Economic Zones as Customary International Law', SDLR 17(3) 661–690 (1979)

Hudson, Manley, 'The Progressive Codification of International Law', AJIL 20(4) 655–669 (1926)

Hulme, Karen, 'Environmental Security: Implications for international law', YbIEL [19] 3–25 (2009)

Infante, María-Teresa, 'The Settlement of Disputes regarding the Law of the Sea and its Bearing on the Legal Nature of the Exclusive Economic Zone', 159–171 in Francisco Orrego-Vicuña (Ed.), *The Exclusive Economic Zone, A Latin American Perspective* (Routledge, 1984)

Infante, María-Teresa, 'South America', 199–234 in James Kraska – Hayoun Ryou-Ellison (Eds), *Cultural Influences on the Law of the Sea: History, legacy, and future prospects* (OUP, 2025)

Irwin, Paul, 'Settlement of Maritime Boundary Disputes: An analysis of the law of the sea negotiations', ODIL 8(2) 105–148 (1980)

Jackson, Andrew, 'The Convention on the Conservation and Management of Fishery Resources in the South East Atlantic Ocean, 2001: An introduction', IJMCL 17(1) 33–77 (2002)

Jacobson, Jon, 'Bridging the Gap to International Fisheries Agreement: A guide for unilateral action', SDLR 9(3) 454–490 (1972)

Jacobson, Jon, 'Managing Marine Living Resources in the Twenty-First Century: The next level of ocean governance?', 311–322 in Myron Nordquist – John Moore (Eds), *Entry into Force of the Law of the Sea Convention* (Nijhoff, 1995)

Jacobson, Jon, 'Conserving and Managing Living Marine Resources: The second story', 17–34 in Myron Nordquist (Ed.), *Implementing the 1982 Law of the Sea Convention* (Center for Oceans Law and Policy, 1996)

Jacovides, Andreas, 'The Law of the Sea – Where Now? A Comment', LCP 46(2) 201–204 (1983)

Jaenicke, Günther, 'Dispute Settlement under the Convention on the Law of the Sea', ZaöRV [43] 813–827 (1983)

Jennings, Robert – Watts, Arthur, *Oppenheim's International Law*, Volume I (Longman, 1992 9th edition)
Jensen, Øystein – Bankes, Nigel, 'Compulsory and Binding Dispute Resolution under the United Nations Convention on the Law of the Sea: Introduction', ODIL 48(3/4) 209–215 (2017)
Jessup, Phillip, 'The Geneva Conference on the Law of the Sea; A study in international law-making', AJIL 52(3) 730–733 (1958)
Jesus, José, 'The Role of ITLOS in the Settlement of Law of the Sea Disputes', 154–164 in Myron Nordquist – John Moore – Alfred Soons – Hak-So Kim (Eds), *The Law of the Sea Convention: US accession and globalization* (Brill, 2012)
Jia, Bing-Bing, 'The Curious Case of Article 281: A *super* provision within UNCLOS?', ODIL 46(4) 266–280 (2015)
Johnson, Barbara, 'A Review of Fisheries Proposals Made at the Caracas Session of LOS III', *Ocean Management* 2(4) 285–313 (1975)
Johnson, Barbara, 'Canadian Foreign Policy and Fisheries', 52–99 in Barbara Johnson – Mark Zacher (Eds), *Canadian Foreign Policy and the Law of the Sea* (University of British Columbia Press, 1977)
Johnson, David, 'The Preparation of the 1958 Geneva Conference on the Law of the Sea', ICLQ 8(1) 122–145 (1959)
Johnston, Douglas, *The International Law of Fisheries: A framework for policy-oriented inquiries* (YUP, 1965)
Johnston, Douglas, 'The Options for LOS III: Appraisal and proposal', 357–372 in Francis Christy – John Gamble – Gary Knight – Edward Miles (Eds), *Law of the Sea: Caracas and beyond* (Ballinger Publishing, 1975)
Joyner, Christopher, 'On the Borderline? Canadian Activism in the Grand Banks', 207–234 in Olav Stokke (Ed.), *Governing High Seas Fisheries: The Interplay of global and regional regimes* (OUP, 2001)
Joyner, Christopher – DeCola, Peter, 'Chile's Presential Sea Proposal Implications for Straddling Stocks and the International Law of Fisheries', ODIL 24(1) 99–121 (1993)
Joyner, Christopher – Martell, Elizabeth, 'Looking Back to See Ahead: UNCLOS/III and lessons for global commons law', ODIL 27(1) 73–95 (1996)
Joyner, Christopher – Von Gustedt, Alejandro, 'The Turbot War of 1995: Lessons for the law of the sea', IJMCL 11(4) 425–458 (1996)
Juda, Lawrence, 'The United Nations Convention on Straddling and Highly Migratory Stocks: Policy problems in implementation', 161–179 in Myron Nordquist (Ed.), *Implementing the 1982 Law of the Sea Convention* (Center for Oceans Law and Policy, 1996)
Juda, Lawrence, *International Law and Ocean Use Management: The evolution of ocean governance* (Routledge, 1996)
Juda, Lawrence, 'The United Nations Agreement on Straddling Fish Stocks and Highly Migratory Fish Stocks: A critique', ODIL 28(2) 147–166 (1997)
Juratowitch, Ben, 'Limitations on Compulsory Settlement of Disputes with Regard to Fisheries in the EEZ Arising from Article 297(3)(a) of UNCLOS', 171–179 in Bjørn Kunoy – Tomas Heidar – Constantinos Yiallourides (Eds), *International Fisheries Law: Persistent and emerging challenges* (Routledge, 2024)
Kaitala, Veijo – Munro, Gordon, 'The Management of High Seas Fisheries', *Marine Resource Economics* 8(4) 313–329 (1993)
Karaman, Igor, *Dispute Resolution in the Law of the Sea* (Brill, 2012)

Katz, Stephen, 'Issues Arising in the Icelandic Fisheries Case', ICLQ 22(1) 83–108 (1973)

Kawano, Mariko, 'Implementation of the Rules of the UNCLOS through Universal and Regional Organizations', 9–35 in Marta Ribeiro – Fernando Bastos – Tore Henriksen (Eds), *Global Challenges and the Law of the Sea* (Springer, 2020)

Kawasaki, Tsuyoshi, 'The 200-Mile Regime and the Management of Transboundary and High Seas Stocks', *Ocean Management* 9(1) 7–20 (1984)

Kedziora, Derrick, 'Gunboat Diplomacy in the Northwest Atlantic: The 1995 Canada-EU fishing dispute and the United Nations Agreement on Straddling and High Migratory Fish Stocks', *Northwestern Journal of International Law & Business* 17(3) 1132–1162 (1996)

Kelly, Christopher, 'Law of the Sea: The jurisdictional dispute over highly migratory species of tuna', *Columbia Journal of Transnational Law* 26(3) 475–513 (1988)

Kennedy, David, *International Legal Structures* (Nomos Verlagsgesellschaft, 1987)

Kesteven, Geoffrey, 'MSY Revisited: A realistic approach to fisheries management and administration', *Marine Policy* 21(1) 73–82 (1997)

Kibel, Paul, 'Alone at Sea: Chile's *presencial* ocean policy', *Journal of Environmental Law* 12(1) 44–63 (2000)

Kildow, Judith, 'The Law of the Sea: Alliances and divisive issues in international ocean negotiations', SDLR 11(3) 558–578 (1974)

Kim, Hyun-Jung, 'Governing Fishing Stocks in Northeast Asia's Disputed Waters: Preventing a "tragedy of the commons"?', IJMCL 33(3) 495–527 (2018)

Kim, Suk-Kyoon, *Maritime Disputes in Northeast Asia: Regional challenges and cooperation* (Brill–Nijhoff, 2017)

Kimball, Lee – Bridgeman, James – Graham, Garry – Schneider, Adolf, 'Balancing the Interests?', *Environmental Policy and Law* 4(2) 69–77 (1978)

Kirk, Elizabeth, 'Maritime Zones and the Ecosystem Approach: A mismatch?', RECIEL 8(1) 67–70 (1999)

Kirk, Elizabeth, 'The Ecosystem Approach and the Search for an Objective and Content for the Concept of Holistic Ocean Governance', ODIL 46(1) 33–49 (2015)

Kiss, Alexandre, *L'Abus de Droit en Droit International* (Pichon & Durand-Auzias, 1953)

Klein, Natalie, *Dispute Settlement in the UN Convention on the Law of the Sea* (CUP, 2005)

Klein, Natalie, 'Expansions and Restrictions in the UNCLOS Dispute Settlement Regime: Lessons from recent decisions', CJIL 15(2) 403–415 (2016)

Klein, Natalie, 'The Vicissitudes of Dispute Settlement under the Law of the Sea Convention', IJMCL 32(2) 332–363 (2017)

Klein, Natalie, 'Provisional Measures and Provisional Arrangements', 117–144 in Alex Oude-Elferink – Tore Henriksen – Signe Busch (Eds), *Maritime Boundary Delimitation: The case law – is it consistent and predictable?* (CUP, 2018)

Klein, Natalie, 'The Many Facets of EEZ Fisheries Disputes and their Resolution under UNCLOS', GöJIL 13(1) 14–25 (2023)

Klein, Natalie – Kroon, Danielle, 'Settlement of International Environmental Law Disputes', 230–258 in Malgosia Fitzmaurice – Marcel Brus – Panos Merkouris (Eds), *Research Handbook on International Environmental Law* (EEP, 2021 2nd edition)

Klein, Natalie – Parlett, Kate, *Judging the Law of the Sea* (OUP, 2022)

Knight, Garry, 'Law of the Sea Negotiations 1971–1972: From internationalism to nationalism', SDLR 9(3) 383–389 (1972)

Koers, Albert, *International Regulation of Marine Fisheries* (Fishing News (Books) Ltd, 1973)
Koh, Tommy, 'The Third United Nations Conference on the Law of the Sea: What was accomplished?', LCP 46(2) 5–10 (1983)
Koh, Tommy, 'The Origins of the 1982 Convention on the Law of the Sea', MLR 29(1) 1–17 (1987)
Koh, Tommy, 'The Exclusive Economic Zone', MLR 30(1) 1–33 (1988)
Koh, Tommy, *Building a New Legal Order for the Oceans* (National University of Singapore Press, 2020)
Kolb, Robert, 'Note sur Certaines Caractéristiques du Différend International', *The Global Community Yearbook of International Law and Jurisprudence* [1] 227–242 (2004)
Kolb, Robert, 'Principles as Sources of International Law (with Special Reference to Good Faith)', *Netherlands International Law Review* 53(1) 1–36 (2006)
Koskenniemi, Martti, *From Apology to Utopia: The structure of international legal argument* (CUP, 2006 2nd edition)
Kraska, James, 'North America', 235–260 in James Kraska – Hayoun Ryou-Ellison (Eds), *Cultural Influences on the Law of the Sea: History, legacy, and future prospects* (OUP, 2025)
Kreća, Milenko, 'A Few Remarks on the Theoretical Basis of the New Law of the Sea', 1207–1216 in Nisuke Ando – Edward McWhinney – Rüdiger Wolfrum (Eds), *Liber Amicorum Judge Shigeru Oda* (Kluwer, 2002)
Kulski, Władysław, 'The Soviet Interpretation of International Law', AJIL 49(2) 518–534 (1955)
Kunoy, Bjørn, 'The Scope of Compulsory Jurisdiction and Exceptions thereto under the United Nations Convention on the Law of the Sea', *Canad. YbIL* [58] 78–141 (2021)
Kunoy, Bjørn, 'Dislodging the Compulsory Dispute Settlement Mechanism: Analysis of article 281 of UNCLOS', LJIL 37(2) 423–439 (2024)
Kwiatkowska, Barbara, *The 200 Mile Exclusive Economic Zone in the New Law of the Sea* (Nijhoff, 1989)
Kwiatkowska, Barbara, 'Creeping Jurisdiction Beyond 200 Miles in the Light of the 1982 Law of the Sea Convention and State Practice', ODIL 22(2) 153–187 (1991)
Kwiatkowska, Barbara, 'The High Seas Fisheries Regime: At a point of no return?', IJMCL 8(3) 327–358 (1993)
Kwiatkowska, Barbara, 'The Southern Bluefin Tuna (New Zealand *v.* Japan; Australia *v.* Japan) Cases', IJMCL 15(1) 1–36 (2000)
Kwiatkowska, Barbara, 'International Decisions – Southern Bluefin Tuna', AJIL 95(1) 162–171 (2001)
Kwiatkowska, Barbara, 'The Australia and New Zealand *v.* Japan – Southern Bluefin Tuna (Jurisdiction and Admissibility) Award of the First Law of the Sea Convention Annex VII Arbitral Tribunal', IJMCL 16(2) 239–294 (2001)
Kwiatkowska, Barbara, *Peaceful Settlement of Oceans and Other Environmental Disputes under International Agreements* (Book World Publications, 2002)
Kwiatkowska, Barbara, 'The 2006 *Barbados/Trinidad and Tobago Award*: A landmark in compulsory jurisdiction and equitable maritime boundary delimitation', IJMCL 22(1) 7–60 (2007)
Kwiatkowska, Barbara, 'The 2006 Barbados/Trinidad and Tobago Maritime Delimitation (Jurisdiction and Merits) Award', 917–988 in Tafsir Ndiaye – Rüdiger Wolfrum

(Eds), *Law of the Sea, Environmental Law, and Settlement of Disputes: Liber amicorum Judge Thomas A. Mensah* (Nijhoff, 2007)
Lachs, Manfred, 'Equity in Arbitration and in Judicial Settlement of Disputes', LJIL 6(2) 323–329 (1993)
Lando, Massimo, *Maritime Delimitation as a Judicial Process* (CUP, 2019)
Lanovoy, Vladyslav, 'Dispute Settlement and Ocean Governance', 72–89 in Simone Borg – Felicity Attard – Patricia Vella-de-Fremeaux (Eds), *Research Handbook on Ocean Governance Law* (EEP, 2023)
Larocque, Emily, 'The Convention on the Conservation and Management of Highly Migratory Fish Stocks in the Western and Central Pacific Ocean: Can tuna promote development of Pacific island nations?', *Asia Pacific* LPJ 4(1) 83–120 (2003)
Larsen, Paul, 'Between Scylla and Charybdis in Treaty Interpretation', AJIL 63(1) 108–110 (1969)
Lauterpacht, Elihu, *Aspects of the Administration of International Justice* (Grotius Publications, 1991)
Lauterpacht, Hersch, *The Function of Law in the International Community* (OUP, 1933; revised and corrected edition reprinted in 2011)
Lauterpacht, Hersch, 'Restrictive Interpretation and the Principle of Effectiveness in the Interpretation of Treaties', BYBIL [26] 48–85 (1949)
Lauterpacht, Hersch, 'Codification and Development of International Law', AJIL 49(1) 16–43 (1955)
Lauterpacht, Hersch, *The Development of International Law by the International Court* (Stevens & Sons, 1958)
Lavalle, Roberto, 'Conciliation under the United Nations Convention on the Law of the Sea: A critical overview', *Austrian Review of International and European Law* [2] 25–48 (1997)
Lear, William, 'History of Fisheries in the Northwest Atlantic: The 500-year perspective', *Journal of Northwest Atlantic Fishery Science* [23] 41–73 (1998)
Lee, Seo-Hang, 'Korea's Claims to Maritime Jurisdiction', *Korean Journal of Comparative Law* [18] 62–89 (1990)
Lee, Seokwoo – Kim, Jeong-Woo, 'UNCLOS and the Obligation to Cooperate: International legal framework for semi-enclosed seas cooperation', 11–29 in Keyuan Zou (Ed.), *Maritime Cooperation in Semi-Enclosed Seas* (Brill–Nijhoff, 2019)
Lehoux, Grégoire, 'La Troisième Conférence sur le Droit de la Mer et le Règlement Obligatoire des Différends', *Canad. YbIL* [18] 31–90 (1981)
Lekkas, Sotirios-Ioannis – Merkouris, Panos – Peat, Daniel, 'The Interpretative Practice of the International Court of Justice', *Yearbook UN Law* 26(1) 316–357 (2023)
Lennan, Mitchell, 'Fisheries Redistribution under Climate Change: Rethinking the law to address the "governance gap"?', 163–177 in Froukje Platjouw – Alla Pozdnakova (Eds), *The Environmental Rule of Law for Oceans: Designing legal solutions* (CUP, 2023)
Lévy, Jean-Pierre – Schram, Gunnar (Eds), *United Nations Conference on Straddling Fish Stocks and Highly Migratory Fish Stocks: Selected documents* (Nijhoff, 1996)
Lewis, Reece, 'The "Constitution for the Oceans"? The Law of the Sea Convention as a Living Treaty', ICLQ 74(1) 1–31 (2025)
Lijnzaad, Liesbeth, 'As Time Goes By: A preliminary inquiry into the "object and purpose" of the Law of the Sea Convention', 264–281 in Kristina Siig – Birgit Feldtmann – Fenella Billing (Eds), *The United Nations Convention on the Law of the Sea: A system of regulation* (Routledge, 2023)

Linderfalk, Ulf, 'Is Treaty Interpretation an Art or a Science? International Law and Rational Decision Making', EJIL 26(1) 169–189 (2015)

Linderfalk, Ulf, 'The Jan Mayen Case (Iceland/Norway): An example of successful conciliation', 193–216 in Christian Tomuschat – Riccardo Pisillo Mazzeschi – Daniel Thürer (Eds), *Conciliation in International Law* (Brill–Nijhoff, 2017)

Lissitzyn, Oliver, *International Law Today and Tomorrow* (Oceana Publications, 1965)

Løbach, Terje – Petersson, Matilda – Haberkon, Eliana – Mannini Piero, *Regional Fisheries Management Organizations and Advisory Bodies: Activities and developments, 2000–2017* (FAO, 2020)

Lodge, Michael, 'The Fisheries Regimes of Enclosed and Semi-Enclosed Seas and High Seas Enclaves', 193–216 in Ellen Hey (Ed.), *Developments in International Fisheries Law* (Kluwer, 1999)

Lodge, Michael, 'The Draft Convention for the Conservation and Management of Highly Migratory Fish Stocks in the Western and Central Pacific Ocean', 19–35 in Myron Nordquist – John Moore (Eds), *Current Fisheries Issues and the Food and Agriculture Organization of the United Nations* (Nijhoff, 2000)

Lodge, Michael – Nandan, Satya, 'Some Suggestions Towards Better Implementation of the United Nations Agreement on Straddling Fish Stocks and Highly Migratory Fish Stocks of 1995', IJMCL 20(3) 345–379 (2005)

Lodge, Michael – Anderson, David – Løbach, Terje – Munro, Gordon – Sainsbury, Keith – Willock, Anna, *Recommended Best Practices for Regional Fisheries Management Organizations* (Chatham House, 2007)

Longhurst, Alan, *Mismanagement of Marine Fisheries* (CUP, 2010)

Lowe, Vaughan, 'The Role of Equity in International Law', *Aust.* YbIL [12] 54–81 (1988)

Lucchini, Laurent, 'La Loi Canadienne du 12 Mai 1994: La logique extrême de la théorie du droit préférentiel de l'Etat côtier en haute mer au titre des stocks chevauchants', *Annuaire Français* [40] 864–875 (1994)

Lugten, Gail, *A Review of Measures Taken by Regional Marine Fishery Bodies to Address Contemporary Fishery Issues* (FAO, 1999)

Macdonald, Ronald, 'The New Canadian Declaration of Acceptance of the Compulsory Jurisdiction of the International Court of Justice', *Canad. YbIL* [8] 3–38 (1970)

Mansfield, Bill, 'Compulsory Dispute Settlement After the Southern Bluefin Tuna Award', 255–272 in Alex Oude-Elferink – Donald Rothwell (Eds), *Ocean Management in the 21st Century: Institutional frameworks and responses* (Nijhoff, 2004)

Markowski, Marion, *The International Law of EEZ Fisheries: Principles and implementation* (Europa Law Publishing, 2010)

Martin, Will, 'Fisheries Conservation and Management of Straddling Stocks and Highly Migratory Stocks under the United Nations Convention on the Law of the Sea', GIELR 7(3) 765–769 (1995)

McDorman, Ted, 'Will Canada Ratify the Law of the Sea Convention?', SDLR 25(3) 535–579 (1988)

McDorman, Ted, 'The Dispute Settlement Regime of the Straddling and Highly Migratory Fish Stocks Convention', *Canad. YbIL* [35] 57–79 (1997)

McDorman, Ted, 'Implementing Existing Tools: Turning words into actions – decision-making processes of RFMOs', IJMCL 20(3) 423–457 (2005)

McDorman, Ted – Bolla, Alexander – Johnston, Douglas – Duff, John, *International Ocean Law Materials and Commentaries* (Carolina Academic Press, 2005)

McDougal, Myres, 'The International Law Commission's Draft Articles Upon Interpretation: Textuality *redivivus*', AJIL 61(4) 992–1000 (1967)
McDougal, Myres – Burke, William, *The Public Order of the Oceans: A contemporary international law of the sea* (YUP, 1962)
McLaughlin-Mitchell, Sara – Owsiak, Andrew, 'Judicialization of the Sea: Bargaining in the shadow of UNCLOS', AJIL 115(4) 579–621 (2021)
McNair, Arnold, *The Law of Treaties* (Clarendon Press, 1961)
McRae, Donald, 'The Applicable Law: The Geneva Convention on the Continental Shelf, the LOSC, and customary international law', 92–116 in Alex Oude-Elferink – Tore Henriksen – Signe Busch (Eds), *Maritime Boundary Delimitation: The case law – is it consistent and predictable?* (CUP, 2018)
Meltzer, Evelyne, *The Quest for Sustainable International Fisheries, Regional Efforts to Implement the 1995 United Nations Fish Stocks Agreement* (National Research Council of Canada, 2009)
Mendenhall, Elizabeth, 'Drivers and Mechanisms of Change in UNCLOS: An international relations research agenda', 135–148 in Justin Alger – Rashid Sumaila (Eds), *A Research Agenda for Sustainable Ocean Governance* (EEP, 2025)
Mensah, Thomas, 'The Role of Peaceful Dispute Settlement in Contemporary Ocean Policy and Law', 81–94 in Davor Vidas – Willy Østreng (Eds), *Order for the Oceans at the Turn of the Century* (Nijhoff, 1999)
Menzel, Anja – Otto, Lisa, 'Understanding Security and Sustainable Development in the Maritime Domain', 7–21 in Lisa Otto – Anja Menzel (Eds), *Global Challenges in Maritime Security: Sustainability and the sea* (Springer, 2024)
Merkouris, Panos, 'In Dubio Mitius', 259–306 in Joseph Klingler – Yuri Parkhomenko – Constantinos Salonidis (Eds), *Between the Lines of the Vienna Convention? Canons and Other Principles of Interpretation in Public International Law* (Kluwer, 2018)
Merrills, John, 'Two Approaches to Treaty Interpretation', *Aust.* YbIL [4] 55–82 (1968)
Meseguer, José, 'Le Régime Juridique de l'Exploitation des Stocks Conjoints de Poissons au-delà des 200 Milles', *Annuaire Français* [28] 885–899 (1982)
Miles, Edward, 'The Structure and Effects of the Decision Process in the Seabed Committee and the Third United Nations Conference on the Law of the Sea', *International Organization* 31(2) 159–234 (1977)
Miles, Edward, *Global Ocean Politics: The decision process at the Third United Nations Conference on the Law of the Sea, 1973–1982* (Nijhoff, 1998)
Miles, Edward – Burke, William, 'Pressures on the United Nations Convention on the Law of the Sea of 1982 Arising from New Fisheries Conflicts: The problem of straddling stocks', ODIL 20(4) 343–357 (1989)
Miller, Denzil – Molenaar, Erik, 'The SEAFC Convention: A comparative analysis in a developing coastal State perspective', *Ocean Yearbook* [20] 305–375 (2006)
Miovski, Lourene, 'Solutions in the Convention on the Law of the Sea to the Problem of Overfishing in the Central Bering Sea: Analysis of the convention, highlighting the provisions concerning fisheries and enclosed and semi-enclosed Seas', SDLR 26(3) 525–574 (1989)
Miron, Alina, 'Compulsory Conciliation Proceedings Regarding Disputes on Conservation and Management of Fish Stocks', 180–192 in Bjørn Kunoy – Tomas Heidar – Constantinos Yiallourides (Eds), *International Fisheries Law: Persistent and emerging challenges* (Routledge, 2024)
Mitchell, Carlyle, 'Fisheries Management in the Grand Banks, 1980–1992 and the Straddling Stock Issue', *Marine Policy* 21(1) 97–109 (1997)

Mizukami, Chiyuki, 'Management of Highly Migratory Species and Fisheries Relations between Japan and South Pacific States', *University of British Columbia Law Review* 24(1) 127–140 (1990)
Molenaar, Erik, 'The Concept of "Real Interest" and Other Aspects of Co-operation through Regional Fisheries Management Mechanisms', IJMCL 15(4) 475–531 (2000)
Molenaar, Erik, 'Addressing Regulatory Gaps in High Seas Fisheries', IJMCL 20(3) 533–570 (2005)
Molenaar, Erik, 'Integrating Climate Change in International Fisheries Law', 263–288 in Elise Johansen – Signe Busch – Ingvild Jakobsen (Eds), *The Law of the Sea and Climate Change: Solutions and constraints* (CUP, 2020)
Molenaar, Erik, 'Regional Fisheries Management Organizations', 81–109 in Marta Ribeiro – Fernando Bastos – Tore Henriksen (Eds), *Global Challenges and the Law of the Sea* (Springer, 2020)
Molenaar, Erik, 'International Cooperation through Regional Fisheries Management Organizations', 23–43 in Olav Stokke – Andreas Østhagen – Andreas Raspotnik (Eds), *Marine Resources, Climate Change and International Management Regimes* (Bloomsbury, 2022)
Molenaar, Erik – Caddell, Richard, 'International Fisheries Law: Achievements, limitations and challenges', 3–10 in Richard Caddell – Erik Molenaar (Eds), *Strengthening International Fisheries Law in an Era of Changing Oceans* (Hart, 2019)
Momtaz, Djamchid, 'L'Accord Relatif à la Conservation et à la Gestion des Stocks de Poissons Chevauchants et de Grands Migrateurs', *Annuaire Français* [41] 676–699 (1995)
Montagu, Basil (Ed.), *The Works of Francis Bacon*, Volume IV (Baldwin, 1826)
Moore, Gerald, 'National Legislation for the Management of Fisheries under Extended Coastal State Jurisdiction', JMLC 11(2) 153–182 (1979)
Moore, John, 'The United Nations Convention on the Law of the Sea: One of the greatest achievements in the international rule of law', 5–23 in Myron Nordquist – John Moore – Ronán Long (Eds), *Legal Order in the World's Oceans* (Brill–Nijhoff, 2017)
Morgan, Edward, *The Aesthetics of International Law* (Toronto University Press, 2007)
Morin, Jacques-Yvan, 'The Quiet Revolution: Canadian approaches to the law of the sea', 11–31 in Ralph Zacklin (Ed.), *The Changing Law of the Sea: Western hemisphere perspectives* (Sijthoff, 1974)
Morin, Michel, '"Creeping Jurisdiction" by the Small Islands of the Pacific Ocean in the Context of Management of the Tuna Fisheries', IJMCL 30(3) 477–511 (2015)
Mossop, Joanna, 'Reimagining the Procedural Aspects of Part XV of the United Nations Convention on the Law of the Sea', IJMCL 38(2) 378–401 (2023)
Mossop, Joanna – Schofield, Clive, 'Adjacency and Due Regard: The role of coastal states in the BBNJ treaty', *Marine Policy* [122] 103877–10394 (2020)
Munro, Gordon, 'Extended Jurisdiction and the Management of Pacific Highly Migratory Species', ODIL 21(3) 289–307 (1990)
Munro, Gordon, 'The United Nations Fish Stocks Agreement of 1995: History and problems of implementation', *Marine Resource Economics* 15(4) 265–280 (1995)
Murphy, Sean, 'Conservation of Fish in the Western and Central Pacific Ocean', AJIL 95(1) 152–155 (2001)

Murphy, Sean, 'Creativity in Dispute Settlement Relating to the Law of the Sea', 195–218 in Charles Brower – Joan Donoghue – Cian Murphy – Cymie Payne – Esmé Shirlow (Eds), *By Peaceful Means* (OUP, 2024)

Murphy, Sean, 'Straddling and Highly Migratory Fish Stocks Compatibility Requirement and Scope of Rights of High Seas Fishing Nations', 222–244 in Bjørn Kunoy – Tomas Heidar – Constantinos Yiallourides (Eds), *International Fisheries Law: Persistent and emerging challenges* (Routledge, 2024)

Nandan, Satya, 'The Draft Agreement on the Conservation and Management of Straddling Fish Stocks and Highly Migratory Fish Stocks', 291–297 in Myron Nordquist – John Moore (Eds), *Entry into Force of the Law of the Sea Convention* (Nijhoff, 1995)

Nandan, Satya, 'An Introduction to the 1982 United Nations Convention on the Law of the Sea', 9–13 in Davor Vidas – Willy Østreng (Eds), *Order for the Oceans at the Turn of the Century* (Nijhoff, 1999)

Nandan, Satya, 'Revisiting the Law of the Sea Negotiations', 16–22 in Myron Nordquist – John Moore – Ronán Long (Eds), *The Marine Environment and United Nations Sustainable Development Goal 14: Life below water* (Brill–Nijhoff, 2018)

Nandan, Satya – Dalaker, Kristine, *Reflections on the Making of the Modern Law of the Sea* (National University of Singapore Press, 2021)

Ndiaye, Tafsir, 'Provisional Measures before the International Tribunal for the Law of the Sea', 95–101 in Myron Nordquist – John Moore (Eds), *Current Marine Environmental Issues and the International Tribunal for the Law of the Sea* (Nijhoff, 2001)

Ndiaye, Tafsir, 'Proceedings before the International Tribunal for the Law of the Sea', *Journal of Law and Judicial System* 1(3) 46–81 (2018)

Nelson, Dolliver, 'The Patrimonial Sea', ICLQ 22(4) 668–686 (1973)

Nelson, Dolliver, 'The Development of the Legal Regime of High Seas Fisheries', 113–134 in Alan Boyle – David Freestone (Eds), *International Law and Sustainable Development: Past achievements and future challenges* (OUP, 1999)

Nguyen, Lan, 'The Chagos Marine Protected Area Arbitration: Has the scope of LOSC compulsory jurisdiction been clarified?', IJMCL 31(1) 120–143 (2016)

Nguyen, Lan, *The Development of the Law of the Sea by UNCLOS Dispute Settlement Bodies* (CUP, 2023)

Nolte, Georg (Ed.), *Treaties and Subsequent Practice* (OUP, 2013)

Nordquist, Myron, 'The Implementation of the 200-mile Exclusive Economic Zone', *International Business Lawyer* 6(2) 170–173 (1978)

Nordquist, Myron – Rosenne, Shabtai – Sohn, Louis, *United Nations Convention on the Law of the Sea 1982: A commentary*, Volume V (Nijhoff, 1989)

Nordquist, Myron – Rosenne, Shabtai – Nandan, Satya – Grandy, Neal, *United Nations Convention on the Law of the Sea 1982: A commentary*, Volume II (Nijhoff, 1993)

Nordquist, Myron – Rosenne, Shabtai – Nandan, Satya – Grandy, Neal, *United Nations Convention on the Law of the Sea 1982: A commentary*, Volume III (Nijhoff, 1995)

Noyes, John, 'Compulsory Third-Party Adjudication and the 1982 United Nations Convention on the Law of the Sea', *Connecticut Journal of International Law* 4(3) 675–696 (1989)

Noyes, John, 'Louis B. Sohn and the Law of the Sea', *Willamette Journal of International Law and Dispute Resolution* 16(2) 238–251 (2008)

Noyes, John, 'Memorializing UNCLOS III, Interpreting the Law of the Sea Convention, and the *Virginia Commentary*', 218–238 in Michael Lodge – Myron Nordquist

(Eds), *Peaceful Order in the World's Oceans: Essays in honor of Satya N. Nandan* (Brill–Nijhoff, 2014)

Ntovas, Alexandros, 'Interpreting the Dispute Settlement Limitation on Fisheries after the Chagos Marine Protected Area Arbitration', 225–253 in Stephen Minas – Jordan Diamond (Eds), *Stress Testing the Law of the Sea* (Brill–Nijhoff, 2018)

O'Connell, Daniel (edited by Ivan Shearer), *The International Law of the Sea* (Clarendon Press, 1982)

O'Reilly-Hinds, Lennox, 'Crisis in Canada's Atlantic Sea Fisheries', *Marine Policy* 19(4) 271–283 (1995)

Oda, Shigeru, 'New Trends in the Regime of the Seas: A consideration of the problems of conservation and distribution of marine resources', ZaöRV [18] 262–286 (1957)

Oda, Shigeru, 'The 1958 Geneva Convention on the Fisheries: Its immaturities', *Die Friedens-Warte* [55] 317–339 (1959)

Oda, Shigeru, 'International Law of the Resources of the Sea', *Recueil des Cours* [127] 355–483 (1969)

Oda, Shigeru, 'Sharing of Ocean Resources – Unresolved Issues in the Law of the Sea', *New York Journal of International and Comparative Law* 3(1) 1–14 (1981)

Oda, Shigeru, 'Fisheries under the United Nations Convention on the Law of the Sea', AJIL 77(4) 739–755 (1983)

Oda, Shigeru, *International Control of Sea Resources* (Nijhoff, 1989 2nd edition)

Oda, Shigeru, 'Dispute Settlement Prospects in the Law of the Sea', ICLQ 44(4) 863–872 (1995)

Oda, Shigeru, *Fifty Years of the Law of the Sea* (Nijhoff, 2003)

Ong, David, 'International Law of the Sea', 710–747 in Michael Bowman – Dino Kritsiotis (Eds), *Conceptual and Contextual Perspectives on the Modern Law of Treaties* (CUP, 2018)

Opsahl, Torkel, 'Towards the Rule of International Law in High Seas Fisheries', NCIL 27(1) 265–322 (1957)

Orakhelashvili, Alexander, 'The Concept of International Judicial Jurisdiction: A reappraisal', *Law and Practice of International Courts and Tribunals* 2(3) 501–550 (2003)

Orakhelashvili, Alexander, *The Interpretation of Acts and Rules in Public International Law* (OUP, 2008)

Oral, Nilufer, 'Protection of Vulnerable Marine Ecosystems in Areas Beyond National Jurisdiction: Can international law meet the challenge?', 85–108 in Anastasia Strati – Maria Gavouneli – Nikolaos Skourtos (Eds), *Unresolved Issues and New Challenges to the Law of the Sea: Time before and time after* (Nijhoff, 2006)

Ørebech, Peter, 'The "Lost Mackerel" of the North East Atlantic – The Flawed System of Trilateral and Bilateral Decision-Making', IJMCL 28(2) 343–373 (2013)

Ørebech, Peter – Sigurjonsson, Ketill – McDorman, Ted, 'The 1995 United Nations Straddling and Highly Migratory Fish Stocks Agreement: Management, enforcement and dispute settlement', IJMCL 13(2) 119–141 (1998)

Orellana, Marcos, 'The Swordfish Dispute between the EU and Chile at the ITLOS and the WTO', NJIL 71(1) 55–81 (2002)

Orellana, Marcos, 'The Law on Highly Migratory Fish Stocks: ITLOS jurisprudence in context', *Golden Gate University Law Review* 34(3) 459–495 (2004)

Orrego-Vicuña, Francisco (Ed.), *The Exclusive Economic Zone, A Latin American Perspective* (Routledge, 1984)

Orrego-Vicuña, Francisco, *The Exclusive Economic Zone: Regime and legal nature under international law* (CUP, 1989)

Orrego-Vicuña, Francisco, 'The "Presential Sea": Defining coastal States' special interest in high seas fisheries and other activities', GYIL [35] 264–292 (1993)
Orrego-Vicuña, Francisco, 'Toward an Effective Management of High Seas Fisheries and the Settlement of the Pending Issues of the Law of the Sea', ODIL 24(1) 81–92 (1993)
Orrego-Vicuña, Francisco, 'Towards an Effective Management of High Seas Fisheries and the Settlement of the Pending Issues of the Law of the Sea: The view of developing countries ten years after the signature of the Law of the Sea Convention', 415–430 in Edward Miles – Tullio Treves (Eds), *The Law of the Sea: New worlds, new discoveries* (LOSI UH, 1993)
Orrego-Vicuña, Francisco, 'Coastal States' Competences over High Seas Fisheries and the Changing Role of International Law', ZaöRV [55] 520–534 (1995)
Orrego-Vicuña, Francisco, *The Changing International Law of High Seas Fisheries* (CUP, 1999)
Orrego-Vicuña, Francisco, 'From the 1893 *Bering Sea Fur-Seals* Case to the 1999 *Southern Bluefin Tuna* Cases: A century of efforts at conservation of the living resources of the high seas', YbIEL [10] 40–47 (2000)
Orrego-Vicuña, Francisco, 'The International Law of High Seas Fisheries: From freedom of fishing to sustainable use', 23–52 in Olav Stokke (Ed.), *Governing High Seas Fisheries: The Interplay of global and regional regimes* (OUP, 2001)
Østhagen, Andreas, 'Fish, Not Oil, at the Heart of (Future) Arctic Resource Conflicts', *Arctic Yearbook* [9] 43–59 (2020)
Østhagen, Andreas – Spijkers, Jessica – Totland, Olav, 'Collapse of Cooperation? The North-Atlantic Mackerel Dispute and Lessons for International Cooperation on Transboundary Fish Stocks', *Maritime Studies* 19(2) 155–165 (2020)
Ostrom, Elinor, *Governing the Commons: The evolution of institutions for collective action* (CUP, 1990)
Oude-Elferink, Alex, *The Law of Maritime Boundary Delimitation: A case study of the Russian Federation* (Nijhoff, 1994)
Oude-Elferink, Alex, 'Fisheries in the Sea of Okhotsk High Seas Enclave – The Russian Federation's Attempts at Coastal State Control', IJMCL 10(1) 1–18 (1995)
Oude-Elferink, Alex, *The Impact of Article 7(2) of the Fish Stocks Agreement on the Formulation of Conservation & Management Measures for Straddling & Highly Migratory Fish Stocks* (FAO, 1999)
Oude-Elferink, Alex, 'The Determination of Compatible Conservation and Management Measures for Straddling and Highly Migratory Fish Stocks', *Yearbook UN Law* [5] 551–607 (2001)
Oude-Elferink, Alex, 'The Sea of Okhotsk Peanut Hole, *De Facto* Extension of Coastal State Control', 179–205 in Olav Stokke (Ed.), *Governing High Seas Fisheries: The Interplay of global and regional regimes* (OUP, 2001)
Oude-Elferink, Alex, 'Coastal States and MPAs in ABNJ: Ensuring consistency with the LOSC', 56–84 in David Freestone (Ed.), *Conserving Biodiversity in Areas Beyond National Jurisdiction* (Brill–Nijhoff, 2019)
Oude-Elferink, Alex, 'What if the United Nations Convention on the Law of the Sea had Entered into Force Unamended: Business as usual or dystopia?', 215–230 in Ingo Venzke – Kevin Heller (Eds), *Contingency in International Law: On the possibility of different legal histories* (OUP, 2021)
Oude-Elferink, Alex – Georgoula, Danae – Nguyen, Lan – Trevisanut, Seline, 'Compulsory Jurisdiction as the DNA of LOSC Dispute Settlement: An evolutionary path', IJMCL 38(2) 185–206 (2023)

Oude-Elferink, Alex – Henriksen, Tore – Busch, Signe, 'The Judiciary and the Law of Maritime Delimitation', 1–32 in Alex Oude-Elferink – Tore Henriksen – Signe Busch (Eds), *Maritime Boundary Delimitation: The case law – is it consistent and predictable?* (CUP, 2018)
Oudendijk, Johanna, *Status and Extent of Adjacent Waters: A historical orientation* (Sijthoff, 1970)
Overfield, Cornell – Yllemo, Jessica, 'Options for Sustainable High Seas Fisheries Management in the Southwest Atlantic', *International Law Studies* [106] 85–100 (2025)
Oxman, Bernard, 'The Third United Nations Conference on the Law of the Sea: The 1976 New York sessions', AJIL 71(2) 247–269 (1977)
Oxman, Bernard, 'The Third United Nation's Conference on the Law of the Sea: The 1977 New York session', AJIL 72(1) 57–83 (1978)
Oxman, Bernard, 'The Third United Nations Conference on the Law of the Sea: The seventh session (1978)', 73(1) AJIL 1–41 (1979)
Oxman, Bernard, 'The Third United Nations Conference on the Law of the Sea: The eighth session (1979)', AJIL 74(1) 1–47 (1980)
Oxman, Bernard, 'The Third United Nations Conference on the Law of the Sea: The ninth session (1980)', AJIL 75(2) 211–256 (1981)
Oxman, Bernard, 'The Third United Nations Conference on the Law of the Sea: The tenth session (1981)', AJIL 76(1) 1–23 (1982)
Oxman, Bernard, 'Coastal States' Competences over High Seas Fisheries and the Changing Role of International Law – Comment', ZaöRV [55] 536–543 (1995)
Oxman, Bernard, 'The Rule of Law and the United Nations Convention on the Law of the Sea', EJIL 7(3) 353–371 (1996)
Oxman, Bernard, 'Complementary Agreements and Compulsory Jurisdiction', AJIL 95(2) 277–312 (2001)
Oxman, Bernard, 'The Territorial Temptation: A siren song at sea', AJIL 100(4) 830–850 (2006)
Oxman, Bernard, 'Courts and Tribunals: The ICJ, ITLOS, and arbitral tribunals', 394–415 in Donald Rothwell – Alex Oude-Elferink – Karen Scott – Tim Stephens (Eds), *Oxford Handbook of the Law of the Sea* (OUP, 2015)
Oxman, Bernard, 'On the Challenges to Stability and to the Rule of Law in Implementation of UNCLOS', 276–284 in Harry Scheiber – Nilufer Oral – Moon-Sang Kwon (Eds), *Ocean Law Debates: The 50-year legacy and emerging issues for the years ahead* (Brill–Nijhoff, 2018)
Oxman, Bernard, 'Conservation and Management of the Living Resources of the Sea: The role of dispute settlement procedures', 419–436 in Charles Brower – Joan Donoghue – Cian Murphy – Cymie Payne – Esmé Shirlow (Eds), *By Peaceful Means* (OUP, 2024)
Paik, Jin-Hyun, 'ITLOS at Twenty: Reflections on its contribution to dispute settlement and the rule of law at sea', 189–209 in Myron Nordquist – John Moore – Ronán Long (Eds), *Legal Order in the World's Oceans* (Brill–Nijhoff, 2017)
Palacios-Abrantes, Juliano – Reygondeau, Gabriel – Wabnitz, Colette – Cheung, William, 'The Transboundary Nature of the World's Exploited Marine Species', *Nature Scientific Reports* 10(1) 17668–17680 (2020)
Paolillo, Felipe, 'The Exclusive Economic Zone in Latin American Practice and Legislation', ODIL 26(2) 105–125 (1995)
Pardo, Arvid, 'Development of Ocean Space – An International Dilemma', *Louisiana Law Review* 45(1) 45–72 (1970)

Pardo, Arvid, 'The Emerging Law of the Sea', 33–76 in Don Walsh (Ed.), *The Law of the Sea: Issues in ocean resource management* (Praeger Publishers, 1977)
Pardo, Arvid, 'The Law of the Sea – Where Now? Before and After', LCP 46(2) 95–105 (1983)
Pardo, Arvid, 'The Convention on the Law of the Sea: A preliminary appraisal', SDLR 20(3) 489–503 (1983)
Pardo, Arvid, 'The Law of the Sea: Its past and its future', OLR 63(1) 7–18 (1984)
Pardo, Arvid – Borgese, Elizabeth, *The New International Economic Order and the Law of the Sea* (International Ocean Institute, 1976)
Paulsson, Jan, *The Unruly Notion of Abuse of Rights* (CUP, 2020)
Pauly, Daniel – Zeller, Dirk, 'Catch Reconstructions Reveal that Global Marine Fisheries Catches are Higher than Reported and Declining', *Nature Communications* 7(1) 10244–10253 (2016)
Payoyo, Peter, *Cries of the Sea: World inequality, sustainable development and the common heritage of humanity* (Brill, 1997)
Pedrozo, Raul, 'Reflecting on UNCLOS Forty Years Later: What worked, what failed', *International Law Studies* [99] 874–943 (2022)
Peterson, Susan – Teal, John, 'Ocean Fisheries as a Factor in Strategic Policy and Action', 114–142 in Arthur Westing (Ed.), *Global Resources and International Conflict: Environmental factors in strategic policy and action* (OUP, 1986)
Phillips, John, 'The Exclusive Economic Zone as a Concept in International Law', ICLQ 26(3) 585–618 (1977)
Pickering, Robert – Taylor, William, 'Recent Developments in the Law of the Sea: A synopsis', SDLR 13(3) 628–666 (1976)
Pierce, George, 'Dispute Settlement Mechanisms in the Draft Convention on the Law of the Sea', *Denver Journal of International Law and Policy* 10(2) 331–354 (1981)
Pinto, Daniela, *Fisheries Management in Areas Beyond National Jurisdiction: The impact of ecosystem based law-making* (Brill, 2012)
Pinto, Moragodage, 'The Duty of Co-Operation and the United Nations Convention on the Law of the Sea', 131–154 in Adriaan Bos – Hugo Siblesz (Eds), *Realism in Law-Making: Essays on international law in honour of Willem Riphagan* (Nijhoff, 1985)
Pollard, Duke, 'The Exclusive Economic Zone: The elusive consensus', SDLR 12(3) 600–623 (1974)
Pontecorvo, Giulio, 'The Impact of the Law of the Sea Treaty on the Organization of World Fisheries: Some preliminary observations on production', 138–156 in Edward Brown – Robin Churchill (Eds), *The UN Convention on the Law of the Sea: Impact and implementation* (LOSI UH, 1987)
Prescott, Victor – Schofield, Clive, *The Maritime Political Boundaries of the World* (Brill, 2005 2nd edition)
Press, Anthony – Hodgson-Johnston, Indi – Constable, Andrew, 'The principles of the Convention on the Conservation of Antarctic Marine Living Resources: Why its commission is not a regional fisheries management organisation', 9–29 in Nengye Liu – Cassandra Brooks – Tianbao Qin (Eds), *Governing Marine Living Resources in the Polar Regions* (EEP, 2019)
Proelss, Alexander, *The United Nations Convention on the Law of the Sea: A commentary* (Beck, 2017)
Pulvenis de Séligny-Maurel, Jean-François, 'The Marine Living Resources and the Evolving Law of the Sea', *Aegean Review of the Law of the Sea and Maritime Law* 1(1) 61–94 (2010)

Pulvenis de Séligny-Maurel, Jean-François, 'Regional Fisheries Bodies and Regional Fisheries Management Organizations and the Settlement of Disputes Concerning Marine Living Resources', 698–712 in Lilian Del Castillo (Ed.), *Law of the Sea, From Grotius to the International Tribunal for the Law of the Sea: Liber amicorum Judge Hugo Caminos* (Brill–Nijhoff, 2015)

Queneudec, Jean-Pierre, 'Coup d'Oeil Retrospectif sur les Origines du Tribunal International du Droit de la Mer', 621–632 in Nisuke Ando – Edward McWhinney – Rüdiger Wolfrum (Eds), *Liber Amicorum Judge Shigeru Oda* (Kluwer, 2002)

Rao, Chandrasekhara, 'Delimitation Disputes under the United Nations Convention on the Law of the Sea: Settlement procedures', 877–897 in Tafsir Ndiaye – Rüdiger Wolfrum (Eds), *Law of the Sea, Environmental Law, and Settlement of Disputes: Liber amicorum Judge Thomas A. Mensah* (Nijhoff, 2007)

Rao, Chandrasekhara – Gautier, Philippe, *The International Tribunal for the Law of the Sea: Law, practice and procedure* (EEP, 2018)

Ray, Hilborn – Ulrike, Hilborn, *Ocean Recovery: A sustainable future for global fisheries?* (OUP, 2019)

Rayfuse, Rosemary, 'The United Nations Agreement on Straddling and Highly Migratory Fish Stocks as an Objective Regime: A case of wishful thinking?', *Aust.* YbIL [20] 253–278 (1999)

Rayfuse, Rosemary, 'The Future of Compulsory Dispute Settlement under the Law of the Sea Convention', VUWLR 36(4) 683–711 (2005)

Rayfuse, Rosemary, 'Some Reflections on What's Wrong with the Law of the Sea', 13–29 in Cedric Ryngaert – Erik Molenaar – Sarah Nouwen (Eds), *What's Wrong with International Law? Liber Amicorum A.H.A. Soons* (Brill–Nijhoff, 2015)

Rayfuse, Rosemary, 'Settling Disputes in Regional Fisheries Management Organisations: Dealing with objections', 240–276 in Hélène Fabri – Erik Franckx – Marco Benatar – Tamar Meshel (Eds), *A Bridge Over Troubled Waters: Dispute resolution in the law of international watercourses and the law of the sea* (Brill–Nijhoff, 2020)

Rembe, Nasila, *Africa and the International Law of the Sea: A study of the contribution of the African States to the Third United Nations Conference on the Law of the Sea* (Brill, 1980)

Rengifo-Lozano, Antonio, *International Regime Theory and the Law of the Sea: A study of fisheries on the high seas* (Editorial Universidad Nacional de Colombia, 2011)

Rhee, Sang-Myon, 'Sea Boundary Delimitation Between States before World War II', AJIL 76(3) 555–588 (1982)

Richardson, Elliot, 'Dispute Settlement under the Convention on the Law of the Sea: A flexible and comprehensive extension of the rule of the law to ocean space', 149–163 in Thomas Buergenthal (Ed.), *Contemporary Issues in International Law: Essays in honor of Louis B. Sohn* (Engel, 1984)

Riphagen, Willem, 'Mechanisms of Supervision in the Future Law of the Sea', 115–148 in Frits Kalshoven (Ed.), *Essays on the Development of the International Legal Order: In memory of Haro Frederik van Panhuys* (Sijthoff, 1980)

Riphagen, Willem, 'Dispute Settlement in the 1982 United Nations Convention on the Law of the Sea', 281–301 in Christos Rozakis – Constantine Stephanou (Eds), *The New Law of the Sea* (North-Holland, 1983)

Röben, Volker, 'The Southern Bluefin Tuna Cases: Re-regionalization of the settlement of law of the sea disputes?', ZaöRV [62] 61–72 (2002)

Roberts, Hayley, 'Identifying "Exclusionary Agreements": Agreement type as a procedural limitation in UNCLOS dispute settlement', ODIL 52(2) 113–142 (2021)

Romano, Cesare, 'The Southern Bluefin Tuna Dispute: Hints of a world to come … like it or not', ODIL 32(4) 313–348 (2001)
Rosenne, Shabtai, 'Relations between Governments and the International Law Commission', *Year Book of World Affairs* [19] 183–198 (1965)
Rosenne, Shabtai, 'Reflections on the Final Clauses in the New Law of the Sea Treaty', VJIL 18(1) 133–146 (1977)
Rosenne, Shabtai, 'Settlement of Fisheries Disputes in the Exclusive Economic Zone', AJIL 73(1) 89–119 (1979)
Rosenne, Shabtai, *Provisional Measures in International Law: The International Court of Justice and the International Tribunal for the Law of the Sea* (OUP, 2005)
Rothwell, Donald, 'The Law of the Sea, International Courts, and Judicialization', AJIL *Unbound* [115] 373–377 (2021)
Rothwell, Donald – Stephens, Tim, 'Dispute Resolution and the Law of the Sea: Reconciling the interaction between the LOS Convention and other environmental instruments', 209–230 in Alex Oude-Elferink – Donald Rothwell (Eds), *Ocean Management in the 21st Century: Institutional frameworks and responses* (Nijhoff, 2004)
Rothwell, Donald – Stephens, Tim, *The International Law of the Sea* (Hart, 2023 3rd edition)
Russell, Denise, *Who Rules the Waves? Piracy, overfishing and mining the oceans* (Pluto Press, 2010)
Saario, Voitto, 'Legal Aspects on the Peaceful Settlement of International Disputes', NJIL 35(1/2) 73–80 (1965)
Safina, Carl, 'Bluefin Tuna: Facing extinction', *Ecologist* 33(8) 47–55 (2003)
Sands, Philippe, 'International Courts and the Application on the Concept of *Sustainable Development*', *Yearbook UN Law* [3] 389–405 (1999)
Sato, Yoichiro, 'Fishy Business: A political-economic analysis of the southern bluefin tuna dispute', *Asian Affairs* 28(4) 217–237 (2002)
Saunders, Phillip, 'Jurisdiction and Principle in the Implementation of the Law of the Sea: The case of straddling stocks', 377–401 in Chi Carmody – Sylvia Rhodes – Yūji Iwasawa (Eds), *Trilateral Perspectives on International Legal Issues: Conflict and coherence* (American Society of International Law, 2003)
Saxena, Jagdish, 'Limits of Compulsory Jurisdiction in Respect of the Law of the Sea Disputes', 328–342 in Ram Anand (Ed.), *Law of the Sea: Caracas and beyond* (Nijhoff, 1980)
Scalieri, Erietta, 'Discretionary Power of Coastal States and the Control of its Compliance with International Law by International Tribunals', 349–382 in Angela Del Vecchio – Roberto Virzo (Eds), *Interpretations of the United Nations Convention on the Law of the Sea by International Courts and Tribunals* (Springer, 2019)
Scanlon, Zoe, 'Incorporating Taiwan in International Fisheries Management: The Southern Indian Ocean Fisheries Agreement experience', ODIL 48(1) 35–51 (2017)
Scanlon, Zoe, 'The Significance of Informal Lawmaking in International Fisheries Law', 210–232 in Natalie Klein (Ed.), *Unconventional Lawmaking in the Law of the Sea* (OUP, 2022)
Scelle, Georges, 'Obsession du Territoire', 347–361 in Frederik van Asbeck (Ed.), *Symbolae Verzijl: Présentées au Professeur J.H.W. Verzijl à l'Occasion de son LXX-ième Anniversaire* (Nijhoff, 1958)
Schachter, Oscar, 'The Nature and Process of Legal Development in International Society', 745–808 in Ronald Macdonald – Douglas Johnston (Eds), *The Structure*

and Process of International Law: Essays in legal philosophy, doctrine and theory (Kluwer, 1983)

Schaefer, Andrew, '1995 Canada–Spain Fishing Dispute', GIELR 10(3) 437–449 (1996)

Schaefer, Milner, 'Some Recent Developments Concerning Fishing and the Conservation of the Living Resources of the High Seas', SDLR 7(3) 371–407 (1970)

Schäli, Judith, 'Intergenerational Justice and the Concept of Common Concern in Marine Resource Allocation and Ocean Governance', 67–89 in Thomas Cottier – Shaheeza Lalani – Clarence Siziba (Eds), *Intergenerational Equity: Environmental and cultural concerns* (Brill–Nijhoff, 2019)

Schane, Sanford, *Language and the Law* (Continuum, 2006)

Schatz, Valentin, 'The Settlement of Disputes concerning Conservation of Fish Stocks in the Arctic and Antarctic High Seas: Towards comprehensive compulsory jurisdiction?', 196–221 in Nengye Liu – Cassandra Brooks – Tianbao Qin (Eds), *Governing Marine Living Resources in the Polar Regions* (EEP, 2019)

Schatz, Valentin, '*Ad Hoc* Expert Panels: RFMOs', *Max Planck Encyclopedia of International Procedural Law* [unpaged entry 3806] (OUP, 2022)

Schatz, Valentin, 'The Settlement of EEZ Fisheries Access Disputes under UNCLOS: Limitations to jurisdiction and compulsory conciliation', GöJIL 13(1) 82–118 (2023)

Schatz, Valentin – Proelss, Alexander – Liu, Nengye, 'The 2018 Agreement to Prevent Unregulated High Seas Fisheries in the Central Arctic Ocean: A critical analysis', IJMCL 34(2) 195–244 (2019)

Scheiber, Harry, 'Origins of the Abstention Doctrine in Ocean Law: Japanese–US relations and the Pacific fisheries, 1937–1958', ELQ 16(1) 23–100 (1989)

Scheiber, Harry, 'Ocean Governance and the Marine Fisheries Crisis: Two decades of innovation – and frustration', *Virginia Environmental Law Journal* 20(1) 119–138 (2001)

Scheiber, Harry, 'The Law of the Sea Institute: A new forum for debate of ocean law in the 1960s "decade of uncertainty"', 9–92 in Harry Scheiber – Nilufer Oral – Moon-Sang Kwon (Eds), *Ocean Law Debates: The 50-year legacy and emerging issues for the years ahead* (Brill–Nijhoff, 2018)

Schiffman, Howard, 'The South Pacific Regional Fisheries Management Organization: An improved model of decision-making for fisheries conservation?', *Journal of Environmental Studies and Sciences* 3(2) 209–216 (2013)

Schram, Gunnar, 'Self-Government and Natural Resources', NJIL 57(3) 331–337 (1988)

Schram, Gunnar – Tahindro, André, 'Developments in Principles for the Adoption of Fisheries Conservation and Management Measures', 251–286 in Ellen Hey (Ed.), *Developments in International Fisheries Law* (Kluwer, 1999)

Schrank, William, 'Extended Fisheries Jurisdiction: Origins of the current crisis in Atlantic Canada's fisheries', *Marine Policy* 19(4) 285–299 (1995)

Schwarzenberger, Georg, *The Dynamics of International Law* (Professional Books Limited, 1976)

Scotcher, Nkeiru, 'Tight Lines at Sea: An interdisciplinary analysis of fisheries, marine biodiversity and institutions', 141–159 in Niels Krabbe – David Langlet (Eds), *Marine Bioprospecting, Biodiversity and Novel Uses of Ocean Resources: New approaches in international law* (Hart, 2024)

Scott, Karen, 'The LOSC: "A constitution for the oceans" in the anthropocene?', *Aust. YbIL* [41] 269–298 (2023)

Scott, Norman, 'Ambiguity *versus* Precision: The changing role of terminology in conference diplomacy', 153–162 in Jovan Kurbalija – Hannah Slavik (Eds), *Language and Diplomacy* (DiploProjects, 2001)

Scott, Shirley, 'Australia's First Tuna Negotiations with Japan', *Marine Policy* 24(4) 309–318 (2000)

Scovazzi, Tullio, 'The Evolution of International Law of the Sea: New issues, new challenges', *Recueil des Cours* [286] 53–233 (2001)

Scovazzi, Tullio, '*Due Regard* Obligations, with Particular Emphasis on Fisheries in the Exclusive Economic Zone', IJMCL 34(1) 56–72 (2019)

Sebenius, James, *Negotiating the Law of the Sea* (Harvard University Press, 1984)

Sellars, Kirsten, *A 'Constitution for the Oceans': The long hard road to the UN Convention on the Law of the Sea* (CUP, 2025)

Serdy, Andrew, *The New Entrants Problem in International Fisheries Law* (CUP, 2016)

Serdy, Andrew, 'Managing Transboundary Fish Stocks for Sustainability', 104–138 in Jensen Øystein (Ed.), *The Development of the Law of the Sea Convention: The role of international courts and tribunals* (EEP, 2021)

Sevaly, Sen, 'The Evolution of High Seas Fisheries Management in the North-East Atlantic', OCM 35(2/3) 85–100 (1997)

Shany, Yuval, *The Competing Jurisdictions of International Courts and Tribunals* (OUP, 2003)

Sharma, Surya, 'Framework of Likely Disputes Under the Law of the Sea Convention – Some Thoughts', ZaöRV [45] 465–496 (1985)

Shearer, Ivan, *Starke's International Law* (Butterworth, 1994 11th edition)

Simma, Bruno, 'Consent: Strains in the treaty system', 485–511 in Ronald Macdonald – Douglas Johnston (Eds), *The Structure and Process of International Law: Essays in legal philosophy, doctrine and theory* (Kluwer, 1983)

Sinan, Hussain – Bailey, Megan – Hanich, Quentin – Azmi, Kamal, 'Common but Differentiated Rights and Responsibilities in Tuna Fisheries Management', *Fish and Fisheries* 23(1) 202–212 (2022)

Sinclair, Ian, *The International Law Commission* (Grotius Publications, 1987)

Singh, Gurdip, *United Nations Convention on the Law of the Sea: Dispute settlement mechanisms* (Academic Publications, 1985)

Sohn, Louis, 'Settlement of Disputes Arising out of the Law of the Sea Convention', SDLR 12(3) 495–517 (1975)

Sohn, Louis, 'Voting Procedures in United Nations Conferences for the Codification of International Law', AJIL 69(2) 310–353 (1975)

Sohn, Louis, 'US Policy Toward the Settlement of the Law of the Sea Disputes', VJIL 17(1) 9–22 (1976)

Sohn, Louis, 'Peaceful Settlement of Disputes in Ocean Conflicts: Does UNCLOS/III point the way?', LCP 46(2) 195–200 (1983)

Sohn, Louis, 'Settlement of Law of the Sea Disputes', IJMCL 10(2) 205–218 (1995)

Sohn, Louis, 'The Importance of the Peaceful Settlement of Disputes Provisions of the United Nations Convention on the Law of the Sea', 265–278 in Myron Nordquist – John Moore (Eds), *Entry into Force of the Law of the Sea Convention* (Nijhoff, 1995)

Sohn, Louis – Noyes, John, *Cases and Materials on the Law of the Sea* (Brill, 2004)

Sohn, Louis – Noyes, John – Franckx, Erik – Juras, Kristen, *Cases and Materials on the Law of the Sea* (Brill–Nijhoff, 2014 2nd edition)

Solomou, Alexia, 'Exceptions to a Rule Must be Narrowly Construed', 359–386 in Joseph Klingler – Yuri Parkhomenko – Constantinos Salonidis (Eds), *Between the Lines of the Vienna Convention? Canons and Other Principles of Interpretation in Public International Law* (Kluwer, 2018)
Sorensen, Max, 'The Law of the Sea', *International Conciliation* [32] 193–256 (1958)
Soroos, Marvin, 'Conflict in the Use and Management of International Commons', 31–43 in Jyrki Käkönen (Ed.), *Perspectives on Environmental Conflict and International Politics* (Pinter Publishers, 1992)
Sovacool, Benjamin, 'A Game of Cat and Fish: How to restore the balance in sustainable fisheries management', ODIL 40(1) 97–125 (2009)
Spiropoulos, Jean, 'The Contribution of the International Law Commission to the Codification of the Law on Fishing and Conservation of the Living Resources of the High Seas', 332–335 in Ricardo Alfaro, [and the other Editors of the Netherlands International Review] (Eds), *Varia Juris Gentium: Questions of international law; Liber amicorum presented to Jean Pierre Adrien François at the occasion of his seventieth birthday* (Sythoff, 1959)
Stephens, Tim, 'A Paper Umbrella which Dissolves in the Rain? Implications of the Southern Bluefin Tuna Case for the Compulsory Resolution of Disputes Concerning the Marine Environment under the 1982 LOS Convention', *Asia Pacific* LPJ 6(3/4) 297–318 (2001)
Stephens, Tim, 'The Limits of International Adjudication in International Environmental Law: Another perspective on the Southern Bluefin Tuna Case', IJMCL 19(2) 173–193 (2004)
Stephens, Tim, 'Fisheries-led Development in the South Pacific: Charting a "Pacific Way" to a sustainable future', ODIL 39(3) 257–286 (2008)
Stevenson, John – Oxman, Bernard, 'The Preparations for the Law of the Sea Conference', AJIL 68(1) 1–32 (1974)
Stevenson, John – Oxman, Bernard, 'The Third United Nations Conference on the Law of the Sea: The 1974 Caracas session', AJIL 69(1) 1–30 (1975)
Stevenson, John – Oxman, Bernard, 'The Third United Nations Conference on the Law of the Sea: The 1975 Geneva session', AJIL 69(4) 763–797 (1975)
Stevenson, John – Oxman, Bernard, 'The Future of the United Nations Convention on the Law of the Sea', AJIL 88(3) 488–499 (1994)
Stokke, Olav, 'Managing Fisheries in the Barents Sea Loophole: Interplay with the UN Fish Stocks Agreement', ODIL 32(3) 241–262 (2001)
Stokke, Olav, 'The Loophole of the Barents Sea Fisheries Regime', 273–302 in Olav Stokke (Ed.), *Governing High Seas Fisheries: The Interplay of global and regional regimes* (OUP, 2001)
Stokke, Olav – Østhagen, Andreas – Raspotnik, Andreas – Stiansen, Erik, 'Climate Change and Resilient Fisheries Management', 3–19 in Olav Stokke – Andreas Østhagen – Andreas Raspotnik (Eds), *Marine Resources, Climate Change and International Management Regimes* (Bloomsbury, 2022)
Stoll, Peter-Tobias – Vöneky, Silja, 'The Swordfish Case: Law of the sea v. trade', ZaöRV [62] 21–35 (2002)
Subedi, Surya, 'An Innovative Solution to an Ambitious Project: Dispute resolution in the 1982 Convention on the Law of the Sea', 163–186 in Charles Jalloh – Olufemi Elias (Eds), *Shielding Humanity: Essays in international law in honour of Judge Abdul G. Koroma* (Brill–Nijhoff, 2015)
Sumaila, Ussif, 'A Review of Game-Theoretic Models of Fishing', *Marine Policy* 23(1) 1–10 (1999)

Sun, Zhen, *Finding a Balance in the Exclusive Economic Zone: Conflict and stability in the law of the sea* (CUP, 2025)

Swan, Judith, *Regional Fishery Bodies and Governance: Issues, actions and future directions* (FAO, 2000)

Sydnes, Are, 'New Regional Fisheries Management Regimes: Establishing the south east Atlantic fisheries organisation', *Marine Policy* 25(5) 353–364 (2001)

Symonides, Janusz, 'The Legal Status of the Enclosed and Semi-Enclosed Seas', GYIL [27] 315–333 (1984)

Székely, Alberto, 'Yellow-Fin Tuna: A transboundary resource of the eastern Pacific', *Natural Resources Journal* 29(4) 1051–1066 (1989)

Tahindro, André, 'Conservation and Management of Transboundary Fish Stocks: Comments in light of the adoption of the 1995 Agreement for the Conservation and Management of Straddling Fish Stocks and Highly Migratory Fish Stocks', ODIL 28(1) 1–58 (1997)

Tahindro, André, 'Sustainable Fisheries: The legal regime of the 1995 United Nations Fish Stocks Agreement and its contribution to subsequent developments promoting sustainable fisheries', 323–369 in Myron Nordquist – John Moore – Ronán Long (Eds), *Legal Order in the World's Oceans* (Brill–Nijhoff, 2017)

Takei, Yoshinobu, *Filling Regulatory Gaps in High Seas Fisheries: Discrete high seas fish stocks, deep-sea fisheries and vulnerable marine ecosystems* (Brill, 2013)

Talmon, Stefan, 'The Chagos Marine Protected Area Arbitration: Expansion of the jurisdiction of UNCLOS Part XV courts and tribunals', ICLQ 65(4) 927–951 (2016)

Tanaka, Norio, 'Some Observations on the Southern Bluefin Arbitration Award', *Japanese Annual of International Law* [44] 9–34 (2001)

Tanaka, Yoshifumi, 'Arbitral Tribunal Award, Annex VII LOSC Barbados v Trinidad and Tobago', IJMCL 21(4) 523–534 (2006)

Tanaka, Yoshifumi, *A Dual Approach to Ocean Governance: The cases of zonal and integrated management in international law of the sea* (Ashgate Publishing, 2008)

Tanaka, Yoshifumi, *The Peaceful Settlement of International Disputes* (CUP, 2018)

Tanaka, Yoshifumi, 'Toward Sustainable Management of Marine Natural Resources', 110–133 in Markus Kotzur – Nele Matz-Lück – Alexander Proelss – Roda Verheyen – Joachim Sanden (Eds), *Sustainable Ocean Resource Governance* (Brill–Nijhoff, 2018)

Tanaka, Yoshifumi, *Predictability and Flexibility in the Law of Maritime Delimitation* (Hart, 2019)

Tanaka, Yoshifumi, 'Reflections on the Implications of Environmental Norms for Fishing: The link between the regulation of fishing and the protection of marine biological diversity', *International Community Law Review* 22(3/4) 389–409 (2020)

Tanaka, Yoshifumi, *The International Law of the Sea* (CUP, 2023 4th edition)

Tanaka, Yoshifumi, 'Jurisdictional Issues in the Settlement of Disputes Concerning Straddling Fish Stocks Under UNCLOS', 193–203 in Bjørn Kunoy – Tomas Heidar – Constantinos Yiallourides (Eds), *International Fisheries Law: Persistent and emerging challenges* (Routledge, 2024)

Tanja, Gerard, *The Legal Determination of International Maritime Boundaries: The progressive development of continental shelf, EFZ and EEZ law* (Kluwer, 1990)

Tarte, Sandra, 'A Duty to Cooperate: Building a regional regime for the conservation and management of highly migratory fish stocks in the western and central Pacific', *Ocean Yearbook* [16] 261–299 (2002)

Taylor, Graham, 'The Content of the Rule against Abuse of Rights in International Law', BYBIL [46] 323–352 (1972)

Taylor, William – Lynch, Abigail – Schechter, Michael (Eds), *Sustainable Fisheries: Multi-level approaches to a global problem* (American Fisheries Society, 2011)

Telesetsky, Anastasia, 'The Sustainable Development Goals and Informal Lawmaking Processes', 271–289 in Natalie Klein (Ed.), *Unconventional Lawmaking in the Law of the Sea* (OUP, 2022)

Titi, Catharine, *The Function of Equity in International Law* (OUP, 2021)

Todd, James, 'The "Law-making" Behavior of States in the United Nations as a Function of their Location within Formal World Regions', *International Studies Quarterly* 15(3) 297–315 (1971)

Townsend-Gault, Ian, 'The "Territorialisation" of the Exclusive Economic Zone: A requiem for the remnants of the freedom of the seas?', 65–76 in Clive Schofield – Seokwoo Lee – Moon-Sang Kwon (Eds), *The Limits of Maritime Jurisdiction* (Brill–Nijhoff, 2014)

Treves, Tullio, 'La Pêche en Haute Mer et l'Avenir de la Convention des Nations Unies sur le Droit de la Mer', *Annuaire Français* [38] 885–904 (1992)

Treves, Tullio, 'The Law of the Sea Tribunal: Its status and scope of jurisdiction after November 16, 1994', ZaöRV [55] 421–451 (1995)

Treves, Tullio, 'The Settlement of Disputes According to the Straddling Stocks Agreement of 1995', 253–269 in Alan Boyle – David Freestone (Eds), *International Law and Sustainable Development: Past achievements and future challenges* (OUP, 1999)

Treves, Tullio, 'New Trends in the Settlement of Disputes and the Law of the Sea Convention', 61–86 in Harry Scheiber (Ed.), *Law of the Sea: The common heritage and emerging challenges* (Nijhoff, 2000)

Treves, Tullio, 'The Jurisdiction of the International Tribunal for the Law of the Sea', 111–131 in Chandrasekhara Rao – Rahmatullah Khan (Eds), *The International Tribunal for the Law of the Sea: Law and practice* (Kluwer, 2001)

Treves, Tullio, 'Preliminary Proceedings in the Settlement of Disputes under the United Nations Law of the Sea Convention: Some observations', 749–761 in Nisuke Ando – Edward McWhinney – Rüdiger Wolfrum (Eds), *Liber Amicorum Judge Shigeru Oda* (Kluwer, 2002)

Treves, Tullio, 'A System for Law of the Sea Dispute Settlement', 417–432 in David Freestone – Richard Barnes – David Ong (Eds), *The Law of the Sea, Progress and Prospects* (OUP, 2006)

Treves, Tullio, 'What Have the United Nations Convention and the International Tribunal for the Law of the Sea to Offer as Regards Maritime Delimitation Disputes?', 71–86 in Rainer Lagoni – Daniel Vignes (Eds), *Maritime Delimitation* (Brill, 2006)

Treves, Tullio, 'Fisheries Disputes: Judicial and arbitral practice since the entry into force of UNCLOS', 328–336 in Rüdiger Wolfrum – Maja Seršić – Trpimir Šošić (Eds), *Contemporary Developments in International Law: Essays in honour of Budislav Vukas* (Brill–Nijhoff, 2016)

Triska, Jan – Slusser, Robert, *The Theory, Law, and Policy of Soviet Treaties* (Stanford University Press, 1962)

Tsamenyi, Martin – Hanich, Quentin, 'Fisheries Jurisdiction under the Law of the Sea Convention: Rights and obligations in maritime zones under the sovereignty of coastal States', IJMCL 27(4) 783–794 (2012)

Tsamenyi, Martin – Manarangi-Trott, Lara, 'Role of Regional Organizations in Meeting LOS Convention Challenges: The Western and Central Pacific Experience',

187–208 in Alex Oude-Elferink – Donald Rothwell (Eds), *Ocean Management in the 21st Century: Institutional frameworks and responses* (Nijhoff, 2004)

Tuerk, Helmut, *Reflections on the Contemporary Law of the Sea* (Brill, 2012)

Ulfstein, Geir, '200 Mile Zone and Fisheries Management', NJIL 52(3) 3–33 (1983)

Urrutia, Osvaldo, 'The Objection Procedure in the South Pacific Regional Fisheries Management Organisation', 106–122 in Bjørn Kunoy – Tomas Heidar – Constantinos Yiallourides (Eds), *International Fisheries Law: Persistent and emerging challenges* (Routledge, 2024)

Valencia, Mark, 'The Yellow Sea: Transnational marine resource management issues', *Marine Policy* 12(4) 382–395 (1988)

Valencia-Rodríguez, Luis, 'The Contributions of Latin America to the Implementation of the UNCLOS', 144–172 in Lilian Del Castillo (Ed.), *Law of the Sea, From Grotius to the International Tribunal for the Law of the Sea: Liber amicorum Judge Hugo Caminos* (Brill–Nijhoff, 2015)

Van Dyke, Jon, 'Modifying the 1982 Law of the Sea Convention, New Initiatives on Governance of High Seas Fisheries Resources: The straddling stocks negotiations', IJMCL 10(2) 219–227 (1995)

Van Dyke, Jon, 'Sharing Ocean Resources – In a Time of Scarcity and Selfishness', 3–36 in Harry Scheiber (Ed.), *Law of the Sea: The common heritage and emerging challenges* (Nijhoff, 2000)

Van Dyke, Jon, 'Allocating Fish Across Jurisdictions', 821–844 in Tafsir Ndiaye – Rüdiger Wolfrum (Eds), *Law of the Sea, Environmental Law, and Settlement of Disputes: Liber amicorum Judge Thomas A. Mensah* (Brill, 2007)

Van Dyke, Jon, 'Allocating Fish Across Jurisdictions', 163–179 in Robin Allen – James Joseph – Dale Squires (Eds), *Conservation and Management of Transnational Tuna Fisheries* (Blackwell, 2010)

Van Logchem, Youri, *The Rights and Obligations of States in Disputed Maritime Areas* (CUP, 2021)

Verzijl, Jan, 'The United Nations Conference on the Law of the Sea, Geneva, 1958', *Nederlands Tijdschrift* 6(1) 115–139 (1959)

Verzijl, Jan, *International Law in Historical Perspective, Part VIII: Inter-State disputes and their settlement* (Sijthoff, 1976)

Vidas, Davor, 'The Delimitation of the Territorial Sea, the Continental Shelf, and the EEZ: A comparative perspective', 33–61 in Alex Oude-Elferink – Tore Henriksen – Signe Busch (Eds), *Maritime Boundary Delimitation: The case law – is it consistent and predictable?* (CUP, 2018)

Vignes, Daniel, 'Will the Third Conference on the Law of the Sea Work According to the Consensus Rule?', AJIL 69(1) 119–129 (1975)

Vignes, Daniel, 'Les Déclarations faites par les États Signataires de la Convention sur le Droit de la Mer sur la Base de l'Article 310 de cette Convention', *Annuaire Français* [29] 715–748 (1983)

Vignes, Daniel – Casado-Raigón, Rafael – Cataldi, Giuseppe, *Le Droit International de la Pêche Maritime* (Bruylant, 2000)

Virzo, Roberto, 'The "General Rule of Interpretation" in the International Jurisprudence Relating to the United Nations Convention on the Law of the Sea', 15–38 in Angela Del Vecchio – Roberto Virzo (Eds), *Interpretations of the United Nations Convention on the Law of the Sea by International Courts and Tribunals* (Springer, 2019)

Von Zharen, Wyndylyn, 'An Ecopolicy Perspective for Sustaining Living Marine Species', ODIL 30(1) 1–41 (1999)

Vrancken, Patrick, *State Ocean Jurisdiction* (Routledge, 2023)
Vukas, Budislav, 'Possible Role of the International Tribunal for the Law of the Sea in Interpretation and Progressive Development of the Law of the Sea', 95–104 in Davor Vidas – Willy Østreng (Eds), *Order for the Oceans at the Turn of the Century* (Nijhoff, 1999)
Vukas, Budislav, *The Law of the Sea: Selected writings* (Brill, 2004)
Vylegzhanin, Alexander – Young, Oran – Berkman, Paul, 'Governing the Barents Sea Region: Current status, emerging issues, and future options', ODIL 49(1) 52–78 (2018)
Watanabe, Hiromoto, 'Current Fisheries Issues: The position of a *fishing nation* and current cases', 305–310 in Myron Nordquist – John Moore (Eds), *Entry into Force of the Law of the Sea Convention* (Nijhoff, 1995)
Watts, Arthur, *International Law Commission 1949–1998, Volume I: The Treaties* (OUP, 1999)
Wespestad, Vidar, 'The Status of Bering Sea Pollock and the Effect of the *Donut Hole* Fishery', *Fisheries* 18(3) 18–24 (1993)
Whiteman, Marjorie, 'Conference on the Law of the Sea: Convention on the continental shelf', AJIL 52(3) 629–659 (1958)
Wiegand, Jane, 'Seizures of United States Fishing Vessels', SDLR 6(3) 428–446 (1969)
Williams, William, 'Reminiscences of the Bering Sea Arbitration', AJIL 37(4) 562–584 (1943)
Wolfrum, Rüdiger, 'Law of the Sea: An example of the progressive development of international law', 309–327 in Christian Tomuschat (Ed.), *The United Nations at Age Fifty: A legal perspective* (Nijhoff, 1995)
Wolfrum, Rüdiger, 'Provisional Measures of the International Tribunal for the Law of the Sea', 173–186 in Chandrasekhara Rao – Rahmatullah Khan (Eds), *The International Tribunal for the Law of the Sea: Law and practice* (Kluwer, 2001)
Wolfrum, Rüdiger, 'Conciliation under the UN Convention on the Law of the Sea', 171–190 in Christian Tomuschat – Riccardo Pisillo Mazzeschi – Daniel Thürer (Eds), *Conciliation in International Law* (Brill–Nijhoff, 2017)
Wolfrum, Rüdiger, 'Realization of Sustainable Management/Development under the Law of the Sea Convention?', 101–109 in Markus Kotzur – Nele Matz-Lück – Alexander Proelss – Roda Verheyen – Joachim Sanden (Eds), *Sustainable Ocean Resource Governance* (Brill–Nijhoff, 2018)
Wolfrum, Rüdiger, 'Implementation and Enforcement of Community Interest-Related Treaties by Judicial Means: Procedural limitations, chances and prospects', IJMCL 38(2) 270–282 (2023)
Xue, Guifang, *China and International Fisheries Law and Policy* (Brill, 2005)
Yamada, Chusei, 'Priority Application of Successive Treaties Relating to the Same Subject Matter: The southern bluefin tuna', 763–771 in Nisuke Ando – Edward McWhinney – Rüdiger Wolfrum (Eds), *Liber Amicorum Judge Shigeru Oda* (Kluwer, 2002)
Yankov, Alexander, 'The Law of the Sea Conference at the Crossroads', VJIL 18(1) 31–41 (1977)
Yee, Sienho, 'Conciliation and the 1982 UN Convention on the Law of the Sea', ODIL 44(4) 315–334 (2013)
Yu, Huming, 'China's Coastal Ocean Uses: Conflicts and impacts', OCM 25(3) 161–178 (1994)
Zalasiewicz, Jan – Williams, Mark, 'The Anthropocene Ocean in Its Deep Time Context', 19–35 in Davor Vidas – Peter Schei (Eds), *The World Ocean in Globalisation:*

Climate change, sustainable fisheries, biodiversity, shipping, regional issues (Brill, 2011)
Zile, Zigurds, 'A Soviet Contribution to International Adjudication: Professor Krylov's jurisprudential legacy', AJIL 58(2) 359–388 (1964)
Zumwalt, Anna, 'Straddling Stock Spawn Fish War on the High Seas', *University of California Davis Journal of International Law and Policy* 3(1) 35–56 (1997)

Index